W0049858

Lecture Notes in Economics and Mathematical Systems

432

Founding Editors:

M. Beckmann
H. P. Künzi

Editorial Board:

H. Albach, M. Beckmann, G. Feichtinger, W. Hildenbrand, W. Krelle
H. P. Künzi, K. Ritter, U. Schittko, P. Schönfeld, R. Selten

Managing Editors:

Prof. Dr. G. Fandel
Fachbereich Wirtschaftswissenschaften
Fernuniversität Hagen
Feithstr. 140/AVZ II, D-58097 Hagen, Germany

Prof. Dr. W. Trockel
Institut für Mathematische Wirtschaftsforschung (IMW)
Universität Bielefeld
Universitätsstr. 25, D-33615 Bielefeld, Germany

Springer
Berlin
Heidelberg
New York
Barcelona
Budapest
Hong Kong
London
Milan
Paris
Santa Clara
Singapore
Tokyo

Mehrdad Tamiz (Ed.)

Multi-Objective Programming and Goal Programming

Theories and Applications

 Springer

Editor

Dr. Mehrdad Tamiz
University of Portsmouth
School of Mathematic Studies
Mercantile House
Hampshire Terrace
Portsmouth P01 2EG, UK

Library of Congress Cataloging-in-Publication Data

Multi-objective programming and goal programming : theories and
applications / Mehrdad Tamiz (ed.)
      p.   cm. -- (Lecture notes in economics and mathematical
systems ; 432)
   "Papers ... presented at the first international conference in
Multi-objective programming and Goal programming theories and
applications (MOPGP94) ... held at the University of Portsmouth,
United Kingdom, from 1 to 3 June 1994"--Pref.
   Includes bibliographical references.
   ISBN 978-3-540-60662-8   ISBN 978-3-642-87561-8 (eBook)
   DOI 10.1007/978-3-642-87561-8

   1. Multiple criteria decision making--Congresses. 2. Programming
(Mathematics)--Congresses.   I. Tamiz, Mehrdad, 1958-
II. Series.
T57.95.M82   1996
519.7--dc20
                                                          95-26207
                                                          CIP

ISBN 978-3-540-60662-8

This work is subject to copyright. All rights are reserved, whether the whole or part of the material is concerned, specifically the rights of translation, reprinting, re-use of illustrations, recitation, broadcasting, reproduction on microfilms or in any other way, and storage in data banks. Duplication of this publication or parts thereof is permitted only under the provisions of the German Copyright Law of September 9, 1965, in its current version, and permission for use must always be obtained from Springer-Verlag. Violations are liable for prosecution under the German Copyright Law.

© Springer-Verlag Berlin Heidelberg 1996

Typesetting: Camera ready by author
SPIN: 10516176   42/3142-543210 - Printed on acid-free paper

Contents

Part 2: Goal Programming

Introduction

Mehrdad Tamiz,
University of Portsmouth, UK.

This book contains a collection of some of the most important and interesting papers that were presented at the first international conference in Multi-objective programming and Goal programming theories and applications (MOPGP94). It is aimed to present some of the latest theoretical and algorithmic developments in the field as well as presenting a very useful selection of application papers.

MOPGP94 was held at the University of Portsmouth, United Kingdom, from 1 to 3 June 1994. The international organising committee consisted of Professors S. Gass, J. Ignizio, G. Mitra, C. Romero, R. Steuer, and myself (Chairman).

This conference was initiated and organised by me and some of my research students (Miss R. Hasham, Mr D. Jones, Mr S. Mardle, and Mr K Mirrazavi) who formed the members of the local organising committee. I am grateful for their help in organising and running the conference. We identified the need for such a specialised conference and believe to date that MOPGP94 was the very first international conference solely devoted to multi-objective and goal programming.

The main aim of MOPGP94 was, therefore, to bring together researchers and practitioners from different disciplines of Optimisation, Operational Research, Mathematical Programming, and Multi Criteria Decision Making, whose common interest was in multi-objective analysis. During the conference delegates unanimously decided that MOPGP should become a regular bi-annual event.

This book is appropriately divided into two parts: 'Multi-Objective Programming and Multi-Criteria Decision Making', and 'Goal Programming'.

Part 1: Multi-Objective Programming and Multi-Criteria Decision Making

Ralph Steuer presented a very interesting plenary talk on Multi-Objective programming which generated a lot of discussions amongst the delegates. Here is a short summary of his talk.

Although most new operational research textbooks contain a chapter on programming with multiple objectives, usually little more than goal programming is discussed. Typically, the concept of nondominance is not mentioned. One reason a more advanced treatment of multiple objective programming is avoided is because of space. Efficient points, criterion space, the size of the nondominated set, payoff tables, aspiration criterion vectors, quasi concave utility functions, Tchebycheff metrics, etc. would have to be discussed. However, by carefully crafting the presentation of these topics, standardizing notation, and exploiting similarities among the procedures of multiple objective programming, one can develop a self-contained, technically rigorous, chapter-length presentation of multiple objective programming problem.

The first paper in this part is on "Dynamic Choices in Economics: A Compromise Approach". The authors formulate a dynamic compromise model where criteria are state variables evolving in time. Motion equations are determined under a few widely accepted assumptions in decision theory. This can apply to several fields such as economics, sociology, ecology, etc. They highlight two applications: a) the choice of consumer goods aggregates and b) the Pigovian-Coasian reparation of negative externality through a negotiation process.

The next paper is on "Application of Multicriteria Analysis to Ranking and Evaluation of Water Development Project (The Case of Jordan). In this paper the authors discuss the water shortage in Jordan. They formulate the problem, present and quantify criteria, and obtain partial and complete ranking of projects by applying the PROMETHEE method to rank a sample of 13 water development projects. They present and discuss some interesting computational results.

The next paper is on "Multi-Criteria Efficiency Profiling". Given a set of organizational units or branches which use multiple inputs (e.g. staff, machinery, materials, money) to generate multiple outputs, the author aims to estimate the relative efficiency with which each input is being utilised at each branch. The techniques presented should be of use in the public sector (e.g. education, health etc.) as well as the private sector. The methods are applied to a benchmark data set and the results are found to be superior to those of the widely investigated technique of data envelopment analysis.

The next paper is on "Nimbus - Interactive Method for Nondifferentiable Multiobjective Optimization Problems". An interactive method, NIMBUS, for nondifferentiable multiobjective optimization problems is introduced. The method starts by moving as far as possible from the starting point into a direction where the values of all the objective functions improve. The decision maker is asked to classify the objectives into five different classes and according to this classification a new problem is formed. The resulting solutions are presented to the decision maker to select the most desirable alternative for continuation. It is guaranteed that only (weakly) Pareto optimal solutions are presented to the decision maker in the selection phase. Then the objective functions are classified at this new solution and so on. Some numerical experiments with test problems are reported.

The next paper is on "A Graphic Search Based on Active Sets For Nonlinear Convex Multiple Objective Programming with Linear Constraints". The authors present an algorithm which makes possible the determination or approximation of the efficient set for a nonlinear convex multiple objective programming problem with linear constraints. The algorithm is based on the idea of the active set methods in nonlinear programming problems.

The next paper is on " A Sequential Network-Based Approach for the Multiobjective Network Flow Problem with Preemptive Priorities". Network flow optimisation is probably one of the most widespread techniques used to model real systems. In many practical problems, where there exist multiple measures of the quality of feasible alternatives, a multiple criteria (associated with preemptive priorities) network flow formulation of the problem is appropriate. A sequential approach for obtaining an optimal solution for this class of problems is given in this paper.

The next paper is on "MOLP Formulation Assistance Using LP Infeasibility Analysis". The correct formulation of a large multiple objective linear program can be very difficult. The problem formulator faces the usual difficulties posed by single objective linear programs, such as isolating and analyzing infeasibility among the constraints. In addition, MOLP problems may raise questions about whether certain relationships should be cast as objectives or constraints, and about how the different objectives interact and interfere with each other. Automated or semi-automated assistance is needed to handle problems of practical scale effectively. The authors demonstrate how the techniques recently developed for the analysis of infeasible LPs can be extended to assist in the formulation of MOLPs.

The next paper is on "Optimizing the Yield of an Extrusion Process in the Aluminum Industry". Aluminum extrusion is a common metallurgical process used to produce elongated components of variable lengths and uniform cross-section. The most common application of extrusion is in the application of aluminum siding. Due to the competitive nature of the industry as well as the variable nature of the price of aluminum, it is important to a manufacturer to operate an efficient extrusion operation. The authors describe a multiple objective model for optimising the efficiency of an aluminum extrusion operation. The model has been successfully applied at Alcan Aluminum's Vancouver Works where it has resulted in annual savings of over one million dollars.

The next paper is on "Projective and Symbolic Degeneracy-reducing Techniques for Multiple Objective Linear Programming". Multiple objective linear programming is typically designed for iterative user assessment that can be crucially enhanced by post-optimal analysis. Generally implemented within a simplex algorithm, post-optimal (sensitivity) analysis is affected by degeneracy. Alternative linear programming methods such as the projective method need not use bases, and are therefore affected differently. Yet, their most common approach to post-optimal analysis has been conservative, e.g. reconstructing a basis in order to use the classical framework. A second, much less investigated approach, is to define post-optimal analysis without bases. Even so, the new methods can be affected by degeneracy, caused by numerical difficulties both in the data representation and the solution process. Symbolic solvers can avoid such difficulties. Viewing MOLP as disjunctive programming, related to discrete optimization and thereby to logic programming, the constraint logic program is selected for its symbolic treatment of algebraic constraints seamlessly embedded in a Prolog syntax.

The next paper is on "Interactive Multiple Criteria Optimization for Capital Budgeting in a Canadian Telecommunications Company". The authors describe DSSORA (Decision Support System Optimal Resource Allocation) which is an interactive system for optimal resource allocation developed to support decisions of investment in capital intensive telecommunications projects. Each manager makes an annual funding recommendation to the budget committee which, through discussions and negotiations, decides what level of funding each program will receive. An analysis of the financial and institutional imperatives, the corporate aspirations and the decision procedures has led to the development of DSSORA. It has been tested by several groups of managers responsible for the management and the implementation of project portfolios with significantly consistent results. The flexibility, user friendliness and quick time response of DSSORA make it an effective negotiation tool in a group setting.

The next paper is on "The Ekeland's Principle and the Pareto ϵ-efficiency". Using a variant of Caristi's fixed point theorem in locally convex spaces. It is well known that the Caristi's fixed point theorem is equivalent to the Ekeland's principle. The author presents a new and interesting variant of Ekeland's principle for vector valued mappings with applications to the study of both, Pareto efficiency and Pareto ϵ-efficiency.

The last, but not least, paper in this part is on "Generation of Pareto Solutions By Entropy-Based Methods". In recent years the Maximum Entropy Principle has been used to develop radically new approaches to various classes of optimisation problems such as those of scaler non-linear constrained optimisation, vector and minimax optimisation. The authors develop two new entropy-based approaches and apply them to the problem of generating Pareto optimal solution sets for general multi-criteria optimisation problems. The algorithms are tested on several test problems and the solutions are reported.

Part 2: Goal Programming

In the light of the recent developments in GP (see the first paper in this part), We believe that it is now becoming a very powerful tool for modelling and solving a wide class of optimisation problems with multiple objectives. We have in recent years identified and addressed various shortcomings in GP. Among these are: normalisation of the objectives prior to optimisation, identifying and generating pareto efficient solutions, and effective modelling of the decision maker's preferences. These ideas as well as various solution speed up techniques are implemented in a new intelligent GP solver, GPSYS, which has been designed and developed by us in the University of Portsmouth.

James Ignizio presented a very interesting plenary talk regarding the past, present, and future status of goal programming. Here is a short summary of his talk:

During the 1960s, the tool that has come to be known as goal programming began to find use, primarily by the practitioner, in the successful (and in some cases often highly successful) solution to a wide variety of real world problems involving multiple, conflicting objectives and both hard and soft constraints; problems that could not be effectively addressed by conventional means without the imposition of numerous simplifying (and questionable) assumptions. Ranging from implementation in engineering design (for the U.S. space program) to that of an alternative to conventional regression (e.g., for the stablishment of predictive models and cost estimates), goal programming proved to be an effective and pragmatic approach to real world problem solving. In the decade of the 1970s, interest in this new tool became more widespread and, for better or worse, attracted considerable attention in the academic community. Unfortunately, some of this interest was generated by individuals who sought nothing more than to exploit the novel nature of this *hot* new tool, rather than attempting to understand its proper role in decision making. This interest was followed by the publication of some rather embarrassing, and certainly naive, papers on the topic. As such, interest in the past decade has seemed to significantly diminish as real and imagined problems with goal programming have been either identified or alleged. Ignizio then described numerous new implementations of goal programming, virtually unknown to those in the multiobjective and multicriteria communities.

The first paper in this part is on "An Overview of Current Solution Methods and Modelling Practices in Goal Programming". We present an overview of the current state-of-the-art methods for modelling and solution of goal programming problems. Strategies for integrating the techniques into an intelligent software package are suggested. Some recent criticisms of goal programming are detailed, together with how these perceived problems can effectively be alleviated by means of such a package.

The next paper is on "Goal Programming in Networks". The methodologies of network analysis and goal programming combine to form a powerful modeling framework that enlarges the application base that can be solved by the powerful computational procedures of networks. The author reviews and discusses goal programming network structures and some applications.

The next paper is on "Solution of Nonlinear Field Problems by Goal Programming". The proposed approach involves approximating the unknown solution by a set of trial functions or modes containing unknown coefficients. It reformulates the field problem as a preemptive goal programming model via the use of hard and soft constraints in linear programming. The advantages of the proposed approach in solving nonlinear problems is outlined. Nonlinear field problems from Fluid Dynamics and Control Theory are employed to illustrate the methodology.

The next paper is on "Incorporating the Decision-Maker's Preferences in the Goal Programming Model with Fuzzy Goal Values: A new Formulation". It is not easy for a decision maker to fix precisely his aspiration levels and his preferences are rarely taken into account within the classical GP model formulation. The authors introduce a concept of decision maker's satisfaction functions even when the goal values are fuzzy. A new formulation of the GP model is proposed in order to explicitly incorporate the decision maker's preference functions.

The next paper is on "A Formulation of a Fuzzy Linear Goal Programming problem with Fuzzy Constraints and Fuzzy Target Values". The authors present solution methods for fuzzy linear goal programming problems. where the aspiration level of the decision maker for each goal-target value is represented by fuzzy numbers and also the constraints of the model define the feasible set as a fuzzy set, described by a membership function. The problem is transformed into a parametric one, and the solution is given in fuzzy terms. A numerical example is provided.

The next paper is on "A Two Staged Goal Programming Model for Portfolio Selection". Many mathematical models have been applied to portfolio selection, however a major drawback of these methods is that a vast majority of input data is needed which requires a large amount of computation. We investigate the multi-objective approach of Goal Programming and its application to portfolio evaluation and selection. A two stage GP model is proposed. The first stage predicts the sensitivity of the shares to specific economic indicators. The second stage of the model selects an optimum portfolio based on the decision maker's priorities and goals together with the information produced by the first stage.

The next paper is on "Application of MOP and GP to Wildlife Management (Deers). A Case of Mediterranean Ecosystem". The case study presented in the paper is about the development of game management operational plan for deer in southern Spain. This paper shows that management models can be improved by using multicriteria approaches, specially when a satisfactory combination of methods is employed. The efficient set is generated for analyzing the conflict between economical and ecological objectives.

The next paper is on "Flight Trajectory Optimisation by Goal Programming with Fuzzy Objectives". A sequential goal programming formulation is proposed for the numerical optimization method of optimal flight trajectory problems. By using a time integration algorithm, trajectory optimization problems are transformed into numerical optimization problems that seek the optimal control variables at discrete time points to minimize objective functions while satisfying the design constraints, e.g. boundary conditions at the terminal time and path constraints for control or state variables. By defining the target values of both the constraints and the objective functions and by prioritizing each goal to its significance, the GP formulation can efficiently solve the optimal solution even if initial design variables are far from optimal.

The next paper is on "An Application of Goal Geometric Programming For Equipment Replacement Under Fuzziness". It is well known that almost all types of equipment are subject to deterioration over time through age and usage and decisions regarding the need for replacement and cost reduction are necessary. In order to support such an analysis the authors use really existing functional relationships between costs (overhaul, replacement, and operating costs) and predictor variables (replacement, overhaul, and inspection intervals). Replacement strategies are directly connected with deterioration and its permanent dynamic changes. It is continuously subjected to influence of many factors that cannot always be predicted by company engineers and experts. This leads to uncertainty of data as well as target values on which optimization model is based. Besides, model parameter values are usually specified by experts, which implies for their subjective character. the authors solve a series of multiobjective linear programming problems in order to build fuzzy regression models for such responses. A real-life example, solved on advanced electronic spreadsheet, illustrates the proposed approach.

The last, but not least, paper is on "An Exploration of Linear and Goal Programming Models in the Downstream Oil Industry". We present a comparison of LP and GP approach as applied to downstream oil industry. Results of some experiments are presented and discussed.

Part 1

Multi-Objective Programming and Multi-Criteria Decision Making

Part 1

Multi-Objective Programming and
Multi-Criteria Decision Making

Dynamic Choices in Economics: A Compromise Approach

Enrique Ballestero[1] and Carlos Romero[2]

[1] E.T.S. Ingenieros Agrónomos, Ciudad Universitaria, 28040 Madrid, SPAIN
[2] E.T.S. Ingenieros de Montes, Avda. Complutense, 28040 Madrid, SPAIN

Abstract. This paper formulates a dynamic compromise model where criteria are state variables evolving in time. Motion equations are determined under a few widely accepted assumptions in decision theory. It is noteworthy to point out that the equilibrium is reached at the L_∞ bound of the compromise set. Applications in several fields (economics, sociology, ecology, etc.) seem to be large. We highlight specially two applications: (a) the choice of consumer goods aggregates and (b) the Pigovian-Coasian reparation of a negative externality. Finally, some properties and tentative extensions are indicated.

Keywords. Compromise programming, consumer theory, environmental economics.

1 Introduction

Multiple Criteria Decision Making (MCDM) represents perhaps the fastest growing area of Decision Analysis in terms of theoretical developments as well as practical applications. It is difficult to question that MCDM is nowdays acknowledged as a logically sound and corroborated decisional paradigm. The different approaches which coexist within this decisional framework have been widely applied in many fields.

Despite this popularity and its wide range of applications, very few efforts connecting MDCM with economic analysis have been undertaken. This is striking since a decision making problem often responding to multi-criteria postulates underlies in every economic problem. Thus, in consumer behaviour we can recognize a multiplicity of criteria when the choice of goods is analysed and this analogously occurs with the enterpreneur's decisions. This paper is a follow-up of our research connections between Compromise Programming (CP) and microeconomics as far as consumer theory and joint production models are concerned.

This foregoing can be summarized as follows. In a previous paper we specified conditions under which traditional utility optimization and CP lead to close solutions (Ballestero & Romero, 1991). We have also shown how these conditions are not strong and are rooted to assumptions commonly accepted in economics such as the law of diminishing marginal rate of substitution (Ballestero & Romero, 1994). In this way, something like a bridge between CP and utility optimization is established. Secondly, within a joint production scenario, we have found (by connecting CP and microeconomic analysis) conditions which guarantee the coincidence of three economic optima (the maximum profit mix, the best technological mix, and the consumer utility optimum). From a mathematical perspective, these conditions lead to a compromise set reduced to a single equilibrium point (Ballestero & Romero, 1993a). Finally, we have proven a general theorem on shadow prices in a joint production framework with m outputs, by connecting CP weighting and internal prices of firms. Thus, the meaning of normalizer weights involved in a compromise model is clarified (Ballestero & Romero, 1993b).

In the present research, we will take a further step in the direction pointed out above. Thus, we will formulate a dynamic compromise model which seeks to capture the basic aspects of any microeconomic problem, that is: (a) the multi-criteria side of the decisional behaviour, (b) its dynamic character, (c) the conflict derived from using scarce means to satisfy alternative ends and (d) the imitative behaviour of individuals.

The paper is organized as follows. After these introductory paragraphs, an Axiom underpinning the model is stated and justified. After that, the analytical structure of the model is developed in an abstract way without any kind of material interpretation or application. After the theoretical approach, the model is interpreted and applied to two relevant microeconomic problems: (1) the choice of consumer goods aggregates, and (2) the Pigovian-Coasian reparation of a negative externality. In the concluding remarks some possible extensions of the model are suggested.

2 Assumptions and Definitions

Henceforward, the following notation will be used:

$[x_1(t), x_2(t) \ldots x_m(t)]$	=	vector of state variables. We will also refer to these variables as $(x_1, x_2, \ldots x_m)$ when there is no risk of misunderstanding.
$\dot{x}_1(t), \dot{x}_2(t) \ldots \dot{x}_m(t)$	=	$dx_i(t)/dt \; \forall \; i$. That is, change of state variables with respect to time.
$x_1^*, x_2^* \ldots x_m^*$	=	standard CP ideal point
$[T(x_1, x_2 \ldots x_m) = K]$	=	opportunity set, trade-off set or efficient frontier.
$(w_1, w_2 \ldots w_m)$	=	vector of weights attached to the state variables.
L_1 and L_∞	=	compromise solutions for metrics 1 and ∞, respectively.
C, C_1, C_2	=	integration constants.
α_1 and α_2	=	$x_1(0)$ and $x_2(0)$, respectively (i.e. initial conditions) in a two-dimensional space.

All variables and parameters are defined as non-negative numbers because of their operational meaning.

The following definition is not essential, but useful to clarify the setting.

Definition In a two dimensional space, a variable $x_i(t)$ is in f position (far from its anchor value) when:

$$w_i [x_i^* - x_i(t)] > w_j [x_j^* - x_j(t)]$$

A variable $x_i(t)$ is in n position (near to its anchor value) when:

$$w_i [x_i^* - x_i(t)] < w_j [x_j^* - x_j(t)]$$

Remark I From the definition, if the variable $x_i(t)$ is in f position, the variable $x_j(t)$ is in n position, and viceversa.

Axiom I In a two dimensional space, the change of x_i with respect to time (i.e. its velocity) is proportional to the difference $[w_i(x_i^* - x_i) - w_j(x_j^* - x_j)]$. If x_i is in f position, the term $w_i(x_i^*-x_i)$ has the meaning of a "rising pull" of x_i towards its anchor value x_i^* and the term $w_j(x_j^* - x_j)$ the meaning of a "downward pull" or resistance of x_j towards its anchor value x_j^*. If x_i is in n position the term $w_j(x_j^* - x_j)$ represents the "rising pull" of x_j towards x_j^* and $w_i(x_i^* - x_i)$ the "downward pull" or resistance of x_i towards its anchor value x_i^*. (*An extension to the m-dimensional space is given in Section 4*).

Justification. In short, the farther x_i (or x_j) is from its anchor value x_i^* (or x_j^*) the quicker x_i (or x_j) must approximate to its anchor value x_i^* (or x_j^*). Indeed, any decision-maker faces a problem of *urgency*, that is, the urgency to correct a big deviation in a variable which remains at the rear, so disequilibrating the harmonic approximation to the anchor values. To stop a gap or to reinforce a weaker side on a "front-line" is an urgent priority for any decision-maker in most scenarios. Axiom I describes the urgency to reinforce the variable x_i by its velocity which is assumed to be proportional to $[w_i(x_i^*-x_i) - w_j(x_j^*-x_j)]$, that is, a measure of its weakness in comparison to the weakness of the other variable x_j.

Common assumptions and definitions Definitions commonly used in CP (Yu 1973, Zeleny 1974) are included in our analysis as well as common assumptions and properties for frontiers. It is important to keep in mind that state variables remain on the frontier (i.e. efficient set in a CP language or trade-off set in an economics language) as this is their path.

Remark II The following lemma is used below:

The bound L_∞ of the compromise set is the intersection point of $w_1(x_1^*-x_1) = w_2(x_2^*-x_2) = \ldots = w_m(x_m^*-x_m)$ with the frontier or opportunity set (Ballestero & Romero, 1991).

Note that $(x_{1\infty}, x_{2\infty}, \ldots, x_{m\infty})$ can be considered as a well-balanced mix since discrepancies $w_1(x_1^*-x_{1\infty})$, $w_2(x_2^*-x_{2\infty})$, $\ldots$, $w_m(x_m^*-x_{m\infty})$ with respect to the ideal are equal.

3 Two-Criteria Model Under Axiom I

The dynamic behaviour underlying Axiom I implies the following measures for velocities (or changes with respect to time) $\dot{x}_1(t)$ or $\dot{x}_2(t)$ of both state variables:

$$\dot{x}_1(t) = w_1(x_1^* - x_1) - w_2(x_2^* - x_2) \tag{1}$$
$$\dot{x}_2(t) = w_2(x_2^* - x_2) - w_1(x_1^* - x_1) \tag{2}$$

From (1) and (2) is obvious that:

$$\dot{x}_1(t) \geq 0 \longleftrightarrow \dot{x}_2(t) \leq 0$$

The state variables will reach their dynamic equilibrium when their velocities are zero ($\dot{x}_1(t) = \dot{x}_2(t) = 0$). From (1) and (2), we get:

$$\dot{x}_1(t) = \dot{x}_2(t) = 0 \longleftrightarrow w_1[x_1^* - x_1(t)] = w_2[x_2^* - x_2(t)] \tag{3}$$

Therefore, the dynamic equilibrium is reached at the L_∞ bound of the compromise set according to (3) and Remark II.

The trajectory followed by the state variables from a certain initial point to the L_∞ bound of the compromise set will be researched. Let us start with a simple case and then go to more general scenarios.

Case 1. The trade-off curve or efficient set is a straight line. Its equation is obviously:

$$x_1/x_1^* + x_2/x_2^* = 1 \tag{4}$$

since max $x_1 = x_1^*$ and max $x_2 = x_2^*$.

Taking out x_2 from (4), introducing its value into (1) and writing x_1 as a function of time $x_1(t)$, the following expression is obtained:

$$\dot{x}_1(t) + \frac{w_1 x_1^* + w_2 x_2^*}{x_1^*} x_1(t) - w_1 x_1^* = 0 \tag{5}$$

Expression (5) is a linear differential equation that can easily be solved. Its general solution (motion equation of x_1) is:

$$x_1(t) = Ce^{-\frac{w_1 x_1^* + w_2 x_2^*}{x_1^*}t} + \frac{w_1 x_1^{*2}}{w_1 x_1^* + w_2 x_2^*} \tag{6}$$

where C is the integration constant. By substituting (6) into (4), the motion equation $x_2(t)$ is obtained:

$$x_2(t) = -\frac{x_2^*}{x_1^*} Ce^{-\frac{w_1 x_1^* + w_2 x_2^*}{x_1^*}t} + \frac{w_2 x_2^{*2}}{w_1 x_1^* + w_2 x_2^*} \tag{7}$$

When $t \to \infty$, expressions (6) and (7) turn into:

$$x_{1\infty} = \frac{w_1 \, x_1^{*2}}{w_1 \, x_1^* + w_2 \, x_2^*} \tag{8}$$

$$x_{2\infty} = \frac{w_2 \, x_2^{*2}}{w_1 \, x_1^* + w_2 \, x_2^*} \tag{9}$$

The terms (8) and (9) correspond to the L_∞ bound of the compromise set as can be easily checked from (3) and (4). Thus, $t \to \infty$ leads also to the equilibrium point (velocities = 0). If an initial condition $x_1(0) = \alpha_1$, $x_2(0) = \alpha_2$ is now considered, the integration constant C can be eliminated, thus obtaining the following motion equations:

$$x_1(t) = \left(\alpha_1 - \frac{w_1 x_1^{*2}}{w_1 x_1^* + w_2 x_2^*} \right) e^{-\frac{w_1 x_1^* + w_2 x_2^*}{x_1^*} t} + \frac{w_1 x_1^{*2}}{w_1 x_1^* + w_2 x_2^*} \tag{10}$$

$$x_2(t) = \left(\alpha_2 - \frac{w_2 x_2^{*2}}{w_1 x_1^* + w_2 x_2^*} \right) e^{-\frac{w_1 x_1^* + w_2 x_2^*}{x_1^*} t} + \frac{w_2 x_2^{*2}}{w_1 x_1^* + w_2 x_2^*} \tag{11}$$

Case 2. The trade-off curve or efficient set is given by a polygonal as usually occurs within a compromise context. If case 1 is implemented iteratively, it is possible to determine the motion equations. In fact, let us use as an initial condition, for example, point $\alpha_1 = 0$, $\alpha_2 = x_2^*$. According to case 1, the motion equations (10) and (11) will be obtained along the first segment of the polygonal, i.e., the segment connecting the extreme efficient point $(0, x_2^*)$ to the adjacent extreme efficient point (x_1^1, x_2^1). If the L_∞ bound belongs to this segment, the process stops, otherwise point (x_1^1, x_2^1) is taken as a next initial condition to obtain the motion equations along the next segment. The process continues until the L_∞ bound lies on a segment.

Case 3. The trade-off curve is a continuous function, which is common in economics. For example, if $T(x_1, x_2) = K$, the x_2 can be expressed as $T^{-1}(x_1, K)$, and (1) becomes:

$$\dot{x}_1(t) + w_1 x_1 - w_2 T^{-1} (x_1, K) + w_2 x_2^* - w_1 x_1^* = 0 \tag{12}$$

The general solution of (12) will provide the motion equation $x_1(t)$. For certain transformation curves, the differential equation (12) cannot be solved by conventional calculus techniques, although its solution can always be approximated by numerical calculus.

4 Extension to m State Variables Under Axiom I

For a general case with m state variables, the main assumption can be formulated as follows:

Changes regarding time (or velocities) of the state variables are governed by the following system of m differential equations:

$$\dot{x}_i (t) = w_i (x_i^* - x_i) - \frac{1}{m-1} \sum_{\forall j \ne i}^{m} w_j (x_j^* - x_j) \quad i = 1, 2, ..., m \tag{13}$$

System (13) can actually be reduced to (m-1) variables and equations by substituting x_m from $T(x_1, x_2, ..., x_m) = K$. When the frontier is linear this system can be solved by resorting to standard calculus techniques as is shown in the section on numerical applications.

The "rising pull" corresponding to ith variable in (13) is: $w_i(x_1^* - x_1)$. The "downward pull" of the ith variable (or resistance to the "rising pull" of variable x_1) is given by the negative term on the right hand side of (13); i.e., the average of the "rising pulls" corresponding to the other (m-1) variables.

Remark III. Equilibrium point. The dynamic equilibrium or convergence of the system is once again given by the bound L_∞. In fact, we have:

$$w_1(x_1^* - x_1) = w_2(x_2^* - x_2) = ... = w_m(x_m^* - x_m) \implies$$

$$\rightarrow w_i (x_i^* - x_i) - \frac{1}{m-1} \sum_{\forall j \ne i}^{m} w_j (x_j^* - x_j) = 0, \forall_i \rightarrow \dot{x}_i (t) = 0, \forall_i \tag{14}$$

hence, every point on the L_∞ path $w_1(x_1^* - x_1) = w_2(x_2^* - x_2) = ... w_m(x_m^* - x_m)$ holds $\dot{x}_1(t) = 0, \forall_i$

Remark IV The L_∞ bound of the compromise set is a particular solution of system (13). This property derives straightforwardly from Remark II.

Numerical example. Let us start with the following data $x_1^* = 10$, $x_2^* = 8$, $x_3^* = 6$, $w_1 = w_2 = w_3 = 1$. As a linear frontier $x_1/10 + x_2/8 + x_3/6 = 1$ is assumed, by introducing these values in (13), making $m=3$ and by taking the frontier equation into account, the following system is obtained.

$$\dot{x}_1 (t) = -1.3x_1 + 0.125x_2 + 6$$

$$\tag{15}$$

$$\dot{x}_2 (t) = 0.2x_1 - 1.375x_2 + 3$$

where the general solution is:

$$x_1 (t) = C_1 e^{-1.18t} - 0.625 C_2 e^{-1.50t} + 4.89$$

$$x_2 (t) = C_1 e^{-1.18t} + C_2 e^{-1.50t} + 2.89 \tag{16}$$

$$x_3 (t) = -1.35 C_1 e^{-1.18t} - 0.375 C_2 e^{-1.50t} + 0.89$$

As usual, the final equilibrium corresponds to the L_∞ bound (4.89, 2.89, 0.89) as can be checked by introducing suitable substitutions in expression (15). By starting in point (10, 0, 0) as the initial condition, the integration constants are eliminated and the following motion equations are obtained:

$$x_1 (t) = 2.03 e^{-1.18t} + 3.075 e^{-1.5t} + 4.89$$

$$x_2 (t) = 2.03 e^{-1.18t} - 4.92 e^{-1.5t} + 2.89 \tag{17}$$

$$x_3 (t) = -2.74 e^{-1.18t} + 1.84 e^{-1.5t} + 0.89$$

System (17) represents the parametric equations defining the path traced by the vector of state variables when moving in the 3-dimensional frontier.

Remark V. Trajectories with some negative coordinates. When $m \geq 3$, some of the coordinates on the L_∞ bound can be negative. For instance, if in our example, the weights change to $w_1 = w_2 = 4$, $w_3 = 1$, then the following coordinates for the L_∞ bound are obtained: 8.75, 5.75, -2.97. This is specially likely when one of terms of $w_i(x_i^* - x_i)$ is very small with respect to the other $w_j(x_j^* - x_j)$ terms. In many scenarios, chiefly in economics, an L_∞ bound with a negative coordinate does not make too much sense.

This problem can be mitigated by resorting to a conventional artifice: calculating the motion equations in the usual way until one of the x_i variables becomes negative. At this moment, x_i is assumed to be zero. For instance, if in a given moment, we find the vector (4, 3, -1, 8); we can start again with (4, 3, 0, 8), by taking this point as the new initial condition and then continue with the analytical procedure on the (4-1)= 3-dimensional space (x_1, x_2, x_4). This type of simplification seems plausible assuming that a small $w_i(x_i^* - x_i)$ implies a little influence of variable x_i in the dynamic process.

5 Economic Applications (the Choice of Consumer Goods Aggregates)

The rationale underlying Axiom I seems sound in applications to several fields. In what follows the applicability of the methodology proposed will be illustrated highlighting two economic approaches. In this section the choice of consumer goods aggregates will be analyzed devoting the next section to the Pigovian-Coasian reparation of a negative externality.

Suppose a consumer who faces the choice of a basket $(x_1, x_2, ..., x_m)$ where x_i is the amount of goods aggregate ith. This individual belongs to a socio-cultural group (e.g. he is a middle-age married lawyer living in Boston) and his consumer behaviour is markedly influenced by pattern in this group as well as by the leaders behaviour, or by the average behaviour of people around him. This socio-cultural influence is likely crucial for the choice of goods aggregates such as clothing, lodging, vacation, food level, etc. Thus, the consumer has a referential point (a pattern or basket to imitate) denoted by $(x_1^*, x_2^*, ..., x_m^*)$.

Therefore, the coordinates of the referential point within a consumer's choice context are not given by the maximum of each variable on the frontier as is usual in CP but is defined exogeneously regardless of the consumer's frontier or opportunity set. The exogenous ideal will be termed *imitative* point or basket. We use this name as it can reflect a social group behaviour to imitate by individuals belonging to this group as will be commented below. It should be noticed that formally a setting referred to an exogeneous ideal can be straightforwardly reduced to a standard CP setting just by a coordinates translation, taking the anti-ideal as a new origin.

We will assume that his imitative behaviour can be described by the dynamic model presented in the above section. More precisely the changes in the amount of goods with respect to time are governed by the system of differential equations given by (13).

The dynamic model interpreted within a consumer choice context has the following meaning. The consumer with a mix $(x_1, x_2, ..., x_m)$ is willing to approximate as close as possible to his imitative point $(x_1^*, x_2^*, ..., x_m^*)$. Generally he cannot reach this imitative point because of his budget restraint. Then the dynamic model assumes he is pursuing a sensible policy such as: the farther is x_i from its anchor value x_i^* the quicker x_i must increase and approximate to x_i^*.

Hence, the change $\dot{x}_i(t)$ regarding time is explained in terms of "rising pull" and "downward pull" as it was described above. Suppose, for example, a lawyer whose consumer imitative point is [x_1^* = decent clothing and footwear (as corresponding to his status); x_2^* = a new car each year; x_3^* = 25 days for vacations]. If his shoes are very old and broken, but his car is only two years old and he plans to spend 20 days in vacations, the "rising pull" of x_1 will likely be extremely marked when compared to the joint "pull" of x_2 and x_3 (in average terms). Therefore, he will probably run to buy footwear before saving for a new car or for longer holidays.

In short, in our consumer choice context first term of (13) can be interpreted as a measure of the urgency for increasing each good i. Weights w_i can play two roles: (a). To normalize heterogeneous goods for computational necessities (quantities of such goods which need to be reduced to standard units for aggregation); then, there are termed as the *normalizer weights*: (b) To express the consumer's preferences with respect to each good; then, there are termed as the *preference weights*. When these preferences do really exist they are the *non-imitative* side of the problem.

When we focus on a *pure* imitative process (i.e. a process where individual preferences are hardly relevant) weights are only normalizers. In this case, the most usual way of normalizing consists on reducing the amounts of goods to percentages over the anchor values x_i^* [see Zeleny (1974), etc as well as an interpretation in terms of shadow prices in Ballestero & Romero (1993b)]. That is, $w_i = 1/x_i^*$. Thus, the percentages x_i/x_i^* are introduced into (14), and we obtain the following equations:

$$\frac{x_1}{x_1^*} = \frac{x_2}{x_2^*} = - - - - = \frac{x_m}{x_m^*} \qquad (18)$$

Taking into account the market prices of goods (P_1, P_2, ..., P_m) expression (18) is tantamount to:

$$\frac{P_1\,x_1}{P_1\,x_1^*} = \frac{P_2\,x_2}{P_2\,x_2^*} = - - - - = \frac{P_m\,x_m}{P_m\,x_m^*} = \frac{\sum\limits_{i=1}^{m} P_i\,x_i}{\sum\limits_{i=1}^{m} P_i\,x_i^*} \qquad (19)$$

From (19) and taking into account the consumer's restraint budget $P_1x_1 + P_2x_2 + \ldots P_mx_m = y$, we obtain the following demand equation.

$$x_i = \frac{y\,x_i^*}{P_1\,x_1^* + P_2\,x_2^* + \ldots + P_m\,x_m^*} \qquad (20)$$

for each good. Therefore, the particularization of the dynamic approach to the consumer choice problem implies functions holding the law of demand (i.e. dx_i/dP_i < 0) and predicts complementary and normal goods (i.e. $dx_i/dP_j < 0$ and dx_i/dy > 0) what seems coherent with a consumption scenario with a high level of aggregation. Concerning to elasticities is easy to obtain that price elasticity is variable and less than one

$$(\text{ i. e.} \mathcal{E}_{x_i}^{P_i} = - P_i \, x_i^{*} / \sum_{i=1}^{m} P_i \, x_i^{*})$$

cross elasticity is also variable and less than one

$$(\text{ i. e.} \mathcal{E}_{x_i}^{P_j} = - P_j \, x_j^{*} / \sum_{i=1}^{m} P_i \, x_i^{*})$$

and income elasticity is constant and equal to one[1].

The above interpretation of the dynamic model for consumer's choices might seem to be rather tentative. It requires further elaborations such as: the clarification of the socio-cultural forces which determine the imitative point; research of the imperfect imitation scenarios (where imitation as well as individual preferences are considered); an establishment of connections between the dynamic approach and the traditional methodology based on the consumer's utility maximization, and finally the testing of the model. These issues are currently being researched.

6 Economic Applications (Level of Negative Externality when Public Opinion is a Referee)

The basic problem in environmental economics can be briefly stated as follows: to determine the optimal level of pollution (externality) produced by an externality generator (e.g. a polluter firm such as a pulp mill operating plant) in an externality sufferer (e.g. people living in the factory neighbourhood or a sufferer firm such as a fish factory). The traditional solution to this problem, since the pioneer work by Pigou (1920), consists in defining the optimal externality as the amount of pollution corresponding to the level of economic activity by which the marginal benefit of the polluter firm equalizes the marginal external cost of the sufferer. The diagrammatic representation of this problem is shown in figure 1 following the classical devise proposed by Turvey (1963).

---

[1] It is also interesting to notice that if the imitative or ideal basket is defined endogeneously, that is $x_i^{*} = y/P_i$ then the demand equation (20) turns into $x_i = y/nP_i$ which coincides with the demand functions deduced from the *impulsive behaviour* commented by Becker (1962).

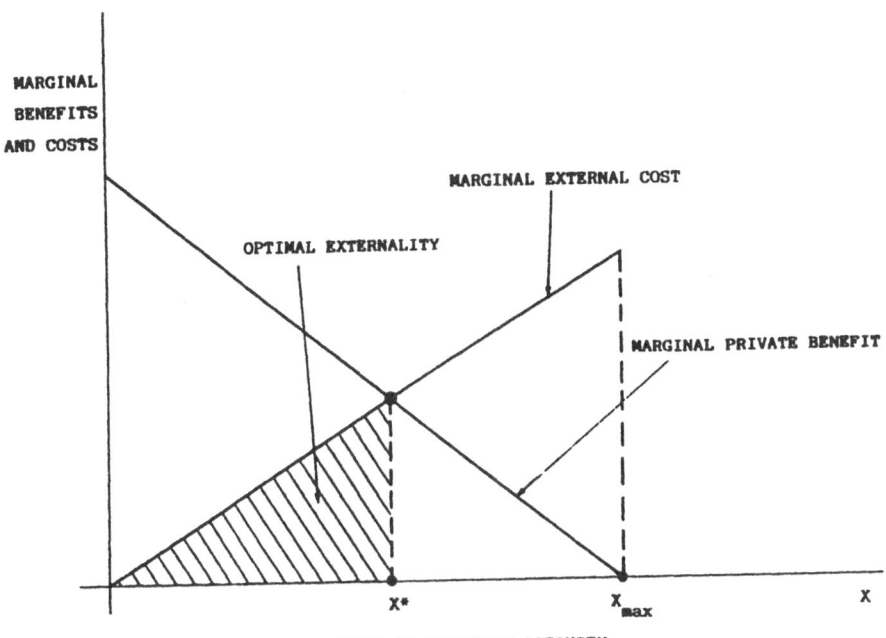

FIGURE 1. DIAGRAMMATIC REPRESENTATION OF THE OPTIMAL EXTERNALITY

The optimal social level of economic activity is given by x^* while the crosshatched area of figure 1 represents the optimal externality or socially optimal level of damage caused by the polluter firm (e.g. the pulp mill plant) to the sufferer (e.g. people in the neighbourhood or the fish factory). The reduction in the level of economic activity from the polluter's (private) optimum x_{max} to the social optimum x^* (mix of interests of polluter and sufferer) can be achieved by governmental intervention.

On the other hand, Coase (1960) has proposed another solution to the externality problem. According to the Coasian approach, anti-pollution regulations become unnecessary as the polluter and the sufferer can negotiate in order to repair the externality. The negotiation process will likely lead to compensatory payment, between sufferer and polluter, until an equilibrium point (the point where the polluter's marginal profit is equal to the sufferer's marginal external cost) is reached.

Indeed, the Coasian Theorem is plausible and realistic when both polluter and sufferer are a few firms (or better, two firms) as they can easily undertake a negotiation. However, a negotiation often reveals itself impossible when the polluter is a firm (or set of firms) while the sufferer are people in the neighbourhood, or perhaps, people living far away. In such frequent cases, there is no negotiation (or this rarely happens) but public campaigns (through newpapers, etc.) are used to stop pollution. In this way, public opinion is raised as a moral authority or moral referee.

An anti-pollution campaign can be considered as a dynamic process in the two-dimensional (B, EC) space, where B is the profit or benefit of the polluter, and EC the external cost affecting the sufferer. We have a classic CP ideal point (B*, EC*) where:

B^* = maximum profit or benefit for the polluter
EC^* = 0 (that is, no pollution)

According to a sensible behaviour of public opinion, the greater $(EC-EC^*)=EC$, the quicker pollution must be stopped (or reduced). In this way, if public opinion has a real influence on the polluter behaviour, the results of the anti-pollution campaign can be described by our dynamic model. If public opinion is not demagogic but strictly impartial and objective, then there are no preferences. Therefore, weights w_1 and w_2 only play a role as normalizer parameters (i.e., they are inversely proportional to the anchor values, as usual).

Coherently with the above argument and according to (13) or to (1) and (2), the changes in both variables B and EC with respect to time are governed by the following two differential equations:

$$\dot{B}(t) = W_1 (B^* - B) - W_2 (EC - EC^*)$$

$$(21)$$

$$\dot{E}C(t) = W_2 (EC - EC^*) - W_1 (B^* - B)$$

Let us now determine the frontier. Assuming that the polluter firm is a price taker its benefit function is given by:

$$B = py - f(y) \qquad f'(y) > 0 \quad f''(y) > 0 \qquad (22)$$

where p is the market price, y the amount of output produced and f(y) the cost function. Let us now represent the external cost of the sufferer firm as:

$$EC = g(y) \qquad g'(y) > 0 \quad g''(y) > 0 \qquad (23)$$

From (23) we obtain:

$$y = g^{-1}(EC) \qquad (24)$$

Substituting (24) in (22) we have:

$$B = pg^{-1}(EC) - f[g^{-1}(EC)] = \Psi(EC) \qquad (25)$$

which represents the production possibility frontier in the B-EC space. Taking into account (25) the motion of the external costs will be governed by the following differential equation:

$$\dot{E}C(t) - W_2 EC - W_1 \Psi(EC) + W_2 C^* + W_1 B^* = 0 \qquad (26)$$

The meaning of (26) will be illustrated with the help of a numerical example. Thus let us assume that the polluter firm can sell its output in a perfectly competitive market at a price of 10 monetary units facing to a cost function such as: $f(y) = 0.5y^2$. The production process of the polluter firm generates an external cost in the sufferer firm ruled by the function $EC = 0.25y^2$. Applying the method exposed above to the particular cost functions of our example the following frontier is obtained:

$$B = 20 \, EC^{0.5} - 2 \, EC \qquad (27)$$

For the data of our example is easy to obtain the following ideal and anti-ideal values:

$B^* = 50$	$B_* = 0$
$EC^* = 0$	$EC_* = 25$

Considering that polluter and sufferer are equal deserving and normalizing as usual by the inverse of the corresponding ranges (i.e. ideal minus anti-ideal values) and taking into account (27) the differential equation (26) becomes:

$$\dot{EC}(t) - 0.4\, EC^{0.5} + 1 = 0 \qquad (28)$$

Implementing the same kind of algebraic manipulation the differential equation governing the motion of the polluter's benefit $\dot{B}(t)$ can be obtained. The equilibrium point of (28) corresponds to the L_∞ bound of the compromise set. For the external costs the value is $EC = 6.35$. Substituting this value of EC in the frontier given by (27) the value $B = 37.5$ as the other coordinate of the L_∞ bound is obtained.

Again as happens with the case of the consumer's choice the interpretation given to our dynamic model within the context of reparation of an externality is tentative and requires further developments which are currently under research.

7 Further Extensions and Concluding Remarks

The dynamic compromise model proposed in this paper has been developed within a framework of differential calculus which is a common point of reference in the field of dynamic models. However it seems possible to formulate the model within a mathematical programming context keeping the basic philosophy underlying Axiom I. This possible extension, which could increase the operational character of the model, is currently being researched.

Conflicting situations in economics, sociology, ecology, etc (involving a trade-off between two or more opposite interests) can be described by the above dynamic model within a compromise context. The model explains the evolution in time of the variables up to the equilibrium point. The examples in the paper illustrate the rationale as well as the operational character of the approach.

Axiom I seems to reflect a general enough behaviour of decision-makers. In many scenarios, the decision-maker has a pattern to imitate (his imitative point or exogenous ideal) as well as an opportunity set. He tries to approximate his position to the imitative point under the following dynamic rule: the farther a variable is from its anchor value, the quicker it evolves toward its anchor value. In real life, this axiom has a variety of meanings, such as the urgency of reinforcing a weak position (for strategic reasons related or not to a balancing policy), the tenacity to recover missing positions, etc.

These kinds of conflict situations are specially relevant in economics. After all, an economic problem involves the following ingredients: (a) a dynamic conflict of uses of scarce means to satisfy alternative ends, (b) patterns to imitate and (c) time as a factor which requires decisions on the urgency to act upon each variable.

The two economic compromise approaches of the dynamic model presented in the paper, although tentative, can highlight the potentiality of Axiom I.

It may be noteworthy that the dynamic equilibrium is always reached at the L_∞ bound. In fact, L_∞ is a significant point, because: (1) is a compromise solution in static CP, (2) it is one of the bounds of the compromise set for bi-criteria problems, and (3) represents a well-balanced weighted mix $(x_{1\infty}, x_{2\infty}, ..., x_{m\infty})$ as $w_1(x_1^*-x_{1\infty})=w_2(x_2^* - x_{2\infty})= ... =w_m(x_m^*-x_{m\infty})$, which turns into $x_1/x_1^* = x_2/x_2^* = ... = x_m/x_m^*$ when only a normalizer weighting is used. This likely brings the L_∞ solution to a seemingly superior place with respect to another best-compromise solutions, chiefly for many situations in real life, where decision-makers prefer an equilibrated mix since harmonic growing and balancing are frequent objectives (or even needs) for average individuals and groups.

References

Ballestero, E. and Romero, C., 1991, A theorem connecting utility function optimization and compromise programming, *Operations Research Letters* 10, 421-427.

Ballestero, E. and C. Romero, 1993a, Economic optimization by compromise programming: the joint production model, *Journal of Multi-criteria Decision Analysis* 2, 65-72

Ballestero, E. and C. Romero, 1993b, Weighting in compromise programming: A theorem on shadow prices, *Operations Research Letters* 13, 325-329.

Ballestero, E. and C. Romero, 1994, Utility optimization when the utility function is virtually unknown, *Theory and Decision* 37, 233-243.

Becker, G.S., 1962, Irrational behavior and economic theory, *The Journal of Political Economy* 70, 1-13.

Coase, R.H., 1960, The problem of social cost. *Journal of Law and Economics* 3, 1-44.

Pigou, A.C., 1932, *The Economics of Welfare*, 4th ed. (Macmillan, London)

Turvey, R., 1963, On the divergence between social and private cost, *Economica* 30, 309-313.

Yu, P.L., 1973, A class of solution for group decision problems, *Management Science* 19, 936-946.

Zeleny, M., 1974, Concept of compromise solutions and the methods of the displaced ideal. *Computers and Operations Research* 1, 479-496.

Application of Multicriteria Analysis to Ranking and Evaluation of Water Development Projects (The Case of Jordan)

B Al-Kloub[1], T T Al-Shemmeri[1], A D Pearman[2], J P Brans[3] and B Mareschal[3]

[1] Staffordshire Univeristy, School of Engineering, Beaconside, Stafford, ST18 0AD, U.K.
[2] University of Leeds, School of Business & Economic Studies, Leeds, LS2 9JT, U.K.
[3] V.U.B and U.L.B Universities of Brussels, Belgium.

Abstract. Water shortage in Jordan is becoming acute, with limited available water and financial resources. With the objective of achieving sustainable development, the Jordan Government is faced with the need to make the best use of these resources. To balance current needs and demands, taking into account the possibility of limited investment in the future, the current Jordanian investment programme (1992-1998) contains various national objectives and constraints to be met. The requirement for this study arises from the need for reliable methodology to select and rank water development projects according to different objectives (criteria). Multicriteria analysis is a realistic approach to this complex problem. In particular, the PROMETHEE method is well adapted to. Its flexibility enables the Decision-Maker to express his preferences more precisely. In this paper the problem is formulated, criteria are presented and quantified, and a partial and complete ranking of projects is obtained by applying the PROMETHEE method to rank a sample of 13 water development projects. The results are discussed and the computational results are given as a step towards the final objective which is a complete ranking of all water sector projects in Jordan taking into account the full range of different constraints, (ie, financial (initial cost, operation and maintenance) and regional development).

Key Words. Water development projects, multicriteria decision making, outranking methods.

1. Introduction

Jordan is located in an arid to semi-arid region. Water availability is becoming a major concern and currently demand exceeds supply. In addition, current practices of over-exploiting and over-pumping have resulted in reducing ground water levels in most wells in the country. Water scarcity is becoming a significant constraint to development and the need for planning the water sector in Jordan is emphasized by the Ministry of Planning and the Ministry of Water and Irrigation. The strategy and planning unit which was recently established at the Ministry of Water and Irrigation responsible for improving and implementing medium and long range water planning is investigating different planning approaches for sustainable water sector development. There is a need for reliable methodologies to address the multiple objective nature of a specific water problem, which is the selection and ranking of water development projects as part of the formulation of a long term, sustainable and environmentally sound water resources development program. This study presents part of this planning approach where decision makers are faced with the problem of the optimal ranking of development projects. A sample of 13 important projects were ranked and analysed as a first step towards ranking all water development projects in the current Jordanian investment plan (1992-1998) subject to different financial and regional development constraints.

2. Multicriteria Analysis

Although there have been many substantial developments in multicriteria methodology, single criterion analysis is still common in many real-world decision making tasks, especially where financial or economic criteria are thought, rightly or wrongly, to be dominant. For single criterion analysis, a number of such criteria have been used to measure, evaluate and rank project performance. A survey carried out by the researchers identified the following well known measures:-

- Net present value criterion
- Net present value ratio
- Internal rate of return
- Pay-back period criterion
- Minimum cost criteria
- Breakeven point analysis
- Benefit/cost criterion
- Effectiveness
- Efficiency
- Quality
- Profitability (ie, liquidity ratios, leverage ratios, activity ratios, profitability ratios, . . . etc)
- Productivity (partial, total and factor)
- Quality of work life
- Innovation

However, the use of multicriteria techniques and decision-making methods is a must where there is limited investment resources and where an effective investment decision must satisfy the broad objectives defined by socio-political conditions which are sometimes non-commensurable and conflicting in nature. Such techniques have been referred to in the literature as multiobjective analysis,

multiple objective optimisation, multiple criteria decision making and v
optimisation.

A single-objective programming problem consists of optimizing one obje
function subjected to constraint set.   On the other hand, a multiobje
programming problem is characterised by a p-dimensional vector of obje
functions as follows:-

$$\text{Max} \quad Z(x) = [Z_1(x), Z_2(x) \ldots Z_p(x)] \quad x \in X$$

where X is a feasible region. However, instead of seeking a single op
solution, a set of "non-dominated" solutions is sought (a subset of the fea
region).   The main characteristic of the non-dominated set (sometimes c
Pareto, non-inferior, efficiency set, transformation set and efficient frontier) i:
for each solution outside the set (but still within the feasible region), there is a
dominated solution for which all objective functions are unchanged or impi
and at least one which is strictly improved. There are several methods to ger
the set of non-dominated solutions (eg, weighting method, constraint me
Philip's linear multi-objective method, and Zeleny's linear multi-obje
methods). For more details see [1].

The need for introducing value judgments into the solution process to orde
alternative solutions in the non-dominated set is central. To aid the decision r
in solving problems and to introduce the value judgments and trade offs,
basic types of method have been developed [2]:-
-       outranking methods.
-       trade-off methods which employ utility functions
-       various interactive methods

Duckstein et al [1] discusses in detail different methods and techniques (dis
continuous and stochastic) to solve the multicriteria problem where the
judgments are introduced according to the following classifications:-
-       continuous methods with prior articulation of preference (methods ·
        the Decision Maker (DM) is asked to articulate his worth or prefe
        structure, and these preferences are then built into the formulation
        mathematical model),
-       discrete methods with prior articulation of preferences (which includ
        out-ranking methods),
-       methods of progressive articulation of preferences.

A tendency has been observed such that the outranking methods are
successful because of their adaptability to real problems and the fact that th
more easily comprehended by decision makers [2]. One of the most used m
is the PROMETHEE method developed by Brans, Mareschal and \
[3,4,5,6,7,8].

ᵋ PROMETHEE Method (Preference Ranking Organization METHod
ɔr Enrichment Evaluation)

basic PROMETHEE method treats criteria problems of the following type:-
  Max $\{f_1(a) \ldots f_K(a) / a \in A\}$
A is a finite set of n possible alternatives which are evaluated on K criteria
  $f_j(o), j = 1, 2, \ldots K$
ᵊn we compare two actions (a,b) we must be able to express the results of this
ɪrison in terms of preference
  (a) dominates (b) if $f_h(a) \geq f_h(b),$ ∀$h = 1, \ldots K$ (with at least one
  >)
  non-dominated alternatives are called efficient (or Pareto-optimal) solutions
  .

ROMETHEE method includes three major steps:-
  ꞉

ᵤment of the preference structure by introducing preference functions. Six
le types of preference function are proposed to the decision makers. In each
ᵊro, one, or two parameters having a clear economic meaning are to be fixed.
  preference functions represent the degree of preference of action (a) with
  to action (b) for each criteria f_j In practice the preference function will be a
  ɔn of the difference between the two evaluations $d = f_j(a) - f_j(b)$.
  ꞉

ᵤment of the dominance relation. A valued outranking relation is built taking
ɔcount all the criteria. For each pair of alternatives the overall degree of
ᵊnce of an alternative over the other one is obtained by
ɣ, a multicriteria preference index is defined as follows:-

$$\pi(a,b) = \sum_{j=1}^{k} \omega_j P_j(a,b) \quad , \qquad \sum_{j=1}^{k} \omega_j = 1$$

ω_j are normalised weights associated with each criterion.
is expresses how and with which degree (a) is preferred to (b). $\pi(a,b)$
res how far (a) is preferred to (b), taking into account all the criteria.
dly, an associated outranking graph and outranking flows are constructed to
ᵥw each alternative (a) comparing with the other k-1 alternatives. The two
ꞌing outranking flows are constructed:-

Leaving flow $\phi^+(a) = \sum \pi(a,x)$ (this measures how much
alternative (a) is outranking all the others (power of (a))).

Entering flow $\phi^-(a) = \sum \pi(x,a)$ (this measures the weakness of
(a)).

Step 3:
Exploitation for decision aid. PROMETHEE I provides a partial ranking, including possible incomparabilities. In this latter case a higher power of one alternative is associated with a lower weakness of the other. This usually happens between alternative (a) and (b) when (a) is good on a set of criteria on which (b) is weak and reciprocally (b) is good on another set of criteria on which (a) is weak. As the information corresponding to the flows is not consistent, it seems natural that the method should not decide which alternative is better. In such a case, it is up to the Decision Maker to take his responsibility and to decide. PROMETHEE II, however, provides a complete ranking. These rankings are deduced from the positive and negative outranking flows. In the partial ranking alternative (a) outranks (b) if it is better than (b) on both flows, while in the complete ranking the following net flow is considered:

$$\phi(a) = \phi^+(a) - \phi^-(a)$$

In both cases the ranking is influenced by the weights associated to the criteria. A special feature of the software, called the walking weights, allows the decision maker to modify the weights and to observe the resulting modifications of the PROMETHEE ranking.

PROMETHEE has been applied in a variety of real-world decision contexts: Briggs et al [9] nuclear waste management site location; Ivan et al [10] production system location; Brans and Mareschal [11] bank industrial evaluation advice; Brans et al [12] computer aided diagnosis; Drago et al [8] communal waste disposal site location; Mladineo et al [2] highway location; Ribarovic et al [14] investment in the ready-mixed concrete industry. There are also papers relating to extensions of the technique. Martel and Aouni [15] incorporated the decision makers' preferences into a goal programming model using PROMETHEE. Mareschal [13] introduced it into stochastic multicriteria decision making. Diakoulaki and Koumoutos [6] introduced cardinal ranking of alternative actions.

4. The GAIA Visual Method

Geometrical information can often provide valuable extra insights. For example, it can allow the decision maker to analyse thoroughly the relationships between alternatives on each criterion, to ease the prediction of results in certain (what if?) situations or to obtain an effective presentation of results. Information can be provided regarding the conflicting nature of criteria and a clear view obtained of whether alternatives perform well or otherwise on different criteria, allowing clusters of similar criteria or of similar alternatives to be identified.

The GAIA program (Geometrical Analysis for Interactive Aid) provides a geometrical presentation of results obtained by PROMETHEE and followed by the PROMCALC method. GAJA is based on the reduction of multi-dimensional problems to two-dimensional planes to allow direct visual presentation. The dimension of any multicriteria analysis is defined by the number of criteria. Reduction of dimensions requires a certain simplification (a loss of information about the problem itself is expected). To ensure that this loss is minimal, by using the Principal Components Analysis technique a plane is obtained in which as little information as possible get lost by projection. The GAIA plane provides reliable information [16]. A decision axis (π-axis) along which the decision maker is invited to select the best alternatives, according to the weights allocated to each criterion is presented. It is then possible to modify interactively the weights and observe the results on the PROMETHEE ranking.

5. PROMETHEE V: "MCDM with Segmentation Constraints"

The PROMETHEE V procedure consists of 2 steps:

1. The multicriteria problem is solved without segmentation constraints (PROMETHEE I & II).
2. The additional segmentation constraints are now integrated subject to:-
    (a) Constraints between the clusters.
    (b) Constraints within the clusters.

The constraints are introduced by considering (0-1) linear integer program ((1) if a_i is selected and (0) if not selected) and applying the Branch and Bound algorithm.

The coefficients of the objective function are the net dominance flows, and the purpose of the maximum problem is to collect as much dominance flow as possible in favour of the subset of alternatives to be selected (see reference 22 for more details).

6. Development of Criteria

In the remainder of the paper, we discuss an initial application of PROMETHEE and GAIA to the question of ranking competing water development projects in Jordan. The situation calls for the definition of a number of decision making criteria, covering economic parameters, market parameters, technical-technological parameters and social and environmental parameters. Since not all the criteria are equally important, weights, reflecting expert judgement, will be allocated.

In Jordan, the national objectives and directions of the country have been developed by the Government as part of the planning process, but these objectives are broad in nature. Generally, we can say that priorities and objectives have changed during the last 20 years. Firstly there was a need to provide safe drinking water and public health became a primary concern, then there was a period of growing emphasis on economic development. Currently, economic and environmental concerns are considered as co-objectives of development [17].

Abu Taleb [18] developed the Jordanian national objectives hierarchy for the water sector. Utilising a Delphi method, he elecited priorities for these objectives (as shown in Figure 1). The criteria and weights developed by Abu Taleb will be used in this preliminary study, as shown in Table 1.

Goals	Objectives	Sub-Objectives	Criteria
Balanced social and economic development	Maximise net economic benefits	Productivity Food Security Foreign Exch Output	Efficiency Irrigated area Energy; cost Output O & M costs
	Maximise health standards	Water Supply Sanitation	Water supply Coverage of sanitation
	Maximise socio-economic benefits	Technology Transfer Employment	Water conservation Foreign labour
	Minimize adverse envir. impacts	Maximise protection of resources	Ground water safe yield Env impacts

Fig. 1. National Objectives Hierarchy [18]

7. Results of Analysis

A complete analysis for a sample of 13 major water projects was carried out, partial and complete ranking was demonstrated (the optimal final complete ranking of projects is shown in Table 2) and sensitivity analysis was undertaken by changing the weights of the criteria and observing the changes in the ranking of the projects.

Table 3 is the spreadsheet of PROMCALC and GAIA showing the generalized criteria. Associated weights at the top of the spreadsheet are displayed, as well as alternatives, type of preference functions for each criteria and its defined parameters, input data for each alternative and the problem type (min or max). Results obtained include descriptive statistics for the criteria as a preliminary analysis (average values, standard deviations and correlation matrix). Partial ranking and final ranking of the selected projects as represented graphically are given in Figures 2 and 3. The actual values of the net outranking flows are represented in "Phi-scale" (as shown in Figure 4) and by looking at this display (and the results shown in Table 4) it is possible to detect clusters of alternatives which are close to each other (eg, alternatives A_7, A_8 (NWW Amman & f, KAC)) and A_{12}, A_{11} (Desal Hisb, Desal Zerqa).

Alternative "further development of Disi Aquifer for agricultural purposes" is the lowest in ranking (the worst) while alternative "improve metering, billing and revenue collection" is the first ranking (the best). Some incomparabilities appear in the PROMETHEE I partial ranking (eg, (A_1, A_2, A_3) are incomparable).

TABLE 1. Jordan National Criteria and Weights as Developed by Abu Taleb
Applying Delphi Technique for Two Rounds [18].

CRITERIA	MEASURE	EVALUATION WEIGHT
C_1 Groundwater Extractions Environmental Impacts:	+/- MCM from safe yield	0.08
C_2 Surface Water Quality	5 point scale	0.06
C_3 Surface Water Quantity	river flow (MCM)	0.06
C_4 Ground Water Quality*	5 point scale	0.12
C_5 Ground Water Quantity	ground water volume (MCM)	0.12
C_6 Sedimentation	million tons of sediment	0.05
C_7 Land Quality*	5 point scale	0.05
C_8 Aesthetics/Recreation*	5 point scale	0.04
C_9 Air Quality	5 point scale	0.04
C_{10} Coverage of Sanitation	+/- % coverage	0.07
C_{11} Supply of Water	MCM	0.08
C_{12} Water Conservation	MCM saved for reallocation	0.08
C_{13} Energy Requirement	Mwh	0.06
C14 O & M Costs*	Scale 1-5	0.06
C_{15} Foreign Laor	number	0.03
C_{16} Irrigated hectares	number	0.06
C_{17} Value of Output	million JD (Jordanian Dinars)	0.07
C_{18} Efficiency	MCM saved	0.08
C_{19} Cost Recovery	proportion of investment to be recovered	0.08

*   Subjective assessment

TABLE 2. Optimal Complete Ranking of the Sample Projects (as abreviated in PROMETHEE)

	PROJECT	As Abbreviated In PROMETHEE
1	Improve metering, billing and revenue collection	(Improv net)
2	Ground water recharge project	(G W recharge)
3	Higher compensation for workers in control of diversions, inspections, etc	(COMP WORK)
4	Construction of Al-Wehdah dam	(Wehdah)
5	Construction of new WW networks and plants in Amman, Fuhis and others	(NWW AMM,f)
6	Maintenance of KAC	(KAC)
7	Rehabilitation of WWTP in the north	(Reh WWTP N)
8	Rehabilitation of Khirbet El Samra WWTP	(SAMRA)
9	Rehabilitation of water networks, Amman	(Reh WNAM)
10	Construction of Karameh dam	(KARAMEH)
11	Desalination of brackish water at Hisban and Kafrain	(Desal Hisb)
12	Desalination of brackish water at Zerka-Mai'n and Zara	(Desal Zerk)
13	Further development of Disi Aquifer for agricultural activities	(Disi)

Checking the weight stability interval indicates how much each criterion weight can be changed without affecting the total ranking. Also if we consider two different sets of weights a different ranking is achieved

8. The GAIA Method Applied on Problem of Projects Ranking

Figures 5 and 6 present the geometric representation of the problem. The alternatives appear to be well clustered in the plane. In addition the conflicting aspects of the criteria appear. Clearly, it is also possible to detect which alternatives are good or bad on the different critiera.

According to the weights associated to the criteria, the PROMETHEE decision axis π is oriented in a direction consistent with the PROMETHEE II complete ranking.

Table 3. The Spread Sheet of Promcalc and Gaia.

Problem loaded : JORDAN1.PR3

Number of actions : 13 (active : 13)
Number of criteria : 19 (active : 19)

Evaluation table :

	C..1 GW Extract min	C..2 SW Quality max	C..3 SW Quant. max	C..4 GW Quality max	C..5 GW Quant. max	C..6 Sediment. min	C..7 Land Qual. max	C..8 Recreation max	C..9 Air Qual. Cov. max	C.10 Sanit max
Type	3	4	3	4	3	3	4	4	4	3
q		0.50		0.50			1.50	0.50	1.50	
p	25.00	1.50	35.00	1.50	25.00	0.60	2.50	1.50	3.50	12.00
s	0.08	0.06	0.06	0.12	0.11	0.05	0.05	0.04	0.04	0.07
	active	active	active	active	active	active	active	active	active	active
A..1 Impr.met. +	40.00	2.00	30.00	3.00	40.00	0.00	3.00	3.00	3.00	0.00
A..2 GW recharg +	0.00	3.00	20.00	4.00	10.00	0.00	5.00	4.00	3.00	0.00
A..3 Comp Work +	0.00	3.00	0.00	3.00	0.00	0.00	3.00	3.00	3.00	0.00
A..4 Wehdah +	0.00	3.00	50.00	3.00	0.00	2.00	4.00	5.00	1.00	0.00
A..5 Reh WNAM +	0.00	3.00	4.00	3.00	4.00	0.00	4.00	4.00	4.00	4.00
A..6 SAMRA +	0.00	3.00	10.00	3.00	0.00	0.00	4.00	4.00	4.00	0.00
A..7 NWW AMM,f +	0.00	3.00	7.00	3.00	0.00	0.00	3.00	4.00	4.00	20.00
A..8 KAC +	0.00	2.00	20.00	2.00	0.00	0.00	3.00	5.00	4.00	6.00
A..9 Reh WWTP N +	0.00	2.00	0.00	3.00	7.00	0.00	3.00	3.00	4.00	0.00
A.10 KARAMEH +	0.00	1.00	0.00	0.00	0.00	0.25	5.00	5.00	1.00	0.00.
A.11 Desal Zerq +	0.00	2.00	0.00	0.00	0.00	0.00	3.00	3.00	2.00	0.00
A.12 Desal Hisb +	0.00	2.00	0.00	1.00	0.00	0.00	3.00	3.00	2.00	0.00
A.13 Disi +	50.00	2.00	0.00	0.00	1.00	0.00	3.00	3.00	3.00	0.00

	C.11 Suppl.Wat. max	C.12 Wat.Cons. max	C.13 Energ.Req. min	C.14 O&M Costs min	C.15 Foreign L. min	C.16 Irrig.Area max	C.17 Val.Output max	C.18 Efficiency Cost max	C.19 REcov max
Type	3	3	3	3	1	1	3	3	3
q									
p	35.00	35.00	375.00	2.00			40.00	30.00	60.00
s	0.08	0.08	0.06	0.06	0.03	0.06	0.07	0.08	0.08
	active	active	active	active	active	active	active	active	active
A..1 Impr.met. +	70.00	0.00	20.00	1.00	0.00	0.00	37.00	70.00	100.00
A..2 GW recharg +	0.00	20.00	5.00	2.00	0.00	1.00	2.00	20.00	0.00
A..3 Comp Work +	10.00	0.00	0.00	1.00	0.00	2.50	12.00	25.00	100.00
A..4 Wehdah +	50.00	0.00	2.00	3.00	50.00	3.50	7.00	0.00	10.00
A..5 Reh WNAM +	0.00	0.00	150.00	3.00	0.00	0.05	1.00	0.00	10.00
A..6 SAMRA +	0.00	0.00	0.00	4.00	0.00	0.10	2.00	0.00	0.00
A..7 NWW AMM,f +	7.00	0.00	200.00	2.00	0.00	0.70	1.30	7.00	10.00
A..8 KAC +	20.00	0.00	0.00	1.00	0.00	2.00	4.00	20.00	10.00
A..9 Reh WWTP N +	7.00	0.00	0.00	3.00	0.00	0.00	1.00	7.00	100.00
A.10 KARAMEH +	30.00	0.00	100.00	3.00	0.00	3.00	6.00	0.00	10.00
A.11 Desal Zerq +	10.00	0.00	200.00	5.00	0.00	0.00	0.00	0.00	20.00
A.12 Desal Hisb +	5.00	0.00	200.00	4.00	0.00	0.00	0.00	0.00	20.00
A.13 Disi +	50.00	0.00	400.00	5.00	50.00	0.00	10.00	0.00	50.00

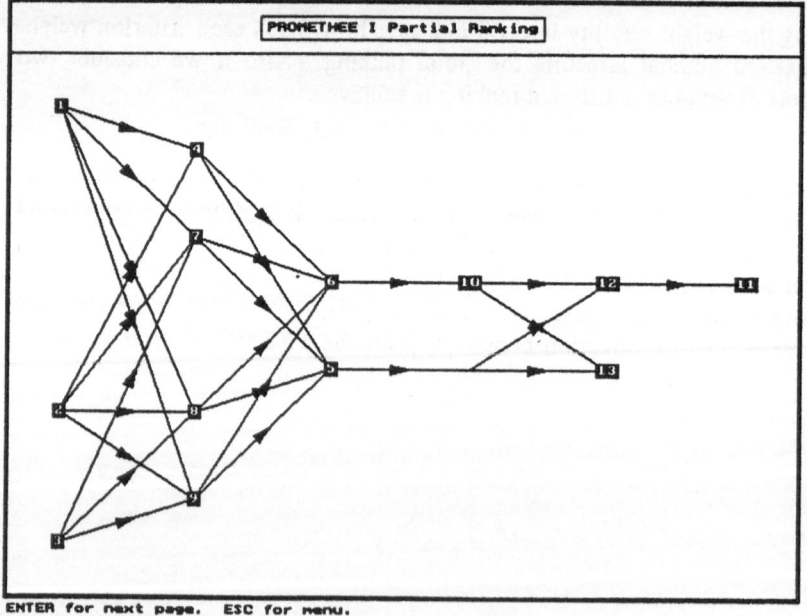

PROMETHEE I Partial Ranking

ENTER for next page.   ESC for menu.

Fig. 2. PROMETHEE I Partial Ranking

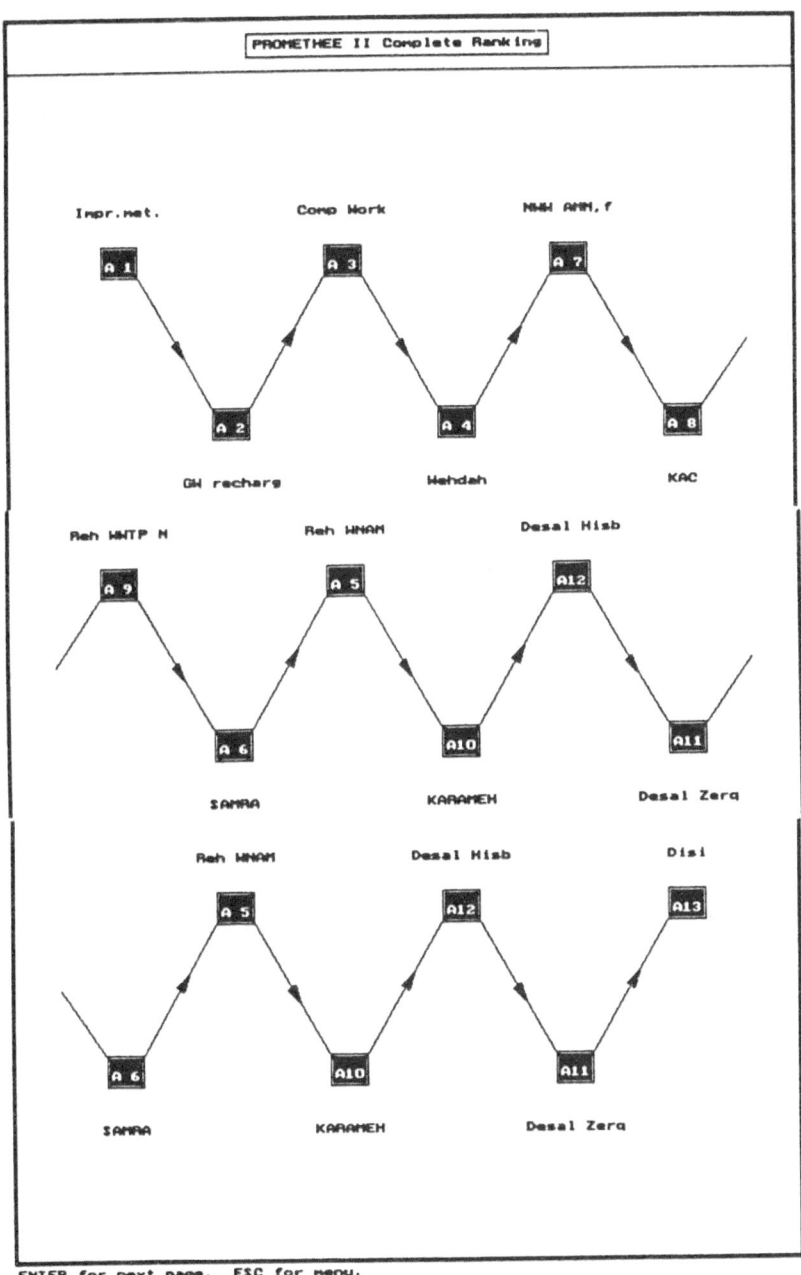

Fig. 3. PROMETHEE Complete Ranking

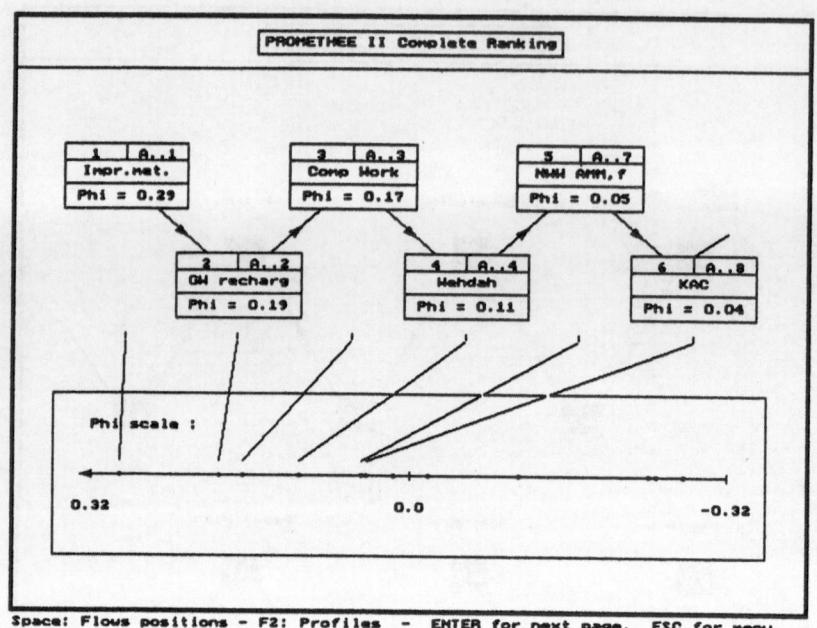

Fig. 4. Complete Ranking, Phi Scale

Table 4. PROMETHEE II Complete Ranking and Net flows.

```
PROMETHEE II Complete Ranking

Rank   Action                   Phi          Phi+          Phi-

    1. A..1  Impr.met.         0.2941       0.4318        0.1378
    2. A..2  GW recharg        0.1913       0.3088        0.1175
    3. A..3  Comp Work         0.1666       0.2654        0.0988
    4. A..4  Wehdah            0.1103       0.2684        0.1581
    5. A..7  NWW AMM,f         0.0484       0.1918        0.1434
    6. A..8  KAC               0.0427       0.2003        0.1576
    7. A..9  Reh WWTP N        0.0423       0.1926        0.1503
    8. A..6  SAMRA             0.0156       0.1868        0.1712
    9. A..5  Reh WNAM         -0.0203       0.1482        0.1685
   10. A.10  KARAMEH          -0.1135       0.1529        0.2664
   11. A.12  Desal Hisb       -0.2442       0.0448        0.2890
   12. A.11  Desal Zerq       -0.2528       0.0447        0.2974
   13. A.13  Disi             -0.2806       0.0982        0.3788
```

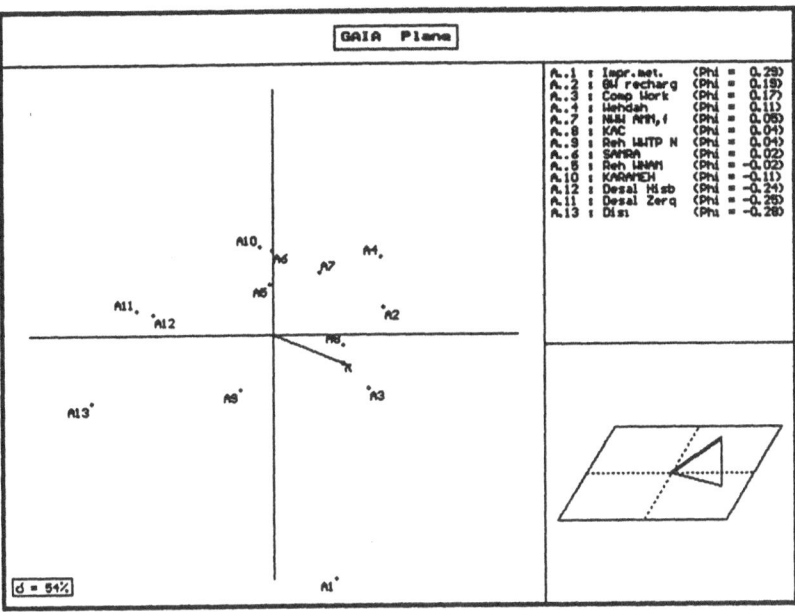

Fig. 5. Gaia Plane (Alternatives)

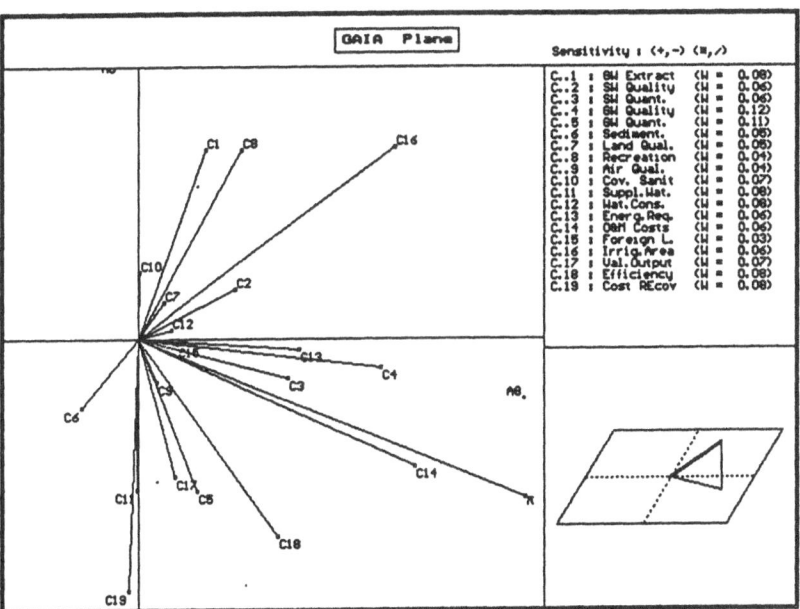

Fig. 6. Gaia Plane (Criteria)

9. Application of PROMETHEE V

Applications of PROMETHEE V were undertaken to introduce constraints to the problem, which are mainly:-

-     financial constraint (initial cost limit of 700,000 JD)
-     incomparability constraint (Al-Wehda X_4 dam and Karameh X_{10} dam were considered as options that cannot coexist with others).
-     regional development constraints (minimum number of projects for each region in the country (R_1 North, R_2 Middle, R_3 South).

The different constraints and results obtained after solving the problem and projects selected based on the constraints results show that Al-Wehdah dam was selected compared to Karameh dam. The first 3 projects kept their ranking as before, and an additional project was selected (NWW Amman & f).

```
Constraints :

K..1 : Exclusion
       active

 X..4 +  X.10  <=  1

K..2 : Cost (00,000 JD)
       active

5 X..1 + 5 X..1 + 100 X..2 + 5 X..3 + 240 X..4 + 120 X..5 + 300 X..6 + 150 X..7

 + 30 X..8 + 80 X..9 + 400 X.10 + 90 X.11 + 20 X.12 + 90 X.13  <=  700

K..3 : Regional develop. R1
       active

 X..2 +  X..4 +  X..8 +  X..9  <=  2

K..4 : Regional develop. R2
       active

 X..2 +  X..5 +  X..6 +  X..7 +  X..8 +  X.10 +  X.11 +  X.12  <=  7

K..5 : Regional develop. R3
       active

 X..2 +  X.11 +  X.12 +  X.13  <=  3

Optimal solution :

Zopt =    0.811

Var.   Action      Value

X..1 : Impr.met.     1    --> Impr.met.  selected
X..2 : GW recharg    1    --> GW recharg selected
X..3 : Comp Work     1    --> Comp Work  selected
X..4 : Wehdah        1    --> Wehdah     selected
X..5 : Reh WNAM      0      -
X..6 : SAMRA         0
X..7 : NWW AMM,f     1    --> NWW AMM,f  selected
X..8 : KAC           0
X..9 : Reh WWTP N    0
X.10 : KARAMEH       0
X.11 : Desal Zerq    0
X.12 : Desal Hisb    0
X.13 : Disi          0
```

10. Conclusions

From the presented information it can be concluded that the PROMETHEE method has potentially a decisive positive contribution both to the process of decision making for selection, and ranking of projects. The geometrical representation of the multicriteria problem represents powerful tools in the hands of the system analysts and a valuable help in solving problems which include conflicting criteria.

As a result of this research a complete study to generate national objectives utilizing the Nominal Group Technique and ranking of all water development projects in the new investment programme (1992-1998) is to be carried out as a future research project, after introducing financial, and regional constraints to the problem.

List of References

1. Ambrose Goicoechea, Don R.Hansen and Lucien Duckstein, "Multiobjective Decision Analysis with Engineering and Business Applications", John Wiley and Sons, 1982.
2. N.Mladineo, I.Lozic and S.Stosic, D.Mlinaric and T.Radica, "An Evaluation of Multicriteria Analysis for DSS in Public Policy Decision", European Journal of Operational Research, 61(1992), 219-229.
3. J.P Brans, Ph.Vincke and B.Mareschal, "How to Select and How to Rank Projects: the PROMETHEE Method", European Journal of Operational Research, 24(1986), 228-238.
4. N.Mladineo and Margeta, J.P Brans and B.Mareschal, "Multicriteria Ranking of Alternative Locations for Small Scale Hydro Plants", European Journal of Operational Research, 31(1987), 215-222.
5. B.Mareschal, "Weight Stability Intervals in Multicriteria Decision Aid", European Journal of Operational Research, 33(1988), 54-64.
6. D.Diakoulaki and N.Koumoutos, "Cardinal Ranking of Alternative Actions Extension of the PROMETHEE Method", European Journal of Operational Research, 53(1991), 337-347.
7. J.P Brans and P.H Vincke, "A Preference Ranking Organisation Method (the PROMETHEE Method for Multiple Criteria Decision-Making)", Management Science, Vol 31, No 6, June 1985.
8. Drago Vuk, Bogomir Kozelj and Nenad Mladineo, "Application of Multicriterional Analysis on the Selection of the Location for Disposal of Communal Waste", European Journal of Operational Research, 55(1991), 211-217.
9. Th.Briggs, P.L Kunsch and B.Marechal, "Nuclear Waste Management; An Application of the Multicriteria PROMETHEE Methods", European Journal of Operational Research, 44(1990), 1-10.

10. Ivan Pavic and Zoran Babic, "The Use of the PROMETHEE Method in the Location Choice of a Production System", International Journal of Production Economics, 23(1991), 165-174.

11. B.Mareschal and J P Brans, "Bank Advisor; An Industrial Evaluation System", European Journal of Operational Reserach, 54(1991), 318-324.

12. Ph.Du Bois, J.P Brans, F.Cantraine and B Mareschal, "Medicis; An Expert System for Computer Aided Diagnosis Using the Promethee Multicriteria Method", European Journal of Operational Research, 39(1989), 284-292.

13. B.Mareschal, "Stochastic Multicriteria Decision Making and Uncertainty", European Journal of Operational Research, 26(1986), 58-64.

14. Zoran Ribarovic and Nenand Mladineo, "Application of Multicriteria Analysis to the Ranking and Evaluation of Investment Programmes in the Ready Mixed Concrete Industry", Engineering Costs and Production Economics, 12(1987), 367-374.

15. Jean-Marc Martel and Belaid Aouni, "Incorporating the Decision-Maker's Preferences in the Goal Programming Model", Journal of Operational Research Society, Vol 41, No 12, pp1121-1132, 1990.

16. Bertrand Mareschal and J.Brans, "Geometrical Representation for MCDA", European Journal of Operational Research", 34(1988), 69-77.

17. Bashar Al-Kloub, "Optimization and Ranking of Water Development Projects Utilizing the Multicriteria Analysis (the Case of Jordan)", Internal Report No/TAS/BK/1/Dec. 1993, School of Engineering, Staffordshire University, England

18. Abu Taleb, "Optimization of Non-renewable Ground Water Development and Related Conveyance Systems Design Using Multicriteria Analysis", PhD Dissertation, Catholic University of America, 1992.

19. Voogd H, "Multicriteria Evaluation for Urban and Regional Planning", Pion Limited, 1983.

20. Jean Brans and Bertrand Mareschal, "The PROMETHEE Method for MCDM, The PROMCALC, GAIA and Bank Advisor Software", 4th International School on Multicriteria Decision Making.

21. Keller H R, Massar D L and Brans J P, "Multicriteria Decision Making: A Case Study, Chemometrics and Intelligent Laboratory Systems, 11(1991) 175-189.

22. Brans J and Mareschal B, "PROMETHEE V: MCDM Problems with Segmentation Constraints", INFOR Vol 30, No 2, May 1992.

MULTI-CRITERIA EFFICIENCY PROFILING

Dr. Christopher Tofallis

The Business School,
Division of Economics, Statistics, and Decision Sciences
University of Hertfordshire,
Mangrove Rd.,
Hertford, SG13 8QF, England
Telephone (U.K.): 01707 285486
e-mail: C.TOFALLIS@HERTS.AC.UK

Abstract. Consider a set of decision-making units (DMUs, e.g. branches, departments, firms) which employ a variety of resources to produce multiple outputs. The units being compared may be in the public sector e.g. health or education, so that the outputs and inputs need not be measured in monetary terms. We present two methods for evaluating the relative efficiency with which each individual resource input is utilised at each organizational unit. We therefore end up with a set of scores for each DMU i.e. a profile rather than a single overall score. Unlike conventional DEA (data envelopment analysis) a large data set is not required and the discriminatory power is shown to be higher.

Keywords: Efficiency, data envelopment analysis, performance measurement, productivity, production.

1 Introduction

Our starting point is the definition of efficiency used in science and engineering, namely, output/input . There the efficiency value naturally lies between zero and unity (100%), but in an organisational context we need to impose constraints to ensure all the scores remain in this range[1] . (Such a score might also be termed a productivity index, but the literature in this area almost always refers to 'efficiency'.) Once an efficiency score , E , is obtained it is possible to calculate target values for inefficient DMUs . This can be done in three ways :--
(i) keep output fixed , in which case the target input is E * (current input)
(ii) keep the input fixed so that the target output is (current output) / E

---

[1] There is no *a priori* upper limit to productivity and we are exploring the removal of this condition in a separate line of research (see section 5). In this paper a figure of 100% merely indicates the highest achieved score relative to which others can be compared.

(iii) change both input and output such that their ratio becomes unity .

If we have multiple inputs and/or outputs we encounter the question of how to aggregate these when there is no pricing system available and the units of measurement differ.

In recent years there has been a great deal of interest in a technique (DEA) which aims to measure the overall technical efficiency of a DMU relative to others which carry out the same type of activities; the efficiency is calculated as the sum of weighted outputs divided by the sum of weighted inputs . DEA applies linear programming to find for *each* DMU its set of weights which will give it the maximum score subject to the conditions that none of the scores exceed unity. Users of DEA sometimes calculate target values as above for individual inputs and outputs, however these are derived from the overall score and so this may not be an appropriate way of proceeding.

Whilst there is a wealth of literature on the estimation of overall technical efficiency, there is very little indeed that deals with a separate efficiency measure for each input when there are multiple outputs; Kumbhakar[4] assumed a particular type of output function (namely Cobb-Douglas: the output is equated to the product of the inputs raised to powers) and also assumed that technical efficiency followed a half-normal distribution and was time-invariant. His stochastic frontier model was applied to U.S. railroad data spanning 25 years and the results indicated that the most inefficient railroad used 40%-50% more labour and 6%-9% more fuel than those on the efficient frontier. Kopp [3] suggested another approach to resource efficiency measurement and he too required the assumption of a known production function relating output to the inputs. Both these papers deal with the case of a single output or require that everything is reducible to this form.

In this paper two efficiency profiling methods are presented which enable the relative (technical) efficiency of *individual* inputs to be evaluated for each decision-making unit. There are a number of advantages that such an approach has over the efficiency score provided by the currently popular DEA models:

(1) It enables the source and extent of inefficiencies in the individual DMU to be more precisely determined: a particular DMU may be efficient in its utilisation of one resource (e.g. sales staff) and inefficient in its utilisation of another (e.g. clerical staff). Such directive information may assist local managers in improving the efficiency of their DMU.

(2) It rules out the possibility of effectively ignoring some inputs at some DMUs by attaching zero or near-zero weights to them. (It would clearly not be sensible to use efficiency scores for input target-setting when no account has been taken of that input in the efficiency assessment.) For the methods of this paper these problems will be avoided by providing a score for each input at every DMU.

(3) Only those outputs to which a given resource acts as an input will be considered in the assessment of that input. As well as being intuitively sensible this rules out the unjustifiable appearance of efficiency by, for instance, placing all the weight on a single input and a single output which are not causally related.

(4) The number of dimensions and 'free parameters' for the new LP model will be fewer than in the equivalent DEA model and so one would expect greater discrimination between the DMUs and a lower proportion appearing to be 100% efficient.

2 Method Using Individual Weights

Suppose that resource I_i acts as an input to s outputs O_r ($r=1,...,s$); this may be a subset of the outputs. Note that a different resource may act as an input to a different set of outputs, possibly fewer or more. The relative efficiency (E_{ik}) with which resource i is being utilised by DMU k is evaluated using the following L.P.(linear program) in which the u-variables are being solved for, and the O's and I's are the observed output and input values.

$$\text{Maximize } E_{ik} = \frac{\sum_{r=1}^{s} u_{irk} O_{rk}}{I_{ik}} \tag{1}$$

$$\text{subject to } \frac{\sum_{r=1}^{s} u_{irk} O_{rj}}{I_{ij}} \leq 1 \quad , \text{ j=1,...,n} \tag{2}$$

$$\text{and } u_{irk} > \varepsilon \quad , \quad r=1,...,s \tag{3}$$

Where ε is a small positive number, n is the number of DMUs and u_{irk} is the weight attached to output r when evaluating the efficiency of input i of DMU k. As with DEA each DMU has its own set of weights which show it in its best light within the constraints of the method (which merely ensure that efficiency scores do not exceed unity). The key difference between this and the DEA formulation is that here each linear program only deals with a single input rather than a weighted sum of all inputs.

The interpretation of the efficiency scores is best understood in terms of target values. Consider a branch which has a score of 0.8 for one of its inputs, this means it could aim to produce the same outputs as before but using only 80% of the current level of that input. This is because there is a combination of the other branches which could achieve the same outputs using only 80% of the input at the branch being studied.

3 Method Using Individual Weights

Once again we study each input resource separately but using a common set of weights to be used on all DMUs. These weights will determine the position of the efficient frontier, hence in estimating them we must first discard any DMUs which are seen to be dominated (by linear combinations of other DMUs) in their usage of the given input; this has already been done for us by our first method since a score below 100% when there is complete flexibility of weights implies that the DMU is dominated by others. We are then left with the non-dominated set (unit efficiency and zero slacks) ; we now try to find a relationship which shows how these DMUs disperse the input amongst their relevant outputs. Hence we might try to fit an expression of the form :

$$I_i = \sum_{r=1}^{s} c_{ir} O_r \qquad (4)$$

The parameters (c) can be interpreted as the amount of input i used per unit of output j (i.e. they are resource consumption rates). Hence each term on the right of (4) estimates the amount of input contributed to that output by an efficient firm with the given output mix. Such an expression is a departure from the usual econometric approach in which aggregate output is expressed as a function of inputs; however we believe that this approach may have the benefits of being easier to comprehend and use. (In econometrics one might try various production functions (Cobb- Douglas,translog,CES,etc) whose form is far from being intuitive, particularly to the average manager.)

   The parameter values c in (4) are determined separately for each input. They can be found using least-squares regression subject to the constraints that they be non-negative and that

$$I_i \geq \sum_{r=1}^{s} c_{ir} O_r \qquad (5)$$

   This says that the input used cannot be lower than the efficient level, i.e. it ensures the efficiency scores do not exceed unity.

   When the values of the parameters are found, the efficiency scores using common weights are simply the ratios of the right hand side of (5) to its left hand side i.e. the ratio of the efficient or target input value to the actual input value. Notice that these "weights" are derived without any subjective choice and are based only on those DMUs which display some indication of good practice in their usage of the given resource. Note that the number of parameters must not exceed the number of non-dominated DMUs arising out of our first method otherwise they are not uniquely determined. In practice this is not much of a restriction by comparison to what is needed for DEA - Charnes and Cooper [2] state that, as a minimum, the number of DMUs should exceed three times the sum of the number of input and output measures. Although the notion of a common set of weights is perhaps foreign to users of DEA , it is likely that in some situations central managers will feel that conditions at each DMU are sufficiently similar for a common basis of comparison to be justifiable. Other methods for generating a common weight set were presented in Tofallis [4],[5]: The former presentation [4] employed a linear program in which the objective was to maximise the minimum efficiency over all the DMUs whereas [5] included an approach which maximised the sum of the efficiencies.

4   Test Results

By using data generated from a known production model it is possible not only to compare our methods with DEA but also to see how closely each method reproduces the true efficiency scores according to the production model. The data set is taken from Bowlin et al.[1] and is reproduced with permission in Table 1.
There are 15 hospitals (H1 to H15) each with three inputs: the number of full-time-equivalent staff, the number of hospital bed-days available in the year, and the expenditure on supplies (I_1 to I_3 respectively). The three outputs are the number of people receiving training at the hospital, the number of regular patients treated in the year, and the number of severe patients treated in the year (O_1 to O_3 respectively). The production model that was used is as follows :

$$\text{Staff:} \quad I_1 = 0.03\, O_1 + 0.004\, O_2 + 0.005\, O_3 \qquad (6)$$
$$\text{Bed-days:} \quad I_2 = (7/0.95)\, O_2 + (9/0.95)\, O_3 \qquad (7)$$
$$\text{Supplies:} \quad I_3 = 200\, O_1 + 20\, O_2 + 30\, O_3 \qquad (8)$$

By inserting values for the outputs into these equations one can find what the efficient input values should be. This was done for the first seven hospitals. Whereas for hospitals 8 to 15 the input levels were chosen to be larger than the efficient levels i.e. these hospitals were set up to be inefficient in at least one input. Dividing the efficient input value from the model by the actual input value one obtains the true efficiency score.
Table 2 compares the true results with those of the methods we are considering. All scores are of the form 'target input divided by actual or observed input'. The DEA results are based on an input minimisation model ; our DEA results are actually different from those in Bowlin et al [1], ours make DEA appear closer to the true scores (we suspect the difference may be due to using different lower bounds on the weights). All three methods obtained the correct scores (to two decimal places) for the first nine hospitals so those results are not shown. However DEA has incorrectly rated two of the inefficient hospitals (H10 and H13) as being efficient in all three inputs; in fact for hospital 10 DEA is out by as much as 31 and 18 percentage points for inputs 1 and 3 respectively. The individual weights method actually deduced precisely the correct score in most cases; in the few remaining cases it is still closer to the true result than DEA. Errors will always be in the direction of an overestimate with this method for the

same reason that this occurs in DEA , namely that an optimal set of weights is being found for that branch. However the amount of over-estimation will not be as great as for DEA because there are fewer weights to be manipulated.

TABLE 1. Data used as observations

	Inputs			Outputs		
	FTE staff	Bed days	Supply $000's	Teaching units	Regular patients	Severe patients
H1	23.5	41 050	130	50	3 000	2 000
H2	24.5	43 160	140	50	2 000	3 000
H3	26.0	43 160	150	100	2 000	3 000
H4	25.0	41 050	140	100	3 000	2 000
H5	28.5	50 530	160	50	3 000	3 000
H6	36.0	62 105	210	100	2 000	5 000
H7	51.5	92 630	270	50	10 000	2 000
H8	25.0	49 475	140	100	3 000	2 000
H9	24.5	43 160	165	50	2 000	3 000
H10	77.0	92 630	340	100	10 000	2 000
H11	44.5	65 260	265	50	5 000	3 000
H12	30.0	60 000	170	100	3 000	3 000
H13	43.5	81 110	245	50	4 000	5 000
H14	30.0	60 000	170	100	3 000	3 000
H15	26.5	47 370	160	50	3 000	2 000

Even more remarkable are the results from our second method. For inputs 1 and 2 the true scores were reproduced exactly for all 15 hospitals. For input 3 the results were always correct to at least two decimal places and the largest error was 0.004 . Hence, as these results were the same as the true scores (to at least two decimal places) there was no need to include another identical column in Table 2. Due to the less restrictive nature of our first method, it will provide scores which are at least as great as those of the method using common weights. If there were a large difference between the scores from our two methods for a given DMU it could be due to the weight flexibility of the first method disguising deficiencies in performance (one would check for near-zero weights). However it may be due to justifiable reasons relating to the particular circumstances at that DMU - further investigation would be called for.

From these comparisons it would seem that the approaches presented show considerable promise and are therefore worthy of further investigation.

TABLE 2. Comparison of efficiency estimates with true values

Hospital	Input	DEA efficiency estimate	True efficiency score *	Efficiency estimate (individual weights method)
10	1	1.0	0.69	0.74
10	2	1.0	1.00	1.00
10	3	1.0	0.82	0.90
11	1	0.89	0.82	0.82
11	2	1.00	1.00	1.00
11	3	0.83	0.75	0.76
12	1	1.00	1.00	1.00
12	2	0.84	0.84	0.84
12	3	1.00	1.00	1.00
13	1	1.0	0.98	1.00
13	2	1.0	0.95	0.95
13	3	1.0	0.98	1.00
14	1	1.00	1.00	1.00
14	2	0.84	0.84	0.84
14	3	1.00	1.00	1.00
15	1	0.89	0.89	0.89
15	2	0.87	0.87	0.87
15	3	0.81	0.81	0.81

*The scores according to the common weights method are the same as those in this column.

5 Further Developments

(i) When two (or more) inputs are known to act as substitutes for each other then it may be desirable to assess their efficiency jointly rather than separately. If the substitution rate is known then the value of the denominator in the objective function (1) and in the constraints can be replaced by the value of the aggregate input. If the substitution rate is not known then one would resort to conventional DEA but using only those two inputs in the formulation. The usual way that DEA is used (i.e. including all inputs in each L.P.) can now be viewed as a special case where all inputs can act as substitutes for each other.

(ii) We have assumed a linear relationship between each input and the outputs , this assumption could be relaxed to take account of factors such as economies of scale. Also, interactive terms (i.e. cross-products in the outputs) could be included to deal with economies of scope. Note that neither of these need involve solving nonlinear programs since the unknowns can appear linearly, the nonlinearity is in the input-output observations which appear as constant coefficients in the mathematical formulation.

(iii) In our method which used a common set of weights, the weights were derived by means of regression (using the non-dominated DMUs only) with the input under investigation being the dependent variable. This gave us a set of scores which in some contexts might be termed partial productivities. To get an overall performance measure based on a common set of weights we would need to have multiple dependent and multiple independent variables in our 'regression'. A computational technique for achieving this would involve the difference between the sum of the weighted inputs and the sum of the weighted outputs for each non-dominated DMU. By minimising the sum of squares of these differences one could obtain the weights using constrained regression, where the constraints would ensure non-negativity of the weights as well as scores not exceeding unity.

(iv) Whereas in the physical sciences and engineering the ratio output/input cannot exceed unity (law of conservation of energy), in a business context one expects to get more out than one puts in - particularly where money is concerned. One might therefore be justified in making this ratio exceed unity in the linear program constraints and have a minimization objective; this would generate weights which showed each DMU in the worst possible light and would thereby complement the results of this paper. Such a method might be called 'Anti-DEA'. (I am grateful to my project student, Jason Hanning , for suggesting this name.)

6 Conclusion

This paper has dealt with the efficiency of utilisation of *individual* resource inputs in organisations with multiple outputs. We have shown how these can be calculated using a simple adaptation of the widely explored DEA technique. For those managers who want an assessment based on a common standard we showed how this could be achieved using a combination of DEA and regression. This involved generating a best-practice model where the common set of weights is based only on the best-practice DMUs.

7 References

[1] Bowlin,W.F., A.Charnes , W.W.Cooper and H.D.Sherman (1985) "Data Envelopment Analysis and Regression Approaches to Efficiency Estimation and Evaluation", *Annals of O.R.*,2,113-138.

[2] Charnes,A.,and W.W.Cooper (1991) "Data Envelopment Analysis" in Operational Research '90 ed. H.E.Bradley ,Pergamon Press, N.Y.

[3] Kopp,R.J. (1981), "The Measurement of Productive Efficiency: A Reconsideration", Quarterly J.Econ.,96,No.3,477-504.

[4] Kumbhakar,S.C. (1988),"Estimation of Input-specific Technical and Allocative Inefficiency in Stochastic Frontier Models", Oxford Econ. Papers,40,535-549.

[5] Tofallis,C. (1992), "Evaluating Relative Efficiency: A Maximin Approach". O.R. Society Annual Conference , Birmingham, U.K. 1992.

[6] Tofallis,C. (1993), "Estimating the Efficiency with which Inputs are Utilised", University of Hertfordshire Business School Working Paper.

NIMBUS – Interactive Method for Nondifferentiable Multiobjective Optimization Problems

Kaisa Miettinen and Marko M. Mäkelä

Department of Mathematics, University of Jyväskylä, P.O. Box 35, FIN-40351 Jyväskylä, Finland

Abstract. An interactive method, NIMBUS, for nondifferentiable multiobjective optimization problems is introduced. We assume that every objective function is to be minimized. The idea of NIMBUS is that the decision maker can easily indicate what kind of improvements are desired and what kind of impairments are tolerable at the point considered.

At each iteration, the decision maker is asked to classify the objectives into up to five different classes: those to be improved, those to be improved down to some aspiration level, those to be accepted as they are, those to be impaired till some upper bound and those allowed to change freely. The aspiration levels and the upper bounds are asked from the decision maker. The decision maker can also attach weighting coefficients to the objective functions. According to this classification, a new (possibly multiobjective) optimization problem is formed.

An MPB (Multiobjective Proximal Bundle) method is employed to solve the new problem. The MPB method is a generalization of the Kiwiel's proximal bundle approach for nondifferentiable single objective optimization into the multiobjective case. The multiple objectives are treated individually without employing any scalarization. The method is capable of handling several nonconvex Lipschitz continuous objective functions subject to nonlinear (possibly nondifferentiable) constraints.

Keywords. Multiobjective optimization, interactive methods, nonsmooth optimization, bundle methods

1 Introduction

Many real world applications have several, often conflicting objectives. In addition, nondifferentiability and all kinds of discontinuities are characteristic of real world problems. For these reasons, it is important to create methods which are able to solve nondifferentiable (or nonsmooth) multiobjective (or multiple criteria, or vector) optimization problems.

In spite of the importance of the area, consideration of nondifferentiable multiobjective optimization problems is less frequent in the literature. One of the exceptions is a so-called subgradient GDF method which we have earlier described in [4]. Our aim here is to develop an interactive multiobjective optimization method for nondifferentiable problems with special interest in user friendliness.

2 Basic Concepts

A (nondifferentiable) multiobjective optimization problem to be solved is of the form

$$(2.1) \quad \begin{aligned} &\text{minimize} \quad \{f_1(\mathbf{x}), \ldots, f_k(\mathbf{x})\} \\ &\text{subject to} \quad \mathbf{g}(\mathbf{x}) = (g_1(\mathbf{x}), \ldots, g_m(\mathbf{x}))^T \leq \mathbf{0}, \end{aligned}$$

where $\mathbf{x} \in \mathbf{R}^n$ is a *decision variable vector* of n components, that is, $\mathbf{x} = (x_1, \ldots, x_n)^T$. We denote by $\mathbf{f}(\mathbf{x})$ a vector of *objective functions*, that is, $\mathbf{f}(\mathbf{x}) = (f_1(\mathbf{x}), \ldots, f_k(\mathbf{x}))^T$ and $k \ (\geq 2)$ is the number of the objective functions. Further, we have m constraint functions $g_j(\mathbf{x})$. The word "minimize" means that we want to minimize all the objective functions simultaneously.

We understand optimality in the sense of Pareto optimality and weak Pareto optimality. A feasible point $\mathbf{x}^*$ is *Pareto optimal* if there does not exist another feasible point $\mathbf{x}$ such that $f_i(\mathbf{x}) \leq f_i(\mathbf{x}^*)$ for all $i = 1, \ldots, k$ and $f_j(\mathbf{x}) < f_j(\mathbf{x}^*)$ for at least one f_j. A feasible point $\mathbf{x}^*$ is *weakly Pareto optimal* if there does not exist another feasible point $\mathbf{x}$ such that $f_i(\mathbf{x}) < f_i(\mathbf{x}^*)$ for all $i = 1, \ldots, k$.

Instead of continuous differentiability, we assume here that all the functions involved are locally Lipschitz continuous.

Definition 2.1. *A function $f_i \colon \mathbf{R}^n \to \mathbf{R}$ is locally Lipschitz continuous at a point $\mathbf{x}^* \in \mathbf{R}^n$ if there exist scalars $K > 0$ and $\delta > 0$ such that*

$$|f_i(\mathbf{x}^1) - f_i(\mathbf{x}^2)| \leq K \|\mathbf{x}^1 - \mathbf{x}^2\| \quad \text{for all} \quad \mathbf{x}^1, \mathbf{x}^2 \in B(\mathbf{x}^*, \delta),$$

where $B(\mathbf{x}^*, \delta)$ is an open ball with centre $\mathbf{x}^*$ and radius δ.

According to Clarke, we define a subdifferential (see [1]).

Definition 2.2. *Let the function $f_i \colon \mathbf{R}^n \to \mathbf{R}$ be locally Lipschitz continuous at a point $\mathbf{x}^* \in \mathbf{R}^n$. The set*

$$\partial f_i(\mathbf{x}^*) = \text{conv}\{\xi \in \mathbf{R}^n \mid \xi = \lim_{l \to \infty} \nabla f_i(\mathbf{x}^l); \ \mathbf{x}^l \to \mathbf{x}^*, \ \mathbf{x}^l \in \mathbf{R}^n \setminus \Omega_{f_i}\},$$

where conv denotes a convex hull and Ω_{f_i} denotes the set where f_i is not differentiable, is called a subdifferential of the function f_i evaluated at the point $\mathbf{x}^*$. In addition, the vectors $\xi \in \partial f_i(\mathbf{x}^*)$ are called subgradients.

Next, we state necessary optimality conditions for nondifferentiable multiobjective optimization problems. The proof can be found, for example, in [3].

Theorem 2.1. *Let the objective and the constraint functions of the problem (2.1) be locally Lipschitz continuous at a feasible point $\mathbf{x}^* \in \mathbf{R}^n$. A necessary condition for the point $\mathbf{x}^*$ to be a (weakly) Pareto optimal solution of the problem (2.1) is that there exist multipliers $0 \leq \boldsymbol{\lambda} \in \mathbf{R}^k$ and $0 \leq \boldsymbol{\mu} \in \mathbf{R}^m$ for which $(\boldsymbol{\lambda}, \boldsymbol{\mu}) \neq (0, 0)$ such that*

$$0 \in \sum_{i=1}^{k} \lambda_i \partial f_i(\mathbf{x}^*) + \sum_{j=1}^{m} \mu_j \partial g_j(\mathbf{x}^*)$$

$$\mu_j g_j(\mathbf{x}^*) = 0 \quad \text{for all} \quad j = 1, \ldots, m.$$

We say that a point is a *substationary point* if it satisfies the (necessary) optimality conditions stated in Theorem 2.1.

If we assume some kind of regularity of the constraints, we can modify the necessary conditions so that $\boldsymbol{\lambda} \neq 0$. If the objective and the constraint functions are convex, then, for example, a so-called Slater's constraint qualification can be used. A convex multiobjective optimization problem satisfies Slater's constraint qualification if there exists some point $\mathbf{x}$ such that $g_j(\mathbf{x}) < 0$ for all $j = 1, \ldots, m$.

3 NIMBUS Method

NIMBUS is based on the classification of the objective functions into up to five classes. At each iteration, the decision maker is asked to classify the objective functions and to specify appropriate information. Then, a new (multiobjective) optimization problem is formed and solved by a black-box optimization routine MPB.

The idea of the MPB routine is to move into a direction where the values of all the objective functions improve (i.e., decrease). The MPB routine is not based on any explicit scalarizing function. We do not here present the MPB routine. We settle for mentioning that it is an extension of the well-known proximal bundle method of single objective optimization into multiobjective optimization (as suggested in [2] and [6]). For details, see [5].

If all the objective and the constraint functions are convex and Slater's constraint qualification is satisfied, then the solution of the MPB routine is guaranteed to be weakly Pareto optimal. If the assumptions are not satisfied, then the solution of the MPB routine is a substationary point. For the proof of these background results, see, for example, [6].

We begin the description of the NIMBUS method itself by stating the assumptions.

1. All the objective and the constraint functions are locally Lipschitz continuous.
2. The feasible region $\{\mathbf{x} \in \mathbf{R}^n \mid \mathbf{g}(\mathbf{x}) \leq 0\}$ is convex.
3. Less is preferred to more to the decision maker.

The five function classes are the following: functions f_i whose values

- o should be decreased ($i \in I^<$),
- o should be decreased down till some aspiration level $\bar{z}_i$ ($i \in I^{\le}$),
- o are satisfactory at the moment ($i \in I^=$),
- o are allowed to increase up till some upper bound ε_i ($i \in I^>$), and
- o can change freely ($i \in I^{\diamond}$).

The difference between the classes $I^<$ and $I^{\le}$ is that the functions in the first class are to be minimized as far as possible, as to the functions in the latter class are to be minimized only till the aspiration level. Also positive weighting coefficients can be connected with the functions in these two classes.

Let $\mathbf{f}(\mathbf{x}^h)$ be the present criterion vector where the classification is performed. After the classification, we form a new problem

$$\text{minimize} \quad \left\{ f_i(\mathbf{x})/w_i \ (i \in I^<), \ \max_{j \in I^{\le}} \left[\max \left[f_j(\mathbf{x})/w_j - \bar{z}_j, \ 0 \right] \right] \right\}$$

(3.1) $\quad$ subject to $f_i(\mathbf{x}) \le f_i(\mathbf{x}^h), \quad i \in I^=$

$\qquad\qquad\qquad f_i(\mathbf{x}) \le \varepsilon_i, \qquad\quad i \in I^>$

$\qquad\qquad\qquad \mathbf{g}(\mathbf{x}) = (g_1(\mathbf{x}), \ldots, g_m(\mathbf{x}))^T \le \mathbf{0},$

where the weighting coefficients w_i, $i \in I^< \cup I^{\le}$, are positive and sum up to one, the aspiration levels must satisfy the condition $\bar{z}_i < f_i(\mathbf{x}^h)$ for $i \in I^{\le}$ and the upper bounds ε_i must satisfy the condition $\varepsilon_i > f_i(\mathbf{x}^h)$ for $i \in I^>$.

If the decision maker does not like the solution of the new problem, (s)he can examine intermediate points between the old and the new solution. It is important to notice that the classification of the functions and the specification of the appropriate parameter information does not necessarily have to succeed as well as in the traditional methods based on the classification. Because of the possibility of considering also intermediate solutions, more information of the problem can be obtained.

Now we can present a detailed NIMBUS algorithm.

3.1 NIMBUS Algorithm

The main interest has been in designing a flexible method where the decision maker can proceed in several different ways. For this reason, the NIMBUS algorithm may seem rather complicated at the first sight. However, the complexity of the algorithm can be hidden by designing a flexible user interface in the implementation phase.

For clarity, we refer to the solutions of the MPB routine as weakly Pareto optimal points, even though they may be substationary, as mentioned earlier.

(1) Choose a feasible starting point $\mathbf{x}^0$ and calculate its weakly Pareto optimal counterpart $\mathbf{x}^1$, employing MPB with $I^< = \{1, \ldots, k\}$. Set $h = 1$.

(2) Ask the decision maker to divide the objective functions into the classes $I^<$, $I^{\le}$, $I^=$, $I^>$, and $I^{\diamond}$ at the point $\mathbf{f}(\mathbf{x}^h)$ such that $I^> \cup I^{\diamond} \ne \emptyset$ and

$I^< \cup I^\leq \neq \emptyset$. If either of the unions is empty, goto step 9. Ask the aspiration levels $\bar{z}_i^h$ for $i \in I^\leq$ and the upper bounds ε_i^h for $i \in I^>$ from the decision maker. Ask also the possible positive weighting coefficients w_i^h for $i \in I^< \cup I^\leq$ summing up to one.

(3) Calculate $\hat{\mathbf{x}}^h$ by solving the problem (3.1) by the MPB routine. If $\hat{\mathbf{x}}^h = \mathbf{x}^h$, ask the decision maker whether (s)he wants to try another classification. If yes, set $\mathbf{x}^{h+1} = \mathbf{x}^h$, $h = h+1$ and go to step 2. If no, go to step 9.

(4) Present $\mathbf{f}(\mathbf{x}^h)$ and $\mathbf{f}(\hat{\mathbf{x}}^h)$ to the decision maker. If the decision maker wants to see different alternatives, set $\mathbf{d}^h = \hat{\mathbf{x}}^h - \mathbf{x}^h$ and go to step 6. If the decision maker prefers $\mathbf{f}(\mathbf{x}^h)$, set $\mathbf{x}^{h+1} = \mathbf{x}^h$, $h = h+1$, and to go step 2.

(5) Now the decision maker wants to continue from $\mathbf{f}(\hat{\mathbf{x}}^h)$. If $I^< \neq \emptyset$ set $\mathbf{x}^{h+1} = \hat{\mathbf{x}}^h$, $h = h+1$ and go to step 2. Otherwise $(I^< = \emptyset)$, the weak Pareto optimality must be checked, setting $I^< = \{1, \ldots, k\}$ and employing MPB. Let the solution be $\check{\mathbf{x}}^h$ and set $\mathbf{x}^{h+1} = \check{\mathbf{x}}^h$. Set $h = h+1$ and go to step 2.

(6) Calculate P different criterion vectors $\mathbf{f}(\mathbf{x}^h + t_j \mathbf{d}^h)$, where $t_j = \left(\frac{j-1}{P-1}\right)$, $j = 1, \ldots, P$.

(7) Produce weakly Pareto optimal solutions from the criterion vectors, employing MPB (with $I^< = \{1, \ldots, k\}$).

(8) Present P alternative criterion vectors to the decision maker and let her or him choose the most preferred one among them. Denote the corresponding decision variable by $\mathbf{x}^{h+1}$. Set $h = h+1$.

(9) Check the Pareto optimality of $\mathbf{x}^h$ by solving the problem (3.2). Let the solution be $(\tilde{\mathbf{x}}, \tilde{\boldsymbol{\gamma}})$. Stop. The final solution is $\tilde{\mathbf{x}}$.

In the last step, the Pareto optimality of the final solution is guaranteed by solving an additional problem

$$(3.2) \qquad \begin{aligned} &\text{maximize} && \sum_{i=1}^k \gamma_i \\ &\text{subject to} && f_i(\mathbf{x}) + \gamma_i \leq f_i(\mathbf{x}^h) \quad \text{for all} \quad i = 1, \ldots, k, \\ &&& \gamma_i \geq 0 \quad \text{for all} \quad i = 1, \ldots, k, \\ &&& \mathbf{g}(\mathbf{x}) = (g_1(\mathbf{x}), \ldots, g_m(\mathbf{x}))^T \leq 0 \end{aligned}$$

with respect to $(\mathbf{x}, \boldsymbol{\gamma})$, applying the MPB routine.

If the solution of the problem (3.2) is equal to $(\mathbf{x}^h, 0)$, then $\mathbf{x}^h$ is guaranteed to be Pareto optimal and it can be appointed a final solution. If $\mathbf{x}^h$ is not Pareto optimal, then the solution $\tilde{\mathbf{x}}$ is. As the decision maker was assumed to prefer less to more, we can presume that the decision maker is satisfied with $\tilde{\mathbf{x}}$ as a final solution.

In the second step of the algorithm, the solution process stops if the decision maker does not want to improve the value of any objective function or if the decision maker is not willing to let any of the objective function values increase.

Because of the structure of the MPB method, its solutions are quite sensitive with respect to the starting point. If the solution obtained is not satisfactory, the solution process can be begun again with some other starting point.

It has been proved in [3] that if in the step 3 of the NIMBUS algorithm the set $I^<$ is nonempty, then the solution of the problem (3.1) is weakly Pareto optimal to the original multiobjective optimization problem. This result is exploited in the step 5.

3.2 User Interface

Because the classification phase is the heart of the NIMBUS method, it is important that the decision maker feels comfortable with it. In Fig. 3.1, we present a snapshot of a user interface of a system implementing NIMBUS. This system is under development.

We assume that as the objective functions are inputted, a symbol is attached to each of them. In this way, we can handle only the symbols in the classification phase. The function descriptions can be seen in the Edit menu, if so desired. The objective function symbols can be seen in the topmost row of Fig. 3.1. Under the symbols are the objective function values to be considered at the present iteration.

The decision maker can perform the classification in different ways. With a mouse (s)he can select a symbol of an objective function in the topmost row, drag it and drop it into some of the five boxes underneath, one for each class. Another way is to employ the symbols of the function classes $(<, \leq, =, >, \diamond)$ and enter appropriate symbols to the third topmost row titled Classes.

The decision maker can also employ indirect classification by specifying aspiration levels, upper bounds and values to be fixed in the fourth row called New levels. This way cannot be used for functions to be minimized and functions whose values can change freely.

No matter which of these three ways is used in the classification, the information specified is updated automatically to all the other appropriate fields. Now the decision maker has alternative ways to end up with the same classification.

The weighting coefficients, the aspiration levels and the upper bounds can be specified with the help of graphical devices by moving bars. If the weighting coefficients are not fixed, they vary as some of them is changed so that their sum stays equal to one. The decision maker can also directly express numerical values.

The decision maker is free to select the order of specifying the classification and connected information. Either all the functions are first classified and then the parameter information is specified, or the information is given in proportion as the classification is carried out.

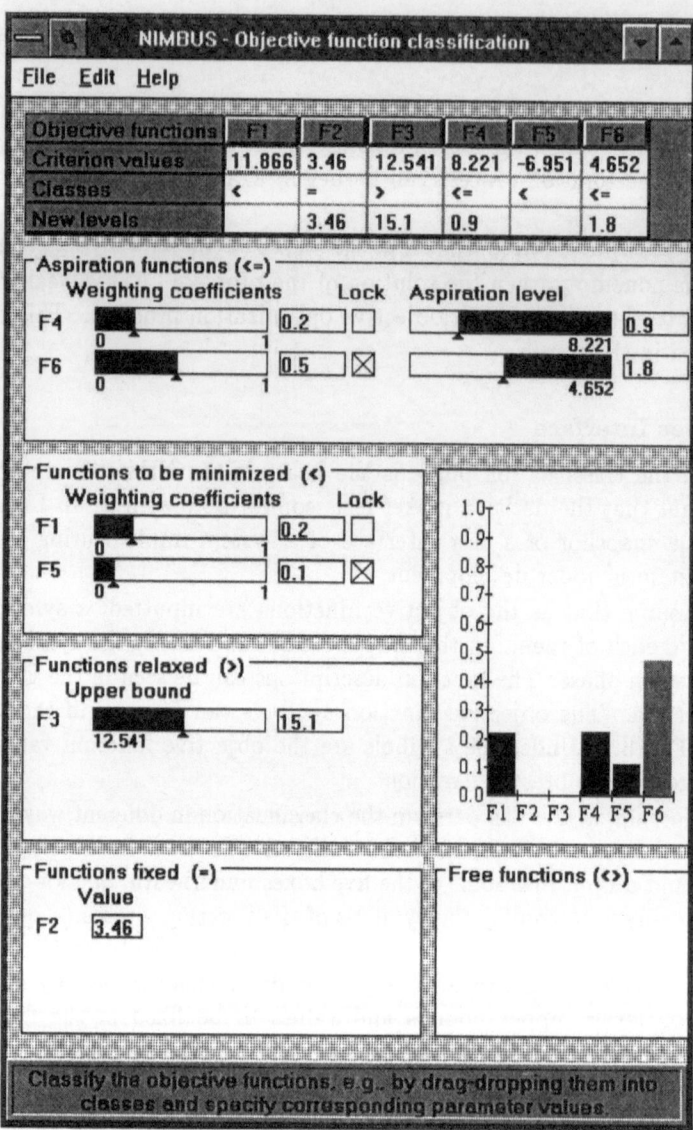

Fig. 3.1. User interface for the classification.

4 Conclusions

We have introduced an interactive multiobjective optimization method which is easy to use and which is able to handle nondifferentiable functions. The role of the decision maker is simple since (s)he only has to indicate what kind of changes (s)he would like to see in the values of the objective functions. The method aims at being flexible so that the decision maker is not forced to adjust oneself into one rigid manner but has different possibilities to proceed. One can select a way that suits one's personal characteristic best.

The decision maker is free to move around the weakly Pareto optimal set (or the set of substationary points) and examine it. (S)he can also change her or his mind if necessary. The decision maker can select to what extent (s)he exploits the versatile possibilities of the method.

The fact that the black-box MPB routine cannot guarantee the Pareto optimality but only substationarity of the solutions can be at least partly compensated by its computational efficiency. The final solution is projected to be Pareto optimal and the intermediate solutions can also be projected if so desired by paying the price of increased computational costs.

The method does not assume the existence of any underlying value function. Thus, no explicit convergence results can be put forward based on the assumptions on the properties of the function. The intention has particularly been to release the decision maker from the presumption of a value function.

The NIMBUS method satisfies two widely agreed desirable properties of interactive methods. It does not place too much demands on the decision maker and it is able to find substationary/weakly Pareto optimal solutions quickly.

It has been described in [3] how the NIMBUS method has been successfully applied to solve several complicated state-constrained nondifferentiable multiobjective optimization problems of optimal control.

References

1. F. H. Clarke, "Optimization and Nonsmooth Analysis," John Wiley & Sons, New York, 1983.
2. K. C. Kiwiel, *A Descent Method for Nonsmooth Convex Multiobjective Minimization*, Large Scale Systems **8** (1985), 119–129.
3. K. Miettinen, "On the Methodology of Multiobjective Optimization with Applications," Doctoral Thesis, University of Jyväskylä, Department of Mathematics, Report 60, 1994.
4. K. Miettinen and M. M. Mäkelä, *An Interactive Method for Nonsmooth Multiobjective Optimization with an Application to Optimal Control*, Optimization Methods and Software **2** (1993), 31–44.
5. M. M. Mäkelä, *Issues of Implementing a Fortran Subroutine Package NSOLIB for Nonsmooth Optimization*, University of Jyväskylä, Department of Mathematics, Laboratory of Scientific Computing, Report 5 (1993).
6. S. Wang, *Algorithms for Multiobjective and Nonsmooth Optimization*, in "Methods of Operations Research 58," Edited by P. Kleinschmidt, F. J. Radermacher, W. Schweitzer, H. Wildermann, Athenäum Verlag, Frankfurt am Main, 1989, 131–142.

A Graphic Search Based on Active Sets for Nonlinear Convex Multiple Objective Programming with Linear Constraints

Rafael Caballero Fernández, Lourdes Rey Borrego, Francisco Ruiz de la Rúa[1]

[1] Departamento de Economía Aplicada (Matemáticas). Facultad de Ciencias Económicas y Empresariales. Campus El Ejido s/n. 29071-Málaga. Spain.

Abstract. The aim of this paper is to present an algorithm which makes possible the determination or approximation of the efficient set for a problem of non-linear convex multiple objective programming, with linear constraints, based on the general idea of the active set methods in non-linear programming. Along the process, the efficient set of the problem is delimited, giving the sections of its contour placed on the boundary of the feasible set, and those which lie in the interior of X. To this end, we calculate the initial and final points of each boundary section, using comparative static techniques, and approximate the interior sections.

Although this algorithm is designed to solve problems with two variables, our aim in the future is to extend these results to the general case, even if this implies the impossibility to obtain a graphic representation of the efficient set.

1 Motivation of the algorithm

The idea of this algorithm is motivated by the structure of the efficient sets that we have observed in the non-linear problems under study. If, for example, we consider two general objective functions with two variables and definite minima, the unconstrained efficient set they determine is a curve in the plane that joins both unconstrained minima (Fig 1).

Under convexity conditions, the Kuhn-Tucker conditions for multiple objective programming (see Sawaragi, Nakayama and Tanino, 1985) assure that efficient points are those where the gradients of both functions go in the same direction, but in opposite sense, that is, they are characterised by the condition:

$$\nabla f_1(\mathbf{x}^*) = -\alpha \nabla f_2(\mathbf{x}^*), \quad \alpha > 0,$$

which is equivalent to $\mathbf{x}^*$ being the solution of the unconstrained weighting problem:

$$\min \ f_1(\mathbf{x}) + \alpha f_2(\mathbf{x}).$$

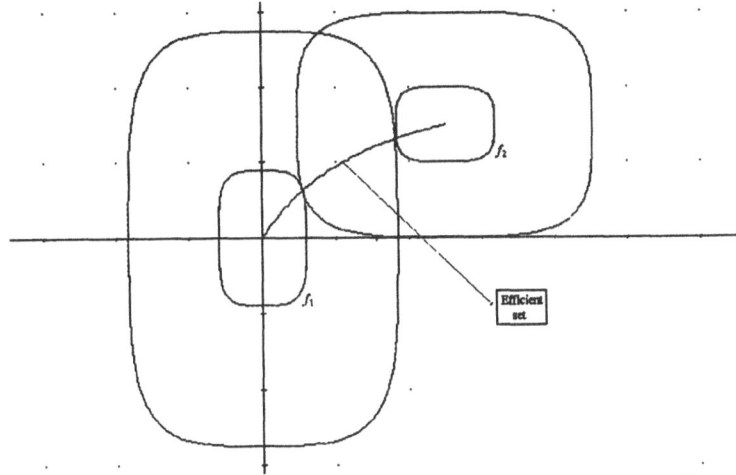

Fig. 1. Efficient set relative to two convex functions. The efficient curve consists of points where the lines of equal value of both functions are tangent.

Let us now suppose that we are working with three convex objective functions. The same conditions assure that in this case the properly efficient points are obtained as the solutions of unconstrained weighting problems of the form:

$$\min \ \mu_1 f_1(\mathbf{x}) + \mu_2 f_2(\mathbf{x}) + \mu_3 f_3(\mathbf{x}), \quad \mu_i > 0.$$

If we take these three functions two by two, each pair will yield the corresponding curve of efficient points relative to them, which joins the two corresponding minima (points which, in general, may not be properly efficient in the original problem, because they have been obtained with zero weights, but are at least weakly efficient).

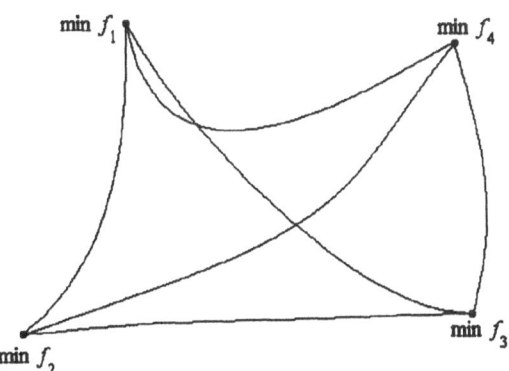

Fig 2. Each pair of functions yields the corresponding efficient curve. The efficient set relative to the four functions is completely surrounded by these curves.

If there are more than three objective functions, there can exist efficient curves associated to two determined functions which are not part of the boundary of the efficient set. Thus, for example, in figure 2, this happens with the curve corresponding to f_2 and f_4.

Let us now consider a linearly constrained problem of the form

$$(P)\begin{cases}\textit{Eff}\,(f_1(\mathbf{x}),\ldots,f_p(\mathbf{x}))\\s.t.\quad A\mathbf{x}\le\mathbf{b}\end{cases}$$

In this case, the efficient set is derived from the unconstrained problem as follows:
- The contour sections of the unconstrained problem which are inside the feasible set X, are still sections of the contour of the efficient set in the constrained problem.
- Those sections which are infeasible, are projected on the boundary of X, yielding a boundary section of the contour of the efficient set.

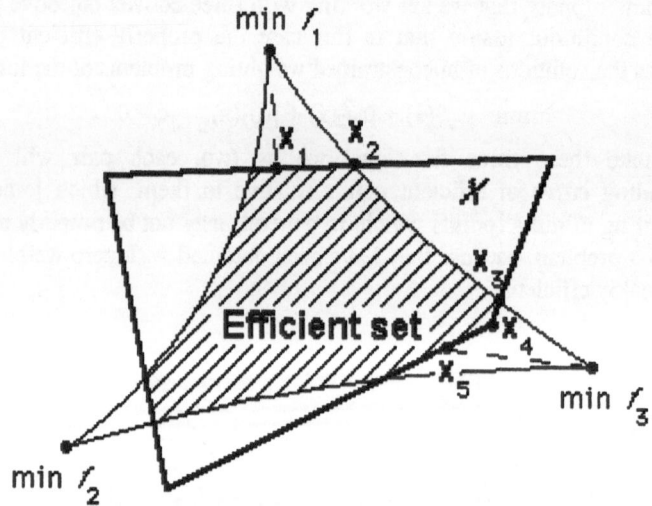

Fig. 3. Efficient set in a problem with three objectives and linear constraints. The curves are the efficient sets relative to each pair of functions. The infeasible sections of the unconstrained efficient set are projected on the boundary of the feasible set.

Taking into account the ideas mentioned above, it seems reasonable to think that the efficient set of the problem may be delimited by the efficient curves corresponding to each pair of objective functions. Ward (1989) confirms this fact. Namely, he proves that for any multi-objective problem with n variables, more than n convex objective functions and convex feasible set X, if each function f_i is continuous over X and E_i (its space of minimising solutions) is nonempty and bounded, then

$$W(P) = U(P, n) \cup S(P, n),$$

where $W(P)$ is the set of weakly efficient solutions for the problem (P), $U(P, n)$ the union of the efficient sets of all subproblems of (P) having n or fewer objectives and $S(P, n)$ the set of all $x \in X$ completely surrounded by $U(P, n)$.[1]

Based on this result, we have built an algorithm to obtain the contour of the efficient set in a convex multiple objective program, in the case where the number of variables is two, and the constraints are linear. Namely, we face the problem (P) with the following conditions:

- $n = 2$
- For each $i = 1, ..., p$, f_i is convex and twice continuously differentiable.
- The feasible set X determined by the linear constraints is closed and bounded.
- There is no redundant constraint, so that only two constraints can pass through each vertex of the feasible set.

Under these conditions, the above result by Ward assures that the contour of the efficient set is included in the union of the efficient sets corresponding to each pair of functions and thus can be obtained as results of weighting problems with two objective functions.

Let us now go back to figure 3, and suppose that we want to obtain the efficient set relative to the functions f_1 and f_3. The points, for example, x_2 and x_3 are just those where the corresponding constraint becomes active or inactive, so that the efficient curve passes from the boundary to the interior of the feasible set, or vice versa. Such points correspond to solutions of weighting problems associated to the two functions which are being considered, for a determined value of the parameter μ, where we suppose that the function to minimise is:

$$(1 - \mu)f_1 + \mu f_3.$$

Thus, the algorithm consists of detecting the extreme points of the efficient set that lie on the boundary of the feasible set, as well as approximating the sections placed in its interior. To this end, we are going to take, as previously mentioned, the functions in pairs, and to make the parameter μ advance from 0 to 1, finding these points where the constraints become active or inactive. To this end, we combine nonlinear programming active set algorithms, with parametric programming and comparative statics procedures. So, at any instant of the procedure, we will be considering only a subset of the objective functions, and a subset of constraints:

- The set of active functions, defined as the two functions which are currently being considered.
- The set of active constraints.

---

[1] See Ward (1989) for precise definitions of these concepts.

2 Development of the algorithm

2.1 Default ordination of the objective functions

Although we know that taking combinatorially all the functions in pairs and carrying out the process we will obtain the contour of the efficient set of the problem, we can also obtain points which are not part of this contour. In order to try to make the process quicker, we are going to consider a default ordination of the objective functions. We do this according to the relative position of their minima in the feasible set, in such a way that they are taken clockwise, as can be seen in figure 4 which corresponds to an imaginary constrained problem, where the resulting ordination is $\{f_4, f_2, f_1, f_3\}$. This ordination is carried out through the consideration of the middle point of the minima, ordering the angles of the resulting vectors:

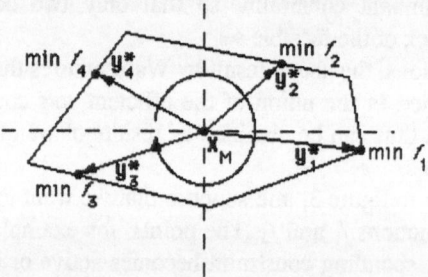

Fig 4. Ordination of the objective functions. The pentagon is the feasible set and x_M is the middle point of the four minima x_i^*. Then we take $y_i^* = x_i^* - x_M$, $i = 1,...,4$, and the angles that these vectors form with the imaginary horizontal axis are ordered clockwise.

This will be taken as the default ordination and will be kept in the algorithm as far as possible. It is only probable that following it (that is, taking only the pairs formed by two consecutive functions according to it) the whole contour of the efficient set will be obtained. During the process, tests are carried out to detect when this ordination has to be left. This will happen when two efficient curves intersect, and the points generated after this moment are in the interior of the efficient set. For the sake of simplicity, we will firstly present the algorithm under the supposition that this ordination can be kept during the whole process, indicating when the mentioned tests are carried out. Later on, we will describe how the process would continue from the moment that this ordination has to be left.

2.2 Initialising the process

We take f_1 and f_2 (already ordered) as the first two active functions. So, the first section of the contour of the efficient set is going to be built making μ grow from 0 to 1, and minimising the resulting functions $(1 - \mu)f_1 + \mu f_2$.

The point which is obtained for $\mu = 0$, x_0^* is thus the result of the problem:

$$\begin{cases} min\ f_1(\mathbf{x}) \\ s.t. \quad A\mathbf{x} \leq \mathbf{b} \end{cases}$$

This is going to be the starting point of process, from which we will begin to build the contour of the efficient set. It will also be the endpoint, because the last pair of functions will be f_n, f_1.

Starting from this point, we determine the first set of active constraint of the process ($J(x_0^*)$ is the corresponding index set). According to what has been previously mentioned, we have to consider three possible cases:

1. $J(x_0^*) = \{k\}$. There is only one active constraint.
2. $J(x_0^*) = \{k, l\}$. There are two active constraints, so x_0^* is a vertex of the feasible set.
3. $J(x_0^*) = \varnothing$. x_0^* is an interior point of the feasible set.

2.3 Boundary section. Main iteration

Let us first suppose that there is only one active constraint (say the k-th). Starting from the point x_0^*, we must generate now a boundary section placed in the constraint $a_k^t\mathbf{x} = b_k$ determined by the minima of the functions $(1 - \mu)f_1 + \mu f_2$ in the active constraint as μ is increased. Our objective will be to determine the endpoint of this segment. Thus, we define a boundary section as a section of the contour of the efficient set placed on a given active constraint, and generated by two given active functions, that is, it is obtained solving the problems

$$\begin{cases} min\ (1 - \mu)f_1(\mathbf{x}) + \mu f_2(\mathbf{x}) \\ s.t. \quad \mathbf{a}_k'\mathbf{x} = b_k \end{cases}$$

as the parameter μ increases. Let us consider the Lagrangian function associated to this problem:

$$\ell(\mathbf{x}, \lambda_k) = (1 - \mu)f_1(\mathbf{x}) + \mu f_2(\mathbf{x}) + \lambda_k(\mathbf{a}_k'\mathbf{x} - b_k).$$

For each value of μ, we can calculate the corresponding optimal solution $\mathbf{x}^*(\mu)$ of the above problem, and the optimal value $\lambda_k^*(\mu)$ of the multiplier. Thus, for example, $x_0^* = \mathbf{x}(0)$. As the k-th constraint is active for $\mu = 0$, then $\lambda_k^*(0) \geq 0$. In order to find the endpoint of the segment, we consider the following property:

Property 1. If $\mathbf{x}^*(\mu_1)$ is the endpoint of a boundary section relative to the active constraint $\mathbf{a}_k^t\mathbf{x} = b_k$ and to the active functions f_i and f_{i+1}, then it verifies one of the following conditions:

i. $\mu_1 = 1$.

ii. $\mathbf{x}^*(\mu_1)$ is a vertex of the feasible set.

iii. $\lambda_k^*(\mu_1) = 0$

Proof. As it can be deduced from the previously mentioned ideas, the endpoint of a boundary section may be reached under one of the following circumstances:

- If, as μ is increased, some constraint is violated before the efficient curve intersects the active constraint, then the corresponding vertex of the opportunity set is the endpoint of the boundary section.

  If not:

- If the corresponding unconstrained efficient curve relative to f_i and f_{i+1} lies outside the feasible set, then the process will reach the value $\mu = 1$, and the corresponding solution will be the endpoint.

- In the remaining case, the endpoint of the boundary section corresponds to a point where the unconstrained efficient curve intersects the active constraint, and enters the interior of X. For some value of μ_1, $\mathbf{x}^*(\mu_1)$ is the optimal solution of

$$\begin{cases} min \ (1 - \mu_1)f_i(\mathbf{x}) + \mu_1 f_{i+1}(\mathbf{x}) \\ s.t. \quad \mathbf{a}_k^t\mathbf{x} = b_k \end{cases}$$

and thus verifies:

$$(1 - \mu_1)\nabla f_i(\mathbf{x}^*(\mu_1)) + \mu_1\nabla f_{i+1}(\mathbf{x}^*(\mu_1)) + \lambda_k^*(\mu_1)\mathbf{a}_k = 0 \qquad [1]$$

$$\mathbf{a}_k^t\mathbf{x}^*(\mu_1) = b_k \qquad [2]$$

But, on the other hand, $\mathbf{x}^*(\mu_1)$ belongs to the unconstrained efficient curve, and thus, it is also the solution of the problem $min \ (1 - \mu_1)f_i(\mathbf{x}) + \mu_1 f_{i+1}(\mathbf{x})$, and verifies:

$$(1 - \mu_1)\nabla f_i(\mathbf{x}^*(\mu_1)) + \mu_1\nabla f_{i+1}(\mathbf{x}^*(\mu_1)) = 0 \qquad [3]$$

As $\mathbf{a}_k \neq 0$, then relations [1], [2] and [3] imply $\lambda_k^*(\mu_1) = 0$.$\square$

Following the previous theorem, we want to approximate the value of μ for which the k-th constraint becomes inactive. That is why we want to express the optimal multiplier λ_k^* associated to this constraint in each problem as a function of μ: $\lambda_k^* = \lambda_k^*(\mu)$. So in this case we are looking for the value of μ_1 that verifies $\lambda_k^*(\mu_1) = 0$. It must be observed that, as the k-th constraint is active in $\mathbf{x}^*(0)$, it is verified that $\lambda_k^*(0) \geq 0$, for the unconstrained minimum violates the constraint.

To estimate this value of μ_1, we are going to carry out an iterative procedure, which is going to generate a sequence of values for this parameter which is expected to converge to the value of μ that annihilates the multiplier. In this process, we are going to use techniques of comparative statics on non-linear programming problems to obtain the derivative of the function $\lambda_k(\mu)$.

Let us suppose given a generic value μ_0. Under the convexity hypothesis that we have supposed in the problem, for any value μ, the corresponding optimal solution $x^*(\mu)$ is a critical point of ℓ, that is, it verifies:

$$\nabla \ell(x_1^*(\mu), x_2^*(\mu), \lambda_k^*(\mu)) = 0$$

Making use of the implicit function theorem, in the point μ_0, these three relations in the variables $x_1^*(\mu)$, $x_2^*(\mu)$, $\lambda_k^*(\mu)$, and μ lets us obtain the numerical value of the derivative $d\lambda_k^*(\mu_0)/d\mu$. Then,

- If $d\lambda_k^*(\mu_0)/d\mu. > 0$, then $\lambda_k^*(\mu)$ increases from μ_0. In this case, we can obtain a first estimation $\mu_{est} = 1$.
- If $d\lambda_k^*(\mu_0)/d\mu. < 0$, then, making use of the Taylor approximation up to the first order, we obtain the estimate:

$$\mu_{est} = min\left\{ \mu_0 - \frac{\lambda_k^*(\mu_0)}{d\lambda_k^*(\mu_0)/d\mu}, 1 \right\}$$

For each value of μ_{est}, we can repeat the process and obtain the corresponding solution $x^*(\mu_{est})$, the optimal multiplier $\lambda_k^*(\mu_{est})$ and the derivative $d\lambda_k^*(\mu_{est})/d\mu$. Therefore, we can keep on iterating until a value of μ_1 is reached for which the corresponding optimal multiplier is small enough, that is, it verifies $|\lambda_k^*(\mu_1)| < \varepsilon$, for a previously fixed value of ε. In the case when the value $\mu_1 = 1$ is taken, it suffices that $\lambda_k^*(1) \geq 0$.

It can be observed that this process is in essence the Newton method[2] applied to find a solution of the equation $\lambda_k^*(\mu) = 0$, provided that its existence is obtained globally for every value of $\mu \in [\mu_0, 1]$. If not, the function can have occasional discontinuities, which do not affect the validity of the process.

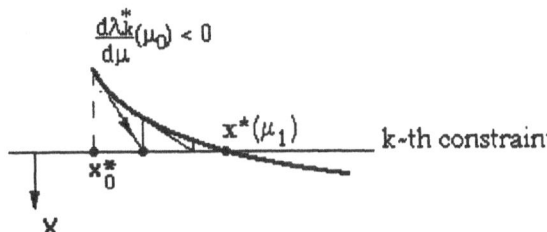

Fig. 5. Graphical interpretation of the iterative procedure. The straight line is the active constraint, and the curve corresponds to the unconstrained efficient set relative to the two active functions.

After this iterative procedure, a value μ_1 is obtained for the parameter μ, where one of the functions becomes inactive, or the current active constraint becomes

---

[2] See Gill et al.(1981), pg. 84

inactive. Nevertheless, before accepting this estimation as a good one, we have to carry out two tests on the solution $\mathbf{x}^*(\mu_1)$:

- A *feasibility test*, to check that it does not violate other constraints. If $\mathbf{x}^*(\mu_1)$ is feasible, and $\mu_1 = 1$, then the active function set is changed (the function f_i becomes inactive and f_{i+1} becomes active), and another boundary section in the same active constraint is generated. If $\mu_1 < 1$, then we have to carry out:
- An *efficiency test*, to check whether the default ordination can be kept or it has to be changed. If it can be kept, then the active constraint becomes inactive, and an interior section follows.

Next, we describe these two tests and how the process continues if they fail.

2.4 Feasibility test

If, $\mathbf{x}^*(\mu_1)$ violates some constraint, the value μ_1 is not admissible, and thus the boundary section ends for a previous value of μ. In this case, we have to find the constraint that was first violated as μ increased from μ_0 (if there is more than one). Let us suppose that this constraint is the one with index j. After that, solving the problem

$$\begin{cases} min \ (1-\mu_1)f_i(\mathbf{x}) + \mu_1 f_{i+1}(\mathbf{x}) \\ s.t. \quad \mathbf{a}'_k \mathbf{x} = b_k \\ \qquad \mathbf{a}'_j \mathbf{x} = b_j \end{cases}$$

the point $\mathbf{x}_1^*$ is obtained, which corresponds to a vertex of X and also of the efficient set, and is the endpoint of the boundary section. The optimal multiplier $\lambda_j^*(\mu_1)$ is also obtained. Following an iterative procedure analogue to that described in 2.3, we can obtain a value $\mu^{(1)}$ for which $\lambda_j^*(\mu^{(1)}) = 0$, that is, the value for which the j-th constraint becomes active. If $\lambda_k^*(\mu^{(1)}) > 0$, then the procedure is again carried out to find a new value $\mu^{(2)}$ for which $\lambda_k^*(\mu^{(2)}) = 0$, that is, the value for which the k-th constraint becomes inactive.

Therefore, two values are obtained, $\mu^{(1)}$ and $\mu^{(2)}$ such that for every value of μ lying between them, the corresponding solution with the two current active functions is always $\mathbf{x}_1^*$, but which define the interval for μ between the activation of the j-th constraint, and the deactivation of the k-th one. Starting with the value $\mu^{(2)}$ another boundary section is generated, on the j-th constraint (see fig. 6).

If the initial point of the procedure is a vertex of the opportunity set, then the initial value $\mu_0 = 0$ is between the values $\mu^{(1)}$ and $\mu^{(2)}$ previously described. So, it is enough to determine which constraint has to become inactive, calculate the value $\mu^{(2)}$, and continue the algorithm with a boundary section on the j-th constraint.

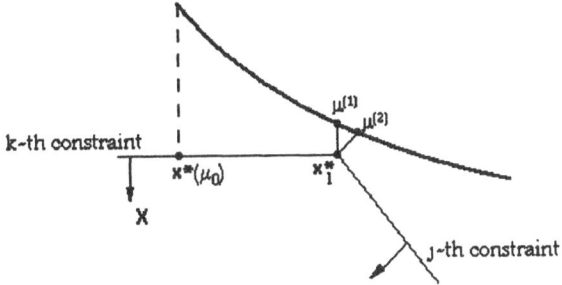

Fig. 6. Change of active constraint. The straight lines are the constraints and the curve is the unconstrained efficient set relative to the two active functions. The j-th constraint becomes active for $\mu = \mu^{(1)}$, and the k-th constraint becomes inactive for $\mu = \mu^{(2)}$. x_1^* is the optimal solution corresponding to both values of μ.

2.5 Efficiency test

Before leaving the active constraint and entering the interior of the feasible set, it is necessary to verify that the boundary section does not have to be extended. This can happen when the intersection of two unconstrained efficient curves causes the existence of sections of the contour of the efficient set determined by pairs of objective functions which are not contiguous with respect to the default ordination.

In this case, an efficiency test is carried out on a point placed on the active constraint immediately after its foreseeable deactivation:

$$x = x^*(\mu_1) + \varepsilon(\ x^*(\mu_1) - x^*(\mu_0)), \quad \varepsilon > 0.^3$$

If the point x is not efficient, then the test is positive, and an interior section follows (see 2.6). If, on the other hand, x is efficient, then the boundary section has to be extended, with a new segment corresponding to another pair of active functions:

Taking the functions in pairs, and carrying out efficiency tests on x with each pair, a new active set of functions is found, $\{f_k, f_i\}$ which generates this new boundary section, which is again a segment on the same active constraint, whose initial point is $x^*(\mu_1)$. The value of μ corresponding to this point and to the new set of active functions is derived from the relations

$$(1 - \mu)\nabla f_k(x^*(\mu_1)) + \mu\nabla f_i(x^*(\mu_1)) + \lambda_k^* a_k = 0.$$

---

[3] The efficiency test we carry out is based on the result by Sawaragi, Nakayama and Tanino (1985) which characterises the efficient points as solutions of several constraint problems.

Thus, a new boundary section follows, according to 2.3, with the same active constraint, the new active functions, and this value of μ.

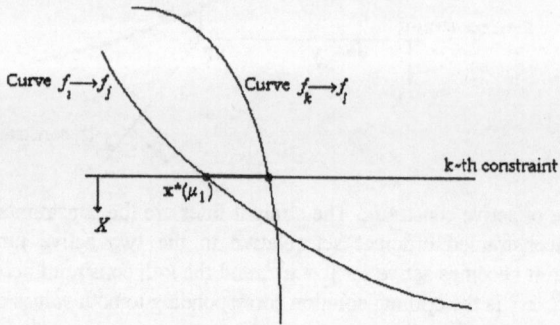

Fig. 7. The efficient curve relative to the functions f_k and f_l originates an extension of the boundary section obtained with the active functions f_i and f_j.

As it can be seen, when this situation takes place, the default ordination has to be left. Therefore, after considering the pair of active functions $\{f_k, f_l\}$, it is not clear which new pair of functions would follow. In section 2.8 we discuss how to keep on the procedure in this case.

2.6 Interior section

If the initial active set of constraints is empty, or if an empty active set is reached along the generating process, we have to determine a section of the boundary of the efficient set which lies in the interior of the feasible set X. Let us suppose that μ_0 is the value for which this happens, and that f_i and f_{i+1} are the current active functions. These interior sections coincide with the unconstrained efficient sets relative to the active functions. Thus, in this case, these sections are not in general straight lines, and so the procedure followed in the boundary sections, consisting of giving the endpoints of the segment is not valid now. To approximate the interior section in the best possible way, μ is increased in steps of previously fixed length, δ, and each of the corresponding unconstrained problem is solved. Therefore, a series of points that lie on the curve is obtained. Similarly to the case of the boundary sections, in each of these points, we have to carry out two tests:

- A *feasibility test*, to detect the value of μ for which the curve leaves the interior of X.
- An *efficiency test* to check whether the default ordination can be kept.
  More precisely, the procedure is as follows:

Given μ_0, we update its value: $\mu_0 = \mu_0 + \delta$, and we solve the unconstrained problem

$$min \ (1 - \mu_0)f_i(\mathbf{x}) + \mu_0 f_{i+1}(\mathbf{x}).$$

After that, a feasibility test is carried out on its solution $\mathbf{x}^*(\mu_0)$. If it is feasible, the process is repeated increasing μ again. If not, the first violated constraint is localised, let us suppose that it is the k-th constraint, and we solve the problem corresponding to μ_0 and with the k-th active constraint:

$$\begin{cases} min \ (1- \mu_0)f_i(\mathbf{x})+ \mu_0 f_{i+1}(\mathbf{x}) \\ s.t. \quad \mathbf{a}_k'\mathbf{x} = b_k \end{cases}$$

Following the iterative procedure described in 2.3, a value μ_1 is calculated such that $\lambda_k^*(\mu_1) = 0$. Starting with this value, a new boundary section on the k-th active constraint is obtained.

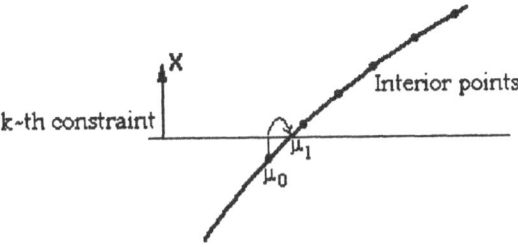

Fig. 8. Interior section. μ is increased in steps of fixed length, and the corresponding optimal solutions constitute an approximation of the interior section. When an infeasible point is reached, the value μ_1 is calculated for which the corresponding constraint becomes active.

If the current iteration $\mathbf{x}^*(\mu_0)$ is feasible, an efficiency test has to be carried out before accepting it. Namely, if the efficient curve that is being considered has previously intersected another one, it is possible that $\mathbf{x}^*(\mu_0)$ belongs to the interior of the efficient set. Therefore, an efficiency set is carried out on a series of points which belong to a neighbourhood of $\mathbf{x}^*(\mu_0)$[4]. If any of them is not efficient, then the test is positive, and the interior section procedure continues. If, on the other hand, they are all efficient, then $\mathbf{x}^*(\mu_0)$ belongs to the interior of the efficient set.

By bisections, the intersection point between the two efficient curves, $\mathbf{x}^*$, is calculated. Then, the new pair of active functions, $\{f_k, f_l\}$ and a new value of μ are derived as described in 2.5, and a new interior section follows.

Again, the default ordination has to be left, and the algorithm continues as described in section 2.8.

---

[4] In our algorithm, this is carried out considering short steps from $\mathbf{x}^*(\mu_0)$ in a fixed number of directions which surround the current point.

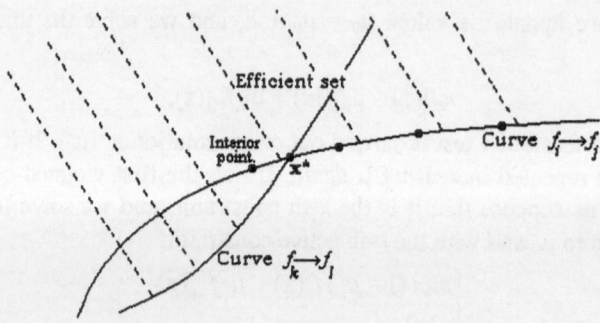

Fig. 9. A point which belongs to the interior of the efficient set is obtained due to the intersection between two efficient curves. In this case, the point **x*** has to be calculated, and the interior section procedure continues with the new set of active functions $\{f_k, f_l\}$.

If along the approximation of the interior section, a value $\mu_0 \geq 1$ is reached, then we take $\mu_0 = 1$ and (if both tests are positive) the active set of functions is updated, leaving f_i and taking f_{i+2}. The interior section procedure continues then with this new active set, from the value $\mu_0 = 0$.

2.7 End of the algorithm

If the default ordination is kept during the whole procedure, the last sections in the generation of the boundary of the efficient set are obtained with the active functions f_p and f_1 (when f_{p-1} becomes inactive, f_1 becomes active again). When the value $\mu = 1$ is reached, we will obtain again the first point of the procedure, and the algorithm ends.

2.8 Leaving the default ordination

If, at any time during the algorithm, the default ordination has to be left, then when the procedure arrives to the value $\mu = 1$, it is not clear which new objective function is to enter the active set.

Nevertheless, it is necessary to point out that the algorithm can work as well taking all the functions combinatorially in pairs, and obtaining the sections corresponding to each pair. A posterior graphical analysis can let us decide which sections belong to the contour of the efficient set.

Therefore, once the default ordination has been abandoned, the algorithm continues in the following way: whenever the value $\mu = 1$ is reached, the function that remains in the active set is combined with all the ones that have not previously shared an active set with it. This branch procedure finishes when all the possible pairs have been examined. We must take into account that the functions which have belonged twice to the active set before leaving the default ordination need not be considered again. For each given pair of functions, the algorithm works exactly as described above, except for the efficiency tests which are not carried out any longer.

This scheme originates an increase in the computing time of the process, and this is why the default ordination is kept as long as it is possible.

3 Example

Next, we show the results obtained when applying the algorithm to a non-linear problem with two variables:

$$
\begin{cases}
Eff\left((x-1)^4 + y^4, \dfrac{1}{x+y}, (x+1)^4 + 50(y-10)^4, 100(x-10)^4 + (y-4)^4\right) \\
s.t. \quad x \geq 0 \qquad\quad (1) \\
\qquad\quad y \geq 0 \qquad\quad (2) \\
\qquad\quad y \leq 8 \qquad\quad (3) \\
\qquad\quad x+y \leq 3 \qquad (4) \\
\qquad\quad -2x+5y \leq 30 \quad (5) \\
\qquad\quad 8x-3y \leq 56 \quad (6)
\end{cases}
$$

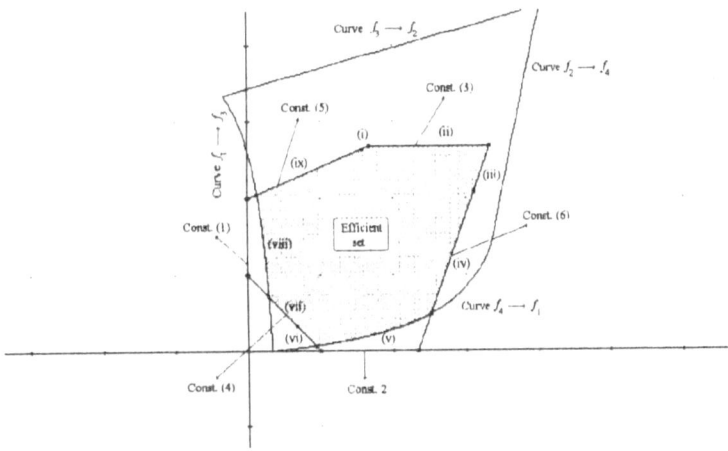

Fig. 10. Each mark on the axis corresponds to 3 units. The hexagon is the feasible set, and the curves are unconstrained efficient sets relative to pairs of functions. The different sections of the contour of the efficient set that are obtained in the algorithm are numbered from (i) to (ix).

After obtaining the minimum of each function in the feasible set, the algorithm puts them in order, with the result in this case: $\{f_3, f_2, f_4, f_1\}$. Figure 10 shows the feasible set, as well as the unconstrained efficient curves which are obtained taking the functions in pairs, according to the above ordination.

Applying the method to this problem, we obtain the following sections of the contour of the efficient set:

i. Active functions: f_3, f_2. Boundary section on the (5)-th constraint, from point (4.726, 7.89) ($\mu = 0$) to point (5, 8) (from $\mu = 0.025$ to $\mu = 0.11$)

ii. Constraint (5) becomes inactive, and (3) becomes active. Boundary section on the third constraint from point (5, 8) ($\mu = 0.11$) to point (10, 8) ($\mu = 1$).

iii. Constraint (3) becomes inactive, and (6) becomes active. At the same time, function f_3 becomes inactive and f_4 becomes active. Boundary section on the (6)-th constraint from point (10, 8) ($\mu = 0$) to (9.335, 6.226) ($\mu = 1$).

iv. Function f_2 becomes inactive and function f_1 becomes active. Boundary section on the (6)-th constraint from point (9.335, 6.226) ($\mu = 0$) to point (7.548, 1.461) ($\mu = 0.12$).

v. Constraint (6) becomes inactive. Interior section up to point (2.796, 0.204) ($\mu = 0.99$).

vi. Constraint (4) becomes active. Boundary section on the (4)-th constraint from point (2.796, 0.204) ($\mu = 0.99$) to point (2, 1) ($\mu = 1$).

vii. Function f_4 becomes inactive and f_3 becomes active. Boundary section on the (4)-th constraint from point (2, 1) ($\mu = 0$) to point (0.861, 2.139) ($\mu = 0.012$)

viii. Constraint (4) becomes inactive. Interior section up to point (0.394, 6.157) ($\mu = 0.72$)

ix. Constraint (5) becomes active. Boundary section on the (5)-th constraint from point (0.394, 6.157) ($\mu = 0.72$) to point (4.726, 7.89) ($\mu = 1$). END OF PROCEDURE.

4 Final remarks

The algorithm has been implemented in FORTRAN language, on a VAX computer, and with the aid of NAG subroutine library. In relation with its performance, we can point out the following two features:

• The average computing C.P.U. time on a series of test problems with a number of functions that ranges between 3 and 6 is about 1 minute. For instance, in the problem solved in section 3, the overall C.P.U. time of the process, with a step of 0.005 for the interior sections, has been 52.24 seconds, so the process can be considered as quick.

• The computational burden of evaluating the derivative of the multiplier with respect to the parameter μ is negligible, for it just amounts a 0'33% of the overall time of the process.

The weights have been normalised, so that the solutions obtained can be related to them, and we avoid bias problems towards certain functions.

Finally, it must be pointed out that this algorithm is just intended for problems with two variables. Its extension to more dimensions is not possible with the same scheme that has been used here. Nevertheless, the authors intend to undertake an extension of this active set approach to higher dimensions, which, of course, requires substantial modifications.

Acknowledgements. The authors wish to express their gratitude to the anonymous referee for his many helpful and valuable suggestions.

References

Fletcher, R. (1987). *Practical Methods of Optimization. Second edition.* J. Wiley & Sons. New York.

Gill, P.E., Murray, W. and Wright, M.H. (1981). *Practical Optimization.* Academic Press. London.

Guddat, J., Guerra Vasquez, F. and Jongen, H.T. (1990). *Parametric Optimization: Singularities, Pathfollowing and Jumps.* J. Wiley & sons. Chichester.

Lowe, T.J., Thisse, J.-F., Ward, J.E., Wendell, R.E. (1984). On efficient solutions to Multiple Objective Mathematical Programs. *Management Science* **30**, pp. 1346-1349

N.A.G. (Numerical Algorithms Group limited, 1991), *The NAG Fortran Library Introductory Guide, Mark 15.*

Sawaragi, Y., Nakayama, H., and Tanino, T. (1985). *Theory of Multiobjective Optimization.* Academic Press. Orlando.

Ward, J. (1989). Structure of efficient sets for convex objectives. *Mathematics of Operations Research* **14 (2)**, pp. 249-257.

A Sequential Network-Based Approach for the Multiobjective Network Flow Problem with Preemptive Priorities

Herminia I. Calvete and Pedro M. Mateo

Dpto. de Métodos Estadísticos, Edificio de Matemáticas, Universidad de Zaragoza, 50009 Zaragoza, Spain

Abstract. This paper presents a sequential approach for the problem of getting an optimal solution to the multiobjective network flow problem with preemptive priorities. This approach considers a priority every time, while maintaining through the sequence the network structure. This allows to use in each iteration of the process any efficient network–based algorithm for the single objective minimum cost flow problem.

Keywords. Multiobjective, network flow optimization, preemptive priorities

1 Introduction

Network flow optimization has been widely used as a tool for modeling real systems. One fact to mention is that models on networks are usually easier to understand by non specialized decision makers. Besides this, probably the most important feature is that algorithms for solving single objective minimum cost flow (SOMCF) problems show, in general, a better computation performance, due to the special structure of the coefficient matrix [4, 6, 7]. Some of these algorithms has been adapted from general purpose ones, but others have been specially designed. Hopefully, all those properties mentioned above should remain when dealing with the multiobjective network flow (MONF) problem.

Given a directed network $G = (\mathcal{N}, \mathcal{A})$, with set of nodes $\mathcal{N} = \{1, \ldots, n\}$ and set of arcs $\mathcal{A} = \{(i,j) : i, j \in \mathcal{N}\}$, denoting by f_{ij} the flow of the arc (i,j), and by f the vector with components $(f_{ij})_{(i,j)\in\mathcal{A}}$, the MONF problem can be stated as follows:

[P1]: *general multiobjective formulation of the network flow problem*

$$\min Z_1(f) = \sum_{(i,j)\in\mathcal{A}} c_{ij}^1 f_{ij}$$

$$\ldots$$

$$\min Z_k(f) = \sum_{(i,j)\in\mathcal{A}} c_{ij}^k f_{ij}$$

subject to:

$$\sum_{\{j:(i,j)\in\mathcal{A}\}} f_{ij} - \sum_{\{j:(j,i)\in\mathcal{A}\}} f_{ji} = 0, \quad \forall i \in \mathcal{N} \tag{1}$$

$$0 \le l_{ij} \le f_{ij} \le u_{ij}, \quad \forall (i,j) \in \mathcal{A} \tag{2}$$

where u_{ij} = upper bound on flow through arc (i,j)

l_{ij} = lower bound on flow through arc (i,j)

c_{ij}^h = unit flowing cost from node i to node j through arc (i,j) in the hth objective function, $h = 1, \ldots, k$.

We will refer to the constraints (1) and (2) as the conservation of flow constraints, and the capacity constraints, respectively. For notational convenience in what follows we will assume that there is at most one arc associated with each ordered pair of nodes (i,j).

In general, there is not a single f which minimizes each objective individually. Usually, trade–off decisions have to be made in order to get the most preferred solution. Since a MONF problem is a special case of a multiple objective linear programming (MOLP) problem, it could be solved by any existing method for MOLP problems [9]. These methods are usually computationally burden, specially in large scale problems. This fact induces to take advantage of the properties of network structure by developing specialized methods for solving MONF problems.

Klingman and Mote [5] have developed a network specialization of the primal simplex multicriteria algorithm, first developed by Yu and Zeleny [10] to generate all nondominated extreme solutions for a general multiobjective linear problem. The specialized algorithm takes advantage of the special basis structure of the network flow problems. Pulat, Huarng and Lee [8] have developed a methodology to generate the set of efficient solutions to a bicriteria network flow problem.

Calvete and Mateo [2] have proposed an approach for the MONF problem using preemptive priorities that are assigned to the objectives in accordance with an order previously established by the decision maker. This approach is most useful when the decision maker is specifically interested in minimizing one objective every time, using the lesser important objectives only to break ties. In that model, it is possible to take advantage of the properties of network flow problems, and to use a specialized version of complementary slackness conditions, in order to develop preemptive multiobjective network specializations of out–of–kilter and primal–dual algorithms, which are extensively used for SOMCF problems.

In this paper a sequential approach for solving the MONF problem with preemptive priorities is developed. After minimizing with respect to the highest priority objective, using any efficient network–based algorithm for the SOMCF, some arcs are *blocked* (i.e. their flow value is fixed) by adding constraints that maintain the network structure. Then, the new SOMCF problem is solved for the second priority objective, and so on. This process is continued until all objectives have been minimized or all arcs have been blocked.

To make the present paper self–contained, we will briefly describe in section 2 the MONF problem with preemptive priorities and discuss the use of complementary slackness conditions to develop algorithms finding an optimal solution. We also give an outline of the algorithms proposed in [2]. Section 3 contains a general description of the sequential approach proposed to solve the problem. It also illustrates the use of two different ways of leaving an arc *blocked* depending on the particular algorithm used to solve in each iteration the SOMCF problem. Section 4 shows the computational performance of the sequential approach, compared with the algorithms developed in [2] and includes a numerical example. Section 5 contains conclusions.

2 Setting the Problem

Given the problem [P1], we will assume that the decision maker is able to rank the objectives $Z_1(f), \ldots, Z_k(f)$ of [P1] according to their importance. Let $Z_1(f)$ be the most important objective, $Z_2(f)$ the second one, and so on. The purpose of the decision maker is in the first place to minimize the highest–level objective; for all alternative optimal solutions to this problem, he/she wants to find those which minimize the second level objective, and so on. Therefore, we propose to assign to the objectives $Z_1(f), \ldots, Z_k(f)$ priority factors $P_1 \gg P_2 \gg \ldots \gg P_k \equiv 1$. In this situation, the lexicographical optimization of the objectives can be stated as follows:

[P2]: *preemptive priority formulation*

$$\min \; Z(f) = \sum_{(i,j) \in \mathcal{A}} \sum_{h=1}^{k} P_h c_{ij}^h f_{ij}$$

subject to: (1), (2).

We will assume, as it is usual in network optimization with a single objective, that bounds on arcs and cost coefficients are integers. We can also select without loss of generality integer priority factors. Under these assumptions, i. e. if problem data are integer, an optimal solution to the multiobjective network flow problem with preemptive priorities can be taken to be integer.

For the SOMCF problem it is a well–known result that if solutions to the primal and dual problems are given, they are optimal for their respective problems if and only if both solutions are feasible and satisfy the following complementary slackness conditions:

$$
\begin{array}{lll}
f_{ij} = l_{ij} & \forall \, (i,j) \text{ such that} & \sum_{h=1}^{k} P_h c_{ij}^h - \pi_i + \pi_j > 0 \\
l_{ij} \leq f_{ij} \leq u_{ij} & \forall \, (i,j) \text{ such that} & \sum_{h=1}^{k} P_h c_{ij}^h - \pi_i + \pi_j = 0 \\
f_{ij} = u_{ij} & \forall \, (i,j) \text{ such that} & \sum_{h=1}^{k} P_h c_{ij}^h - \pi_i + \pi_j < 0
\end{array}
$$

where the π variables are associated with the conservation of flow constraints of the primal problem. In order to circumvent the problem of choosing and managing numerical values of P_h that should be large enough to assure that the minimum of $Z(\boldsymbol{f})$ is equivalent to the lexicographical minimum of $Z_1(\boldsymbol{f}), \ldots, Z_k(\boldsymbol{f})$, we propose to use the following equivalent formulation of the conditions for optimality, based on the lexicographical character of vectors $\tilde{c}_{ij}$:

$$
\begin{aligned}
f_{ij} &= l_{ij} && \forall\,(i,j) \text{ such that } \tilde{c}_{ij} \succ \boldsymbol{0} \ (\text{i.e. } \tilde{c}_{ij}^{r(i,j)} > 0) \\
l_{ij} &\le f_{ij} \le u_{ij} && \forall\,(i,j) \text{ such that } \tilde{c}_{ij} = \boldsymbol{0} \\
f_{ij} &= u_{ij} && \forall\,(i,j) \text{ such that } \tilde{c}_{ij} \prec \boldsymbol{0} \ (\text{i.e. } \tilde{c}_{ij}^{r(i,j)} < 0)
\end{aligned}
\tag{3}
$$

where $\tilde{c}_{ij} = c_{ij} - \pi_i + \pi_j$

$c_{ij} = (c_{ij}^1, \ldots, c_{ij}^k)^t$

$\pi_i = (\pi_i^1, \ldots, \pi_i^k)^t$, the dual variable is $\pi_i = \pi_i^t P$, $P = (P_1, \ldots, P_k)^t$

$r(i,j) = $ the index of the first non–zero component of $\tilde{c}_{ij}$, i.e. $\tilde{c}_{ij}^{r(i,j)} \ne 0$, and $\tilde{c}_{ij}^s = 0, \forall s < r(i,j)$.

In what follows, we will refer to vector π_i as the dual variable.

In [2] this version of complementary slackness conditions is used to develop two algorithms that are specialized versions of out–of–kilter and primal–dual type ones. Among others, advantages of these methods are the extremely simple way to obtain a starting flow and dual variables, and the integrality of solutions obtained without any additional effort of computing. Both algorithms admit similar implementations. The major differences are the initial solution and the levels of achievement of conservation of flow constraints, capacity constraints and complementary slackness conditions during the process.

In each iteration preemptive priority out–of–kilter algorithm (POOK) preserves the conservation of flow constraints, but it does not preserve either the capacity constraints or the complementary slackness conditions. The basic idea of this algorithm is to modify flow along cycles bringing the current solution closer to an optimal solution. These cycles are obtained by a general labeling process. When it is not possible to find such cycles, dual variables are changed trying to get a new labeling that allows to go on searching for a cycle. Perhaps one of the most important practical features of this method in actual applications is its freedom to begin with any initial solution. This property is very convenient when one is interested in seeing, for instance, the consequences of changes in the preferences of the decision maker, since the old optimal primal and dual solutions can be used to start the new problem.

Unlike the previous one, preemptive priority primal–dual algorithm (PPD) preserves during the iterations the capacity constraints and the complementary slackness conditions, but it does not satisfy the conservation of flow constraints. Thus, a surplus is assigned to every node, defined as the difference between the total flow arriving at this node minus the total flow departing from it. When all surpluses are zero, the solution is optimal. The basic idea of this algorithm is to modify flow along paths from a node with surplus greater than zero to a node with surplus less than zero, bringing both surpluses closer to zero. These paths are obtained by a general labeling process. When it is not possible to find such paths, dual variables are changed trying to get a new labeling that allows to go on searching for a path.

It is in the process of changing dual variables where both algorithms differ essentially from classical out–of–kilter and primal–dual algorithms for the SOMCF problem, since now it is necessary to compute the minimum of vectors, so lexicographical rules have to be applied.

3 Sequential Approach

In this section we present a sequential approach for solving the preemptive priority network flow problem. Unlike the algorithms POOK and PPD, the main idea of the sequential approach is to consider an objective every time, but successively adding constraints in such way that the new problems maintain network structure. In general, when a sequential approach is applied, after solving the problem with the highest level objective, the first objective is set equal to its optimal value and it is added to the problem as a constraint. Then the extended problem is optimized with respect to the second objective function, and the procedure continues in the same way until the entire set of objectives have been minimized. If this process was directly applied to a network flow problem, network structure would be lost from the second iteration of the procedure, preventing the use of network–based algorithms.

What we propose is to determine in each iteration which arcs can not change their flow in successive iterations of the process without the detriment of the satisfaction of higher priority objectives. Next, the current flow through these arcs is fixed from then on by adding constraints so that the network structure of the problem will be preserved, and these arcs become *blocked* for the rest of the sequential process. After blocking those arcs, the new problem is solved for the next priority objective. The process finishes when either every objective function is minimized or every arc is blocked.

Assuming that the original problem is feasible, a general description of the sequential process is given below:

Sequential process:

Initialization:

  Set $h = 1$.

Set

$$A^h = A.$$
$$s_i^h = 0, \forall i \in \mathcal{N}.$$
$$l_{ij}^h = l_{ij}, u_{ij}^h = u_{ij}, \forall (i,j) \in A^h.$$
$$B = \emptyset.$$

Go to the typical iteration.

<u>Typical Iteration:</u>

Step 1. Given the network $G^h = (\mathcal{N}, A^h)$, solve the following SOMCF problem:

$[SP^h]$:

$$\min Z_h(\boldsymbol{f}) = \sum_{(i,j) \in A^h} c_{ij}^h f_{ij}$$

subject to:

$$\sum_{\{j:(i,j) \in A^h\}} f_{ij} - \sum_{\{j:(j,i) \in A^h\}} f_{ji} = s_i^h, \quad \forall i \in \mathcal{N}$$

$$0 \le l_{ij}^h \le f_{ij} \le u_{ij}^h, \quad \forall (i,j) \in A^h.$$

Let $f_{ij}^h, (i,j) \in A^h$, and $\pi_i^h, i \in \mathcal{N}$, be the optimal primal and dual solutions.

If $h = k$ terminate, the current solution $\boldsymbol{f}$ is an optimal solution.

Otherwise, go to Step 2.

Step 2. Let

$$A^* = \{(i,j) \in A^h : c_{ij}^h - \pi_i^h + \pi_j^h \ne 0\}.$$

Set $B = B \bigcup A^*$. If $B = A$ terminate, the current solution $\boldsymbol{f}$ is optimal.

Set $h = h + 1$.

Construct the new network $G^h = (\mathcal{N}, A^h)$, $A^h \subseteq A$, blocking arcs $(i,j) \in A^*$ while leaving all others unaffected.

Go to Step 1.

We will explain further on two different ways of blocking arcs. Before that we will prove the optimality of the solution found at the end of the process.

Theorem. Sequential process terminates providing an optimal solution for the problem [P2].

Proof. Obviously, if the original prcblem is feasible every problem constructed in successive iterations of the process will be feasible. Indeed, the only modification in Step 2 is to fix the current value of some flows. Therefore the new problem has at least the previous solution as a feasible solution.

Note that the infeasibility of the original problem would be detected in Step 1 for $h = 1$ since problem $[SP^1]$ would be also infeasible.

Let $\bar{h}$ be the last iteration carried out by the process. We consider the vector $\pi_i = (\pi_i^1, \ldots, \pi_i^k)^t$, $i \in \mathcal{N}$, where π_i^r are obtained successively every time that Step 1 is applied. If $\bar{h} < k$, we will take $\pi_i^r = 0$, $r = \bar{h} + 1, \ldots, k$.

We will next prove that $f = (f_{ij}^{\bar{h}})_{(i,j) \in \mathcal{A}}$ found at the termination of the process, and $\tilde{c}_{ij} = c_{ij} - \pi_i + \pi_j$ satisfy the conditions for optimality (3), thus proving the theorem.

Let (i, j) be an arc such that $\tilde{c}_{ij} = \mathbf{0}$, so this arc has never been blocked. Then, because of the feasibility of the flow $l_{ij} = l_{ij}^{\bar{h}} \leq f_{ij}^{\bar{h}} \leq u_{ij}^{\bar{h}} = u_{ij}$.

Let (i, j) be an arc such that $\tilde{c}_{ij} \succ \mathbf{0}$, then there exists an index r such that $c_{ij}^r - \pi_i^r + \pi_j^r > 0$ and $c_{ij}^s - \pi_i^s + \pi_j^s = 0, \forall s < r$. Consequently, at the end of the rth iteration the flow through that arc must be $f_{ij}^r = l_{ij}$ because of the conditions for optimality for the SOMCF problem. Moreover, at that moment the arc was blocked fixing its flow to this value from then on. Thus, $f_{ij}^{\bar{h}} = l_{ij}$.

Similarly, it can be proved that $f_{ij}^{\bar{h}} = u_{ij}$ when $\tilde{c}_{ij} \prec \mathbf{0}$, so the conditions for optimality are satisfied and $f_{ij}^{\bar{h}}$ is optimal.   Q.E.D.

Although the algorithm is correct, it does not indicate how to implement Step 2. We propose two different ways of blocking arcs, depending on the algorithm used to solve the SOMCF problem in Step 1.

Implementation 1:

One way of blocking arcs is to maintain through all iterations the same network, that is to say, the same nodes and arcs. Then, given the arcs to be blocked, we fix the value of their current flow by making their upper and lower bounds equal to it. Thus Step 2 would be:

Step 2. Let

$$\mathcal{A}^* = \{(i, j) \in \mathcal{A}^h : c_{ij}^h - \pi_i^h + \pi_j^h \neq 0\}.$$

Set $B = B \bigcup \mathcal{A}^*$. If $B = \mathcal{A}$ terminate, the current solution f is optimal.
Set $h = h + 1$.

Set

$$G^h = (\mathcal{N}, \mathcal{A}^h) = (\mathcal{N}, \mathcal{A}).$$
$$s_i^h = 0, i \in \mathcal{N}.$$
$$l_{ij}^h = u_{ij}^h = f_{ij}^{h-1}, \quad \forall (i,j) \in \mathcal{A}^*.$$
$$l_{ij}^h = l_{ij}^{h-1}, u_{ij}^h = u_{ij}^{h-1}, \quad \forall (i,j) \in \mathcal{A}^h - \mathcal{A}^*.$$

Go to Step 1.

Implementation 2:

Another way of blocking arcs is to reduce the network by removing from it those arcs that become blocked. In this case, their flow must be incorporated into the supplies of corresponding start and end nodes. This implementation can be applied when the algorithm used to solve every SOMCF problem in Step 1 allows to use the balance of mass constraints

$$\sum_{\{j:(i,j)\in\mathcal{A}\}} f_{ij} - \sum_{\{j:(j,i)\in\mathcal{A}\}} f_{ji} = s_i, \quad \forall i \in \mathcal{N}.$$

where s_i is the supply of node i. In this case Step 2 can be stated as follows:

Step 2. Let

$$\mathcal{A}^* = \{(i,j) \in \mathcal{A}^h : c_{ij}^h - \pi_i^h + \pi_j^h \neq 0\}.$$

Set $\mathcal{B} = \mathcal{B} \bigcup \mathcal{A}^*$. If $\mathcal{B} = \mathcal{A}$ terminate, the current solution f is optimal.

Set $h = h + 1$.

Set

$$G^h = (\mathcal{N}, \mathcal{A}^h), \text{ where } \mathcal{A}^h = \mathcal{A}^{h-1} - \mathcal{A}^*.$$
$$s_i^h = s_i^{h-1} - \sum_{\{j:(i,j)\in\mathcal{A}^*\}} f_{ij}^{h-1} + \sum_{\{j:(j,i)\in\mathcal{A}^*\}} f_{ji}^{h-1}, \quad \forall i \in \mathcal{N}.$$
$$l_{ij}^h = l_{ij}^{h-1}, u_{ij}^h = u_{ij}^{h-1}, \quad \forall (i,j) \in \mathcal{A}^h.$$

Go to Step 1.

Note that, under the assumption of integrality of problem data, optimal primal and dual solutions found in Step 1 can be chosen to be integer whatever implementation is used. Therefore an optimal solution to [P2] would be integer.

4 Computational Experience

In this section we show the results of a study of the computational performance of the sequential approach. Three different versions were compared. The first one, entitled PS–OOK, uses an out–of–kilter algorithm [4] to solve every SOMCF problem in Step 1, together with the first implementation in Step 2. The second and third versions, entitled PS–PD.1 and PS–PD.2 respectively, use both a primal–dual algorithm [1] in Step 1, and first and second implementation, respectively, in Step 2.

All algorithms were tested on a set of MONF problems in order to compare their performance. Test problems were generated using a modified version of GRIDGEN program [1, 3] which enables to consider multiple criteria. Five groups of problems were generated for testing purposes. The number of nodes and arcs of each group is specified in Table 4.1. For every group, ten distinct test problems were generated and used with 2, 3, 5 and 15 criteria.

Table 4.1. Groups for test problems

	No. of nodes	No. of arcs
Group 1	27	250
Group 2	27	500
Group 3	52	600
Group 4	52	1000
Group 5	102	1000

All codes are in Borland $C++$ version 3.0. The simulations were run on a PC 80486DX/33MHz under DOS. The average CPU time, and its standard deviation are presented in Tables 4.2. through 4.4. They show that algorithm PS–PD.2 gives the best performance, and it is only lightly dependent of the number of criteria. Moreover, when these algorithms and those developed in [2] are compared this sequential version still gives the best performance. Algorithms PS–PD.1, PPD and POOK give comparable computational results and, finally, algorithm PS–OOK gives the worst performance.

In order to illustrate the application of the sequential approach a small eight nodes and sixteen arcs network flow problem with five criteria functions is solved using Algorithm PS–PD.2. Fig. 4.1 displays the network configuration of the problem, where numbers on arcs indicate lower and upper bounds. The coefficients in the objective functions are shown in Table 4.5.

In the first place, the original network is solved with respect to the first objective function (Step 1 of the algorithm). So, we solve

$$\min Z_1(\boldsymbol{f}) = f_{23} + 2f_{35} + 4f_{45} + 2f_{56} + 2f_{68}$$

subject to:

$$\sum_{\{j:(i,j)\in A\}} f_{ij} - \sum_{\{j:(j,i)\in A\}} f_{ji} = 0, \quad \forall i \in \mathcal{N}$$

$$0 \le l_{ij} \le f_{ij} \le u_{ij}, \quad \forall(i,j) \in \mathcal{A}.$$

where lower and upper bounds of each arc and the supply of each node are indicated in Fig. 4.1. Fig. 4.2 displays an optimal solution to the problem, where

Table 4.2. PS-OOK test results (CPU time)

No. of crit.		Group 1	Group 2	Group 3	Group 4	Group 5
2	Mean	0.268	0.957	1.885	4.030	5.730
	Standard Deviation	0.053	0.092	0.201	0.419	0.615
3	Mean	0.369	1.352	2.421	5.519	7.326
	Standard Deviation	0.059	0.233	0.187	0.886	0.404
5	Mean	0.509	2.367	3.796	9.154	10.101
	Standard Deviation	0.083	0.409	0.338	1.089	1.153
15	Mean	1.444	6.958	9.467	25.690	26.193
	Standard Deviation	0.239	0.832	1.008	3.357	1.987

Table 4.3. PS-PD.1 test results (CPU time)

No. of crit.		Group 1	Group 2	Group 3	Group 4	Group 5
2	Mean	0.116	0.265	0.655	1.105	1.994
	Standard Deviation	0.034	0.035	0.037	0.095	0.191
3	Mean	0.161	0.380	0.853	1.515	2.475
	Standard Deviation	0.050	0.032	0.075	0.118	0.220
5	Mean	0.215	0.556	1.165	2.184	3.344
	Standard Deviation	0.032	0.040	0.111	0.124	0.196
15	Mean	0.472	1.346	2.713	5.746	7.969
	Standard Deviation	0.060	0.152	0.212	0.211	0.539

Table 4.4. PS-PD.2 test results (CPU time)

No. of crit.		Group 1	Group 2	Group 3	Group 4	Group 5
2	Mean	0.100	0.208	0.517	0.811	1.630
	Standard Deviation	0.032	0.035	0.048	0.166	0.175
3	Mean	0.105	0.237	0.566	0.829	1.721
	Standard Deviation	0.040	0.028	0.062	0.041	0.112
5	Mean	0.121	0.249	0.605	0.913	1.858
	Standard Deviation	0.023	0.031	0.072	0.093	0.097
15	Mean	0.187	0.329	0.916	1.301	2.973
	Standard Deviation	0.029	0.045	0.054	0.104	0.156

arcs to be blocked are indicated by dashed lines. Because of the implementation chosen these arcs will be removed from the network when applying Step 2 of the algorithm.

The network to be solved with respect to the second objective function (column 3 of Table 4.5) is shown in Fig. 4.3, together with the optimal solution found at the end of the iteration. Arcs (2,6) and (2,4) have now to be blocked, so removed from the network.

Table 4.5. Coefficients in the objective functions for the example

Arc	c_{ij}^1	c_{ij}^2	c_{ij}^3	c_{ij}^4	c_{ij}^5	Arc	c_{ij}^1	c_{ij}^2	c_{ij}^3	c_{ij}^4	c_{ij}^5
(1,2)	0	0	0	1	9	(4,5)	4	0	2	2	4
(1,3)	0	2	0	2	1	(4,6)	0	0	4	0	0
(2,3)	1	0	1	0	0	(5,6)	2	1	0	7	0
(2,4)	0	0	0	0	10	(5,7)	0	0	3	1	2
(2,5)	0	0	2	3	5	(6,8)	2	3	0	0	1
(2,6)	0	2	0	0	0	(7,6)	0	1	3	5	6
(3,5)	2	0	1	1	0	(7,8)	0	0	0	1	0
(3,7)	0	0	4	1	5	(8,1)	0	0	0	0	0

In the third iteration, the network to be solved with respect to the third objective function (column 4 of Table 4.5) is shown in Fig. 4.4, together with the optimal solution found at the end of the iteration. Arc (2,5) has now to be blocked, and therefore removed from the network. Forth and fifth iterations do not block any arc nor modify the flow through the arcs either. An optimal solution to the example is given in Fig. 4.5.

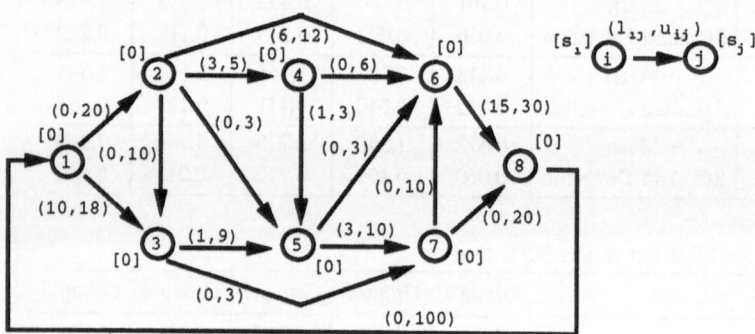

Fig. 4.1. An example network

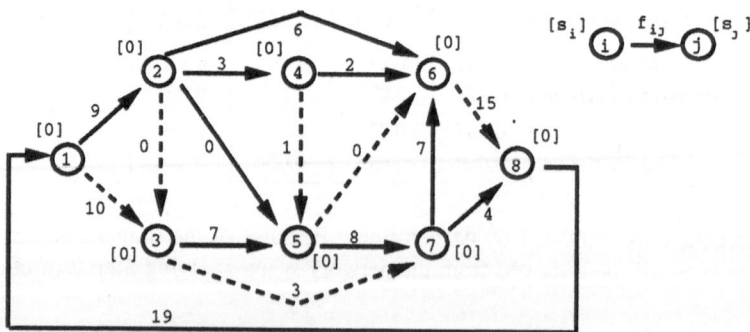

Fig. 4.2. First optimal solution

85

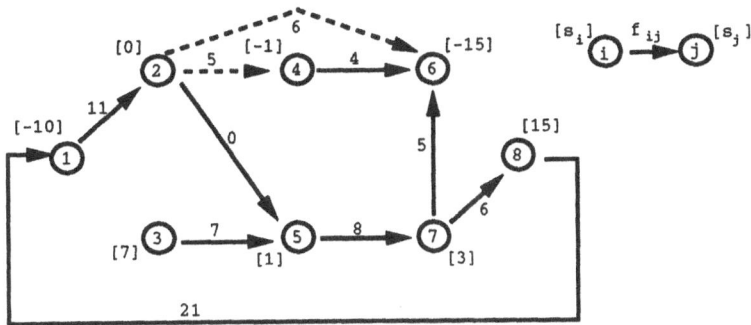

Fig. 4.3. Second optimal solution

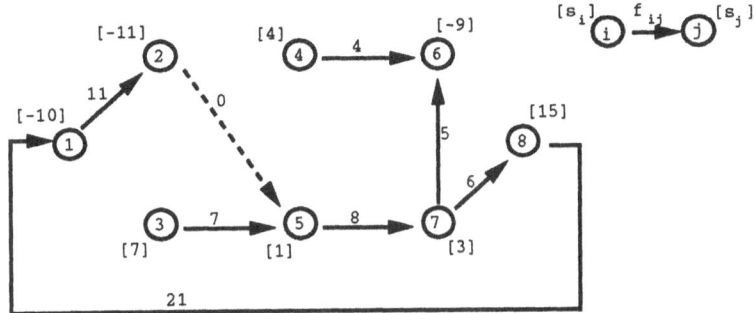

Fig. 4.4. Third optimal solution

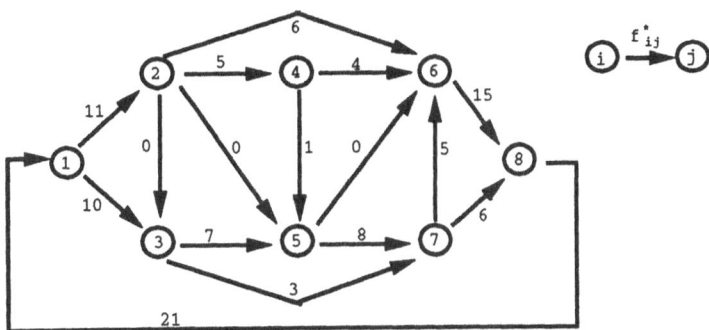

Fig. 4.5. Optimal solution

The same sequence of optimal solutions and blocked arcs is obtained when Algorithm PS–PD.1 is applied. The only difference is that, in this case, the network structure is always the same (that of the Fig. 4.1) and the way of blocking arcs is to equal the lower and upper bounds with the current flow. Thus, for instance in iteration 2 the lower and upper bounds of arcs (1,3), (2,3), (3,7), (4,5), (5,6) and (6,8) would be 10, 0, 3, 1, 0 and 15, respectively.

5 Summary

A sequential approach for the multiobjective network flow problem with preemptive priorities is introduced in this paper. Its main idea is to consider an objective every time, but successively adding constraints in such a way that the resulting problem maintains network structure. This is accomplished by blocking arcs that satisfy a particular condition related to the complementary slackness conditions for optimality. As the network structure is not affected, the procedure can use any network–based algorithm for solving the corresponding single objective network flow problem.

The correctness of the sequential approach is proved and three different versions of it are proposed. They differ in the algorithm used to solve each SOMCF problem (an out–of–kilter or a primal–dual one) and in the way of blocking arcs (modifying lower and upper bounds on arcs or removing arcs). A comparative study of the computational performance is also given.

6 References

1. D. P. Bertsekas, *Linear Network Optimization. Algorithms and Codes.* MIT Press, Cambridge, MA (1991).
2. H. I. Calvete and P. M. Mateo, An Approach for the Network Flow Problem with Multiple Objectives. To appear in *Computers and Operations Research.*
3. H. I. Calvete and P. M. Mateo, PGRIDGEN, una modificación del generador de redes GRIDGEN. *Publi. del Seminario García Galdeano*, Serie II, sección 1, **1**, 1-11 (1994).
4. L. R. Ford and D. R. Fulkerson, *Flows in Networks.* Princeton University Press, Princeton, NJ (1962).
5. D. Klingman and J. Mote, Solution Approaches for Network Flow Problems with Multiple Criteria. *Advances in Management Studies*, **1**(1), 1-30 (1982).
6. E. Lawler, *Combinatorial Optimization: Networks and Matroids.* Holt, Reinhart and Winston, New York (1976).
7. C. H. Papadimitriou and K. Steiglitz, *Combinatorial Optimization. Algorithms and Complexity.* Prentice–Hall, Englewood Cliffs, NJ (1982).
8. P. S. Pulat, F. Huarng and H. Lee, Efficient Solutions for the Bicriteria Network Flow Problem. *Computers and Operations Research*, **19**(7), 649-655 (1992).
9. M. Zeleny, *Multiple Criteria Decision Making.* McGraw-Hill, New York (1982).
10. P. L. Yu and M. Zeleny, The Set of All Nondominated Solutions in Linear Cases and a Multicriteria Simplex Method. *J. of Mathematical Analysis and Applications*, **49**, 430-468 (1975).

MOLP Formulation Assistance Using LP Infeasibility Analysis

John W. Chinneck
Systems and Computer Engineering
email: chinneck@sce.carleton.ca

Wojtek Michalowski
School of Business
email: wojtek@business.carleton.ca

Carleton University
Ottawa, Ontario K1S 5B6
CANADA

ABSTRACT

The correct formulation of a large multiple objective linear program (MOLP) can be very difficult. The problem formulator faces the usual difficulties posed by single objective linear programs (LPs), such as isolating and analyzing infeasibility among the constraints. In addition, MOLP problems may raise questions about whether certain relationships should be cast as objectives or constraints, and about how the different objectives interact and interfere with each other. Automated or semi-automated assistance is needed to handle problems of practical scale effectively. We demonstrate how techniques recently developed for the analysis of infeasible LPs can be extended to assist in the formulation of MOLPs, both by analyzing infeasible constraint sets and by illuminating the interactions among objectives and constraints.

1. INTRODUCTION

Multiple Objective Linear Programming (MOLP) emerged in the 1960s as a way of dealing with the multiple criteria character of many real-life Linear Programming (LP) applications. Most MOLP research since that time has been directed towards devising more effective and efficient solvers for practical problems and towards tools for improving the interpretation of results. Relatively little effort has been put into tools which assist in the formulation and "debugging" of MOLPs, though this is now a major research theme for general LPs. Formulation assistance for MOLPs is increasingly necessary as the models grow in size and complexity. This paper describes tools which assist in MOLP formulation. When the formulation is complete, the final version of the model can be solved in any manner, including lexicographic goal programming or the Multiplex approach [Ignizio 1985].

The rapid increase in computing power in recent years has shifted the focus of general LP research from model solution to model formulation and management. Questions include the kind of interface which makes complex LP models most easily understood (e.g. graphical, algebraic, matrix, spreadsheet etc.), and the kinds of assistance that users need when the model is infeasible or otherwise incorrect [Geoffrion 1987; Greenberg 1983]. This line of research is exemplified by the Intelligent Mathematical Programming System project [Greenberg 1990] whose aim is to develop a tool kit for effective model management. One product of this project, Greenberg's ANALYZE software [Greenberg 1983,1987,1993], includes tools to assist in the manipulation and interpretation of LP solutions, and to perform postoptimality analysis and infeasibility studies.

The research described in this paper is in the same vein. The methods developed here assist in the formulation of LP problems with multiple objectives. Assistance is provided in (i) analyzing constraint infeasibility as for single-objective LPs, (ii) deciding whether relationships should be represented as objectives or constraints, and (iii) simplifying the model by eliminating objectives, or assigning lexicographic ordering or weights to the objectives.

This is also the first study to consider a MOLP as a model with inherent structural inconsistencies. We show how the inability to reach a single solution optimizing all of the objective functions at the same time can be converted to a problem of constraint infeasibility. Following Chinneck's work [Chinneck and Dravnieks 1991, Chinneck 1994] on the infeasibility of LPs, we demonstrate how to identify sources of MOLP inconsistency and how to analyze the options for reducing or removing them. This analysis may reduce the degree of conflict among MOLP objectives, and the resulting problem should be easier to solve by means of, for example, interactive methods (see Michalowski and Szapiro [1992], Steuer [1986], or Zionts and Wallenius [1983] for a discussion of interactive programming methods).

The paper is organized as follows. Section 2 presents our views on the problems of formulating and solving MOLPs. Section 3 briefly reviews the tools of LP infeasibility analysis which will be used in the remainder of the paper. Section 4 develops the formulation assistance algorithms for MOLPs. Section 5 presents an example, and Section 6 provides some brief conclusions.

2. FORMULATING AND SOLVING MOLPs

2.1 Solving MOLPs

The classical formulation of a Multiple Objective Linear Program (MOLP) assumes that a decision alternative and its outcome are represented by vectors. A decision alternative is described by a vector x, $x \in \mathbb{R}^n$. The set X of all admissible decisions is:

$$X = \{x \in \mathbb{R}^n \mid Ax \leq b\}, \quad b \in \mathbb{R}^k, \quad A \in M_{kxn}$$

where M_{kxn} is a set of kxn matrices. It is further assumed that X is nonempty and bounded, therefore it is compact and connected.

An outcome of a decision alternative is described by a vector y, $y \in \mathbb{R}^m$. The correspondence of decision alternative and an outcome is given by a linear mapping $f: X \to \mathbb{R}^m$; $f(x) = Cx$, where $C \in M_{mxn}$ and $f(x) = [f^1(x), \cdots, f^m(x)]^T$. The components $f^i: X \to \mathbb{R}$, $i \in \{1, 2, \cdots, m\} = I$ of a mapping f are called *objective functions*. Given that $y = f(x) = [f^1(x), \cdots, f^m(x)]^T$, we say that a decision alternative x *realizes* a decision maker's objectives on levels $y^1 = f^1(x), \cdots, y^m = f^m(x)$. The direct image $Y = f(X)$ of the set X is called the set of *admissible outcomes*. The set Y is compact and connected as a continuous image of compact and connected set X.

The ordering of the outcomes associated with all admissible decisions is accomplished through the relation "$\geq$" called the *Pareto-preference*:

$$y \geq y' \Leftrightarrow y \neq y' \wedge \forall j \in I \; y^j \geq y'^j.$$

If $y \geq y'$, then it is said that y dominates y'. Given the set Y and relation "$\geq$", the set Y_{ND} of *nondominated* outcomes is defined as: $y \in Y_{ND} \Leftrightarrow (\nexists y' \in Y \; y' \geq y)$. Decisions with nondominated outcomes are called *efficient* ones.

With the introduction of notions of Pareto-preference and dominance, the MOLP problem now becomes one of determining the set Y_{ND} for a given tuple (X, f). For practical purposes such a formulation needs further refinement due to the size of a set of nondominated outcomes. Usually this is accomplished by acquiring additional information about the decision maker's preferences which is then used to generate a single outcome $y \in Y_{ND}$.

It is well known in the literature that the unique optimal solution $y^{iopt} = f^i(x_k^{opt})$ of a single objective LP max $f^i(x)$ s.t. $x \in X$ is a nondominated outcome. For the sake of further analysis we define a *true MOLP* as a problem where $\exists i, j \in I$ such that $y^{iopt} \neq y^{jopt}$. Thus, at least two of the objective functions of a true MOLP are in conflict and their respective optima are achieved at different efficient extreme points.

2.2 Formulating MOLPs

A major difficulty in formulating MOLPs is in the classification of mathematical relationships as constraints or objectives. For example, it is sometimes difficult to determine whether a cost expression should be an objective ("minimize cost"), or a constraint ("cost must not exceed $100,000"). We define four classes of mathematical relationships below; a similar distinction has been used by Ignizio and Cavalier [1994]:

Hard constraint: a mathematical relationship that is definitely classed as a constraint. These are often basic physical relationships such as conservation of flow in a network.

Soft constraint: a mathematical relationship that is currently classed as a constraint, but which could be considered an objective by dropping the right hand side and adding an objective sense (maximize or minimize).

Hard objective: a mathematical relationship that is definitely classed as an objective (maximize or minimize).

Soft objective: a mathematical relationship that is currently classed as an objective, but which could be considered as a constraint if an appropriate right hand side and relationship sense were added.

Constraint to objective conversions are straightforward: drop the right hand side and the relationship symbol, and add the appropriate optimization sense: $\geq$ constraints become maximizations, $\leq$ constraints become minimizations, $=$ constraints can become either (though $=$ constraints are very unlikely candidates for conversion to objectives). Objective to constraint conversions are more complicated. First the constraint sense must be determined. Generally maximization objectives become $\geq$ constraints, and minimization objectives become $\leq$ constraints, but either could be converted to an $=$ constraint and used as a goal constraint with the addition of appropriate deviational variables. Second, the right hand side value must be set; this requires some knowledge of the application.

The value assigned to the right hand side in a soft constraint is called the *aspiration level* of the constraint. Similarly, when a soft objective is converted to a constraint, an aspiration level must be assigned.

We formalize current practice by assuming that the modeller can make an initial assignment of every relationship to one of the four classes described above. Given this initial classification of the relationships, the modeller faces two main MOLP formulation problems, in addition to the usual LP formulation problems:

1. Final Classification: arriving at a final classification of the soft constraints and objectives. Should a soft constraint be converted to an objective? Should a soft objective be converted to a constraint, and if so, what should the aspiration value be? This produces the final form of the MOLP, which can then be solved.

2. Simplification: elimination of constraints and objectives, rewriting of constraints, resetting of aspiration values etc. to arrive at a simpler or clearer formulation.

Our approach to these two formulation problems is to provide algorithmic tools for *Interaction Analysis*, the process of analyzing the interactions and interferences between objectives and between objectives and constraints. Interaction Analysis may also assist in the solution of the MOLP, as described above. The main ingredient of Interaction Analysis is recently-developed techniques for the analysis of infeasible LPs, as reviewed briefly below.

3. A BRIEF REVIEW OF LP INFEASIBILITY ANALYSIS

There are two main approaches to analyzing infeasible LPs. *Bound tightening* is used in most LP presolve routines and operates by tightening row bounds by substituting in the variable bounds. This occasionally detects infeasibility and provides a chain of operations which identifies the cause, but is not reliable. More recently, methods to isolate *Irreducible Infeasible Systems (IISs)* of constraints have been developed. Chinneck's "filtering" methods of IIS isolation [Chinneck and Dravnieks 1991] are robust and effective, and have been incorporated in several well-known LP solvers. The filtering algorithms are briefly reviewed below.

The *deletion filter* (see Algorithm 1) guarantees that only the members of a single IIS are isolated. It requires the solution of a phase 1 LP for each finite

---

INPUT: an infeasible set of linear constraints.
FOR each constraint in the set:
      1. temporarily drop the constraint from the set.
      2. test the feasibility of the reduced set:
            IF feasible THEN
                        return dropped constraint to the set.
            ELSE (infeasible)
                      drop the constraint permanently.
END FOR.
OUTPUT: constraints constituting a single IIS.

Algorithm 1: The Deletion Filter.

---

bound in the model, including variable bounds. Because the final basis of the previous deletion iteration is used as the initial basis for the next iteration, this is not as slow in practice as might be expected.

The *sensitivity filter* is applied to the final basis of an infeasible phase 1 LP to accelerate the isolation. Any nonbasic variable having a reduced cost or shadow price of zero identifies a constraint which does not contribute to the phase 1 objective function, i.e. does not contribute to the infeasibility. The indicated constraints are then eliminated because they are not part of the IIS being isolated. The sensitivity filter is highly effective in rapidly eliminating numerous constraints, but the deletion filter is still needed to provide the positive identification of a single IIS.

A good acceleration of the IIS-isolation process is achieved by the *deletion/sensitivity filter* which combines the two by applying a sensitivity filter to the infeasible deletion filter iterations, as shown in Algorithm 2.

The *reciprocal filter* applies where rows or variables have distinct upper and lower bounds. In the absence of simple upper and lower bound reversal, if a row or column has distinct upper and lower bounds and one of the bounds is involved in an IIS, then the other bound cannot be involved in the same IIS.

INPUT: an infeasible set of linear constraints.
FOR each constraint in the set:
        1. temporarily drop the constraint from the set.
        2. test the feasibility of the reduced set:
            IF feasible THEN
                    return dropped constraint to the set.
            ELSE (infeasible)
                i. drop the constraint permanently.
                ii. apply the sensitivity filter.
END FOR.
OUTPUT: constraints constituting a single IIS.

Algorithm 2: The Deletion/Sensitivity Filter.

Thus as soon as one of the bounds is identified as belonging to an IIS, then the other bound can be discarded.

In the *elastic filter*, nonnegative "elastic variables", e_i are added to the constraints to provide elasticity, as shown in Eqn. 1. Constraints stretch against

$$\begin{array}{cc} \textit{Nonelastic} & \textit{Elastic} \\[6pt] \sum_i a_i x_i \geq b & \sum_i a_i x_i + e \geq b \\[6pt] \sum_i a_i x_i \leq b & \sum_i a_i x_i - e \leq b \\[6pt] \sum_i a_i x_i = b & \sum_i a_i x_i + e_1 - e_2 = b \end{array}$$

Equation 1: Elasticizing Constraints.

the resistance provided by the elastic objective of minimizing the sum of the nonnegative elastic variables, which replaces the original objective function. As shown in Algorithm 3, the elastic filter solves the elastic LP (which will of course be feasible), and then removes the elastic variables from any constraint in which the elastic variables are positive (i.e. *enforces* the constraint) in a process of gradually "tightening" the LP. The cycle of solution and enforcement

INPUT: an infeasible set of linear constraints.
1. Make all constraints elastic by adding nonnegative elastic variables.
2. Solve LP using elastic objective function.
3. IF feasible THEN

Enforce the constraints in which any $e_i > 0$ by permanently removing their elastic variable(s).

Go to step 2.

ELSE (infeasible)

Exit.

OUTPUT: the set of de-elasticized enforced constraints   contains at least one IIS.

Algorithm 3: The Elastic Filter.

continues until the LP becomes infeasible, and the algorithm terminates. The output set of enforced constraints contains at least one IIS.

3.1 SOFTWARE FOR ISOLATING IISs

MINOS(IIS) is a version of MINOS 5.4 [Murtagh and Saunders 1987] modified at Carleton University to incorporate the filtering algorithms for isolating IISs. See [Chinneck 1993] for details. MINOS(IIS) operates exactly like MINOS 5.4, unless infeasibility is detected, in which case the IIS-isolating routines are called.

Version 4.2.1 of MINOS(IIS) incorporates the deletion, sensitivity, deletion/sensitivity, reciprocal and elastic filters as well as various heuristics for finding IISs having few rows, which are generally easier to interpret. A default method is provided, but the others can be selected under user control. The user can also specify *guide codes* for influencing the IIS search by placing constraints in categories, notably "inclusion in IIS encouraged" and "exclusion from IIS encouraged".

MINOS(IIS) is used in analyzing the example in Section 5, but other solvers capable of IIS isolation could also be used. Two commercial solvers have recently added IIS filtering routines: LINDO 5.3 (deletion filter only) and CPLEX 3.0 (deletion/sensitivity and elastic filters). CLAUDIA, proprietary software developed by Roger Main at BP Oil, also uses variations of the deletion and elastic filters.

4. ALGORITHMIC TOOLS FOR MOLP FORMULATION ASSISTANCE

The most direct use of IIS isolation applies the methods directly to the constraint set in the MOLP, as described in Section 4.1. A simple problem conversion also allows IIS isolation to analyze objective interactions, as described in Section 4.2.

4.1 Interaction Analysis of the Constraint Set

If the MOLP constraint set is infeasible, then an IIS can be isolated, which assists in the diagnosis of the problem as it does for any infeasible LP. However, rather than proceeding directly to the isolation of an IIS if the overall constraint set (both hard and soft constraints) is infeasible, it is better to apply a two-step analysis: first test only the hard constraints, then test the entire constraint set including both the hard and soft constraints.

Infeasibility of the set of hard constraints reveals a basic LP formulation error which must be repaired before the multiple objective aspects of the formulation can be addressed. Subsequent infeasibility of the combined set of hard and soft constraints shows that the aspiration levels of the soft constraints are unrealistic. The IIS isolated by the infeasibility analysis in the second case is especially revealing because it shows the *set* of constraints which interact to cause infeasibility. There are two cases to consider.

Case 1: only one soft constraint and a set of hard constraints in the IIS. This shows that the aspiration level of the soft constraint is unrealistic and conflicts with basic hard constraints, such as physical limitations of equipment. This implies that either (i) the aspiration level of the soft constraint must be relaxed to a feasible level, or (ii) the soft constraint should be converted to an objective.

Case 2: more than one soft constraint in the IIS. In this case, the aspiration levels of the soft constraints interact with each other, and any hard constraints in the IIS, to create the infeasibility. The changes needed to repair the model are similar to the first case: (i) change the aspiration level of one or more of the soft constraints, or (ii) convert one or more of the soft constraints to objectives. In addition, we know that if more than one of the soft constraints is converted to an objective, then these converted objectives will interfere with each other. The kind of analysis needed at this point is then similar to the methods described below.

4.2 Interaction Analysis of the Objectives

Infeasibility analysis deals only with constraints, so if we are to use IIS-isolation to analyze the interaction of the objectives in the MOLP, all objectives (both soft and hard) must first be converted to constraints. This is easily accomplished via a two-step process which produces the *converted MOLP*.

Step 1: Find the extreme feasible aspiration level of each objective. This is done by creating and solving one LP for each objective, in which only that objective appears while all of the other objectives are temporarily removed. Note the optimum terminating point(s) (this is the extreme feasible aspiration point, x_A^{opt}, for some objective A) and the extreme feasible aspiration level ($y_A = f_A(x_A^{opt})$ for some objective A) for each LP.

Step 2: Convert each objective to a constraint. This is done by rewriting each objective as a constraint, with the relationship sense determined appropriately (maximization objectives become $\geq$ constraints, minimization objectives become $\leq$ constraints), and the right hand side set at the extreme feasible aspiration level determined in Step 1. For numerical reasons, it is better to relax the right hand side from the extreme feasible aspiration level by a small epsilon amount.

Observation 1: As a side effect of Step 1 we are able to identify sets of objectives which are not in conflict. Any objectives whose Step 1 LPs terminate at the same extreme point (or who have the same extreme points as alternative optima) are not in conflict. Consider Figure 1 for example.

Observation 2: If all of the Step 1 LPs terminate at the same extreme point

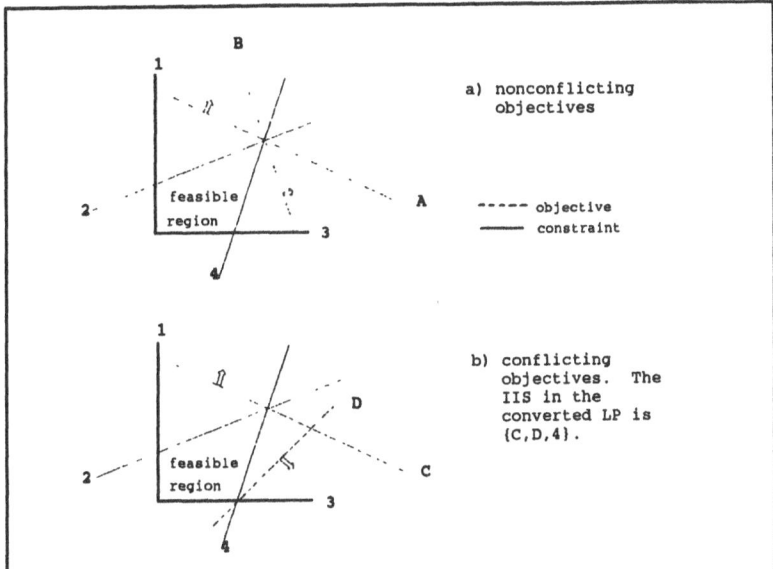

Figure 1: Nonconflicting objectives terminate at the same extreme point (or have the same extreme points as alternative optima).

(or have the same exteme points as alternative optima), then none of the objectives are in conflict, and the model is not a true MOLP.

If the condition described in Observation 2 holds, then the converted MOLP is feasible because the common extreme point exists which satisfies all of the original (hard and soft) constraints, and all of the converted objectives. This leads directly to Theorem 1.

Theorem 1: If the original constraint set (hard and soft) is feasible, then the converted MOLP is infeasible if and only if the original model is a true MOLP.

Proof: Because the constraint set is feasible, infeasibility can happen only due to an interaction involving one or more of the converted objectives. If the converted objectives are not in conflict, then a feasible point exists, as discussed above. Thus an infeasible converted LP can be constructed only if one or more of the converted objectives are in conflict. This is the definition of a true MOLP. QED

Observation 3: Given that the original constraint set is feasible and that the model is a true MOLP, then any IIS in the converted MOLP will involve more than one converted objective. The extreme feasible aspiration levels are defined such that each converted objective is feasible with respect to the constraint set. Infeasibility then requires at least two converted objectives.

Observation 4: Any converted objectives appearing together in an IIS are in conflict. If they are not in conflict, then the infeasibility is not irreducible, since both would restrict the converted model to the same extreme point.

These observations can be used to provide insight on the objective behaviour and to suggest model reformulations and simplifications.

There are two cases to consider:

Case 1: The IIS includes only converted hard objectives, and possibly hard constraints. The IIS isolates sets of conflicting hard objectives. This information can be used to reformulate the problem by abandoning an objective that is clearly of lesser importance among the converted objectives in the IIS, if this is appropriate. In a similar vein, it can be used to guide the setting of the lexicographic order or weights on the objectives once the conflict sets are known.

Case 2: The IIS includes at least one converted soft objective or soft constraint. Here we have more reformulation options, since at least one of the members of the IIS is soft. If the IIS includes a soft constraint, a reformulation of the soft constraint may permit the objectives to achieve their extreme feasible aspiration levels. For example, assume that constraint 4 in Figure 1b is a soft constraint. Objectives C and D can simultaneously achieve their extreme feasible aspiration levels if constraint 4 is relaxed back to the intersection of C and D.

Constraint 4 can be relaxed back to the appropriate point in two ways: (i) it can be converted to an objective instead of a constraint, or (ii) its aspiration level can be adjusted. Further, we can discover exactly how much to relax constraint 4 by constructing and solving a small elastic program (see Eqn. 1) from the IIS: simply elasticize constraint 4 and solve the LP consisting of the IIS constraints only. The value of the elastic objective gives the amount of relaxation in constraint 4 that is needed to allow the other objectives to achieve their extreme feasible aspiration levels.

If the IIS includes a soft objective, then similar considerations apply: if it is important that the other objectives achieve their extreme aspiration levels, then ignore the soft objective, or convert it to a soft constraint whose aspiration level is set appropriately.

4.2.1 Generating Different Interacting Sets of Objectives

We deal with a single IIS at a time when considering only constraints: each IIS found must be repaired before analysis can proceed. When considering objectives as well, the IISs in the converted MOLP simply provide information about interacting sets of relationships, so we wish to be able to shift our focus from one interacting set to another (i.e. from one IIS to another) as the analysis proceeds, while all IISs remain intact in the converted MOLP. Because the IIS algorithms are designed to isolate a single IIS, this causes some difficulty, but three techniques can be applied, as described below.

1. Eliminate a converted objective from the current IIS. One method of generating a different IIS, and hence a different interacting set of relationships, is to eliminate one of the converted objectives from the current IIS. When this is done, restarting the IIS isolation algorithm will isolate a different IIS if one exists. Converted objectives can be eliminated by either (i) actual elimination from the model, or (ii) elasticization.

To examine how much the aspiration level of a particular objective must be adjusted to accommodate the other objectives in a particular IIS, the following procedure can be used: (1) choose the constraint set consisting only of the constraints and converted objectives in the IIS, (2) change the converted objective in question back to an objective, (3) solve the now-feasible LP to optimality, (4) calculate the difference between the original aspiration level and the optimum value found in step (3).

The difference found in step (4) shows whether the conflicts between the objectives in the IIS are serious. A small difference indicates that the conflict is not serious, perhaps resolvable by converting one objective to a constraint with a slightly relaxed aspiration level, or by eliminating one objective entirely.

2. Apply the IIS search guide codes. Different IISs are often found by setting the IIS guide codes so as to encourage or discourage the inclusion of specific constraints or converted objectives. One approach is to discourage the inclusion of all of the members of the current IIS.

A particularly useful technique when examining the objective interactions is to run the IIS analysis once for each objective with that objective encouraged to stay in the IIS, and the other objectives encouraged to drop out of the IIS. If a certain relationship appears in all or most of the IISs generated this way, then that relationship is particularly conflictive and is a good candidate for change in some manner (convert relationship type if soft, adjust right hand side if constraint, etc.).

3. Use an IIS-enumerating algorithm. The IIS-enumerating algorithm described by Gleeson and Ryan [1990] could, in principle, be used to find all of the IISs in the converted model, but it has not been implemented in practice.

Option 2 is used in the complete method described in Section 4.3.

4.2.2 Which Objectives Conflict With a Particular Objective?

One common MOLP formulation question is to find all of the other objectives which conflict with a particular objective. This is easily determined by examining the converted MOLP: all objectives which do not share the same extreme aspiration point x_k^{opt} (or an alternative extreme aspiration point) are in conflict.

4.2.3 Evaluating the Relative Amount of Objective Interference

The degree of interference between some objective A and some objective B is determined by (i) substituting x_A^{opt} into objective B and x_B^{opt} into objective A, then (ii) determining how much each objective moves away from its extreme aspiration level at the new point. The absolute difference found in step (ii) can be used directly, or it can be normalized by dividing by the extreme aspiration level if appropriate.

We say that objective A *interferes strongly* with objective B if the value of objective B at x_A^{opt} is greatly different (absolute or normalized difference as appropriate) from the extreme aspiration level of objective B; otherwise objective A *interferes weakly*. Objective interference is *relative* because it is possible that objective A interferes strongly with objective B while objective B interferes weakly with objective A.

This analysis may suggest model simplifications. For example, if objective B interferes weakly with objective A, then the model can perhaps be simplified by considering objective B superior to objective A and, for example, using lexicographic ordering of the objectives.

Where the tradeoffs among several objectives must be analyzed simultaneously, an *objective interference table* can be constructed. First create a table having columns for the objectives and rows for the extreme aspiration points for the objectives, e.g. x_A^{opt}, x_B^{opt}, etc. The element of the table for the row x_A^{opt} and the column for objective B is then $[y^{Bopt} - f^B(x_A^{opt})]$ or $[y^{Bopt} - f^B(x_A^{opt})]/y^{Bopt}$ if normalized. The normalized table shows the fractional loss in the objective function value relative to the extreme aspiration level when the objective function is evaluated at the extreme aspiration point for a different objective. Examples of how it can be used are given in Section 5.

4.3 Summary of the Method

The steps in the method are summarized below. Note that each test is re-applied until passed because there may be more than one IIS in the model, perhaps unrelated. A brief discussion follows the statement of the method.

Analyze the constraint interactions:

1. Select the set of hard constraints and apply a phase 1 feasibility test. If feasible, go to Step 2, else (infeasible) identify an IIS and repair the basic LP formulation error. Go to Step 1.

2. Select the entire set of constraints (hard and soft) and apply a phase 1 feasibility test. If feasible, go to Step 3, else (infeasible) identify an IIS and proceed as follows:
    Case 1: only one soft constraint in the IIS. Either (i) relax aspiration level of the soft constraint, or (ii) convert the soft constraint to an objective.
    Case 2: more than one soft constraint in the IIS. Either (i) relax the aspiration level(s) of one or more of the soft constraints, or (ii) convert one or more of the soft constraints to objectives.
    Go to Step 2.

Create the converted MOLP:

3. Find the extreme feasible aspiration level and extreme feasible aspiration point (and any alternative extreme feasible aspiration points) of each objective by selecting the entire set of constraints and only one objective at a time and solving to optimality.

4. Group objectives having the same extreme feasible aspiration points (or alternative extreme feasible aspiration points): members of each group are non-conflicting objectives. If there is only one group of objectives, then the model is not a true MOLP, so exit.

5. Convert each objective to a constraint by appending the extreme feasible aspiration level as the right hand side and using the constraint sense appropriate to the objective ($\geq$ for maximize and $\leq$ for minimize).

Analyze the objective interactions:

6. Identify a set of IISs as follows. For each objective, set the guide codes to encourage the inclusion of that objective in the IIS and to encourage the exclusion of the other objectives. Identify frequently occuring elements in the IISs and proceed as follows.
    Case 1: frequent element is a hard objective or constraint. Either (i) abandon objective(s) in the set that are of lesser importance, or (ii) use the objective inteference table to set the lexicographic order or weights on the objectives appearing in the IIS.
    Case 2: frequent element is a converted soft objective or soft

constraint. Either (i) soft constraint: adjust the aspiration level or convert to an objective, or (ii) soft objective: use the objective interference table to decide whether to ignore or convert to a soft constraint with appropriate aspiration level.

7. If analysis complete then construct objective interference table for final setting of lexicographic order or objective weights and exit. Else (analysis not complete) go to Step 3.

Steps 1-5 of the method are straightforward. Step 6 allows a great deal of flexibility. The modeller could, for example, choose to generate and analyze a single IIS in Step 6. Decisions on how to simply the model (e.g. convert a soft constraint to an objective or just modify it's right hand side?) require domain knowledge which cannot be hard-coded into the algorithm. Similarly, only the modeller, applying domain knowledge, can determine when the model is sufficiently well formulated to proceed to the solution stage.

5. EXAMPLE

The example is an adapted and simplified version of a land-use problem developed by Steuer and Schuler [1981]. Details of the relationships are given in the Appendix. The seven objectives relate to, in order, pasturage, dispersed recreation, timber production, and populations of deer, rabbits, squirrels, and quail. The first three objectives are soft (i.e. they could be considered as constraints), while the last four are hard (they are definitely objectives), since this is basically a forestry problem. Soft constraint s1 is a budget limitation. The steps of the analysis method are applied below.

Steps 1 and 2. The complete set of constraints is feasible, so the tests in Steps 1 and 2 are passed.

Step 3. The extreme feasible aspiration levels of the various objectives are summarized below:

objective	extreme feasible aspiration level
zs1	1577.59
zs2	164.33
zs3	7437.00
zh4	28.96
zh5	81.07
zh6	40.28
zh7	100.57

Step 4. Every extreme aspiration point is different, so no groupings can be made. All objectives are in conflict and the original problem is a true MOLP.

Step 5. All objectives are converted to constraints. For example, maximize zs1 is converted to $zs1 \geq 1577.59 - \epsilon$, where ϵ is about 0.01.

Step 6. The set of IISs identified by using the guide codes follows. Notice how each objective appears in it's respective IIS.

$$\{zs1, zs2, s1, h2, h3\}$$
$$\{zs1, zs2, s1, h2, h3\}$$
$$\{zs1, zs3, s1, h2, h6, h8, h12, h16\}$$
$$\{zs1, zh4, s1\}$$
$$\{zs1, zh5, s1\}$$
$$\{zs1, zh6, s1\}$$
$$\{zs1, zh7, s1, h2, h3, h6, h8, h12\}$$

Frequently occuring elements of the IISs are $zs1$ and $s1$ which are members of all seven IISs. We elect to remove the budget constraint $s1$ entirely, which is equivalent to adjusting the right hand side to a high value.

Step 7. Analysis is not complete, so proceed to Step 3.

Step 3. The new extreme feasible aspiration levels are shown below. Notice that all of the extreme feasible aspiration levels have increased (except $zs3$) with the removal of $s1$.

objective	extreme feasible aspiration level
zs1	1853.88
zs2	185.65
zs3	7437.00
zh4	31.87
zh5	95.58
zh6	44.01
zh7	134.88

Step 4. Every extreme aspiration point is different, so no groupings can be made. All objectives are in conflict and we are still dealing with a true MOLP.

Step 5. All objectives are converted to constraints. For example, maximize $zs1$ is converted to $zs1 \geq 1853.88 - \epsilon$.

Step 6. The set of IISs identified by using the guide codes follows:

$$\{zs1, zs2, s2\text{-}s7, h1, h3\text{-}h5, h8\text{-}h9, h12\text{-}h13, h16\}$$
$$\{zs1, zs2, s2\text{-}s7, h1, h3\text{-}h5, h8\text{-}h9, h12\text{-}h13, h16\}$$
$$\{zs1, zs2, s2\text{-}s7, h1, h3\text{-}h5, h8\text{-}h9, h12\text{-}h13, h16\}$$
$$\{zs1, zh4, s2\text{-}s7, h1\text{-}h5, h8\text{-}h9, h13\}$$
$$\{zs2, zh5, s2\text{-}s7, h1, h3\text{-}h5, h8\text{-}h9, h12\text{-}h13, h16\}$$
$$\{zs2, zh6, s2\text{-}s7, h1, h3\text{-}h5, h9, h13\}$$
$$\{zs1, zh7, s2\text{-}s7, h1, h3\text{-}h5, h9, h13\}$$

The results are less conclusive this time. The IISs contain many more elements, $zs3$ does not appear in any of the IISs generated, there are several common soft constraints ($s2$-$s7$), and all of the IISs contain a lengthy list of hard constraints. Because there is no clear-cut conclusion to be drawn, we construct a normalized objective interference table, as shown in Table 1. Rows in the table are for the extreme aspiration points for the named objective, columns are for the objectives, and the elements are the fractional decreases in the objectives at the points (i.e. $[y^{Bopt} - f^B(x_A^{opt})]/y^{Bopt}$ for table element AB).

Note that the diagonal elements of the table are zero, because each objective

Table 1: Normalized objective interference table.

	zs1	zs2	zs3	zh4	zh5	zh6	zh7
zs1	0.000	0.426	1.000	0.550	0.490	0.435	0.154
zs2	0.103	0.000	0.000	0.063	0.014	0.219	0.016
zs3	0.827	0.816	0.000	0.670	0.795	0.583	0.753
zh4	0.250	0.132	0.000	0.000	0.046	0.258	0.056
zh5	0.000	0.006	0.000	0.092	0.000	0.201	0.023
zh6	0.000	0.357	0.000	0.267	0.423	0.000	0.021
zh7	0.183	0.449	0.000	0.207	0.430	0.053	0.000

reaches it's extreme aspiration level at it's extreme aspiration point.

Since no particular relationship is clearly identified as a candidate for change, the final formulation phase activity is to make recommendations about lexicographic ordering or weighting of the objectives. We choose to use the objective interference table to examine how best to set the lexicographic order.

One way to set the order is to look at which extreme aspiration points have the least negative impact on the other objectives. This can be done in a many ways: average deterioration in objective value, smallest maximum decrease in objective value, etc. For illustrative purposes, we choose to look at the average decline in objective value by calculating the average of the off-diagonal elements in each row of the table, with results as follows:

zs1	zs2	zs3	zh4	zh5	zh6	zh7
.509	.069	.741	.125	.054	.178	.220

With the idea of ordering the objectives so that a greater degree of satisfaction of one objective early has the least negative impact on the later objectives, we can order the objectives based on increasing values of the average decline in objective values: zh5, zs2, zh4, zh6, zh7, zs1, zs3. The reverse of this ordering shows the objectives having the most to the least impact on the others. It is not surprising to see that timber production (zs3) and pasturage (zs1), both land-intensive activities, have the most impact on the other recreation and wildlife objectives. This sort of insight can be used to prepare the final model formulation for solution.

Step 7. Finished formulating, so exit and go to solver. Choosing a lexicographic ordering approach, our final model is thus LEXMAX(zh5,zs2,zh4,zh6,zh7,zs1,zs3) s.t. $x \in X$, with constraint s1 removed.

6. CONCLUSIONS

This paper describes algorithmic tools which assist in formulating MOLPs. Such semi-automated assistance is necessary when the models are large and complex. The principal ingredients of the tools described here are the IIS-isolation approach to LP infeasibility analysis, and the Objective Interference Table. Assistance is provided in analyzing constraint infeasibility as for single-objective LPs, in deciding whether relationships should be represented as objectives or constraints, and in simplifying the model by eliminating objectives, adjusting constraints, or assigning lexicographic ordering or weights to the objectives.

Human input guided by knowledge of the application is needed in deciding the kinds of changes to make to the original model, and in determining when the model is sufficiently well-formulated to present to a solver (either lexicographic goal programming or multiplex).

Empirical testing on a set of real-life problems is needed to confirm the value of these tools.

ACKNOWLEDGEMENTS

This research was partially supported by research grants to the authors provided by the Natural Sciences and Engineering Research Council of Canada.

REFERENCES

J.W. Chinneck (1994). "MINOS(IIS): Infeasibility Analysis Using MINOS", **Computers and Operations Research**, Vol. 21, no. 1, pp. 1-9.

J.W. Chinneck and E.W. Dravnieks (1991). "Locating Minimal Infeasible Constraint Sets in Linear Programs", **ORSA Journal on Computing**, vol. 3, no. 2, pp. 157-168.

A.M. Geoffrion (1987). "An Introduction to Structured Modelling", **Management Science**, vol. 33, no. 5, pp. 547-588.

J. Gleeson and J. Ryan (1990). "Identifying Minimally Infeasible Subsystems of Equations", **ORSA Journal on Computing**, vol. 2, no. 1, pp. 61-63.

H.J. Greenberg (1983). "A Functional Description of ANALYZE: A Computer-Assisted Analysis System for Linear Programming Models", **ACM Transactions on Mathematical Software**, vol. 9, no. 1, pp. 18-56.

H.J. Greenberg (1987). "ANALYZE: A Computer-Assisted Analysis System for Linear Programming Models", **Operations Research Letters**, vol. 6, no. 5, pp. 249-255.

H.J. Greenberg (1990). "An Industrial Consortium to Sponsor the Development of an Intelligent Mathematical Programming System", **Interfaces**, vol. 20, no. 6, pp. 88-93.

H.J. Greenberg (1993). **A Computer-Assisted Analysis System for Mathematical Programming Models and Solutions: A User's Guide for**

ANALYZE, Kluwer Academic Publishers.

J.P. Ignizio (1985). **Introduction to Linear Goal Programming**, Sage Publications, Beverley Hills.

J.P. Ignizio and T.M. Cavalier (1994). **Linear Programming**, Prentice Hall, Englewood Cliffs.

W. Michalowski and T. Szapiro (1992). "A Bi-reference Procedure for Interactive Multiple Criteria Programming", **Operations Research**, vol. 40, no. 2, pp. 247-258.

B.A. Murtagh and M.A. Saunders (1987). **MINOS 5.1 User's Guide**, technical report SOL 83-20R, Systems Optimization Laboratory, Department of Operations Research, Stanford University.

R.E. Steuer (1986). **Multiple Criteria Optimization: Theory, Computation and Application**, Wiley, New York.

R.E. Steuer and A.T. Schuler (1981). "Interactive Multiple Objective Linear Programming Applied to Multiple Use Forestry Planning", publication FWS-1-81, School of Forestry and Wildlife Resources, Virginia Polytechnic Institute and State University.

S. Zionts and J. Wallenius (1983). "An Interactive Multiple Objective Linear Programming Method for a Class of Underlying Nonlinear Utility Functions", **Management Science**, vol. 29, no. 5, pp. 519-529.

APPENDIX

All objectives are to be MAXIMIZED. We have classified objectives as soft or hard based on the problem description given by Steuer and Schuler [1981].

SOFT OBJECTIVES:

$zs1 = 1.67x11 + 2.32x12 + 1.28x13 + 2.68x14 + 2.04x15 + 0.34x41 + 0.44x42 + 0.43x51$

$zs2 = 0.008x11 + 0.057x12 + 0.007x13 + 0.056x14 + 0.024x15 + 0.023x41 + 0.17x42 + 0.169x51 + 0.171x61 + 0.024x62 + 0.025x63 + 0.023x64 + 0.024x66 + 0.024x67 + 0.17x71 + 0.024x72 + 0.023x73 + 0.025x74 + 0.024x76 + 0.024x77 + 0.17x81 + 0.024x82 + 0.023x83 + 0.025x84 + 0.024x86 + 0.024x87$

$zs3 = 95x66 + 47x67 + 46x76 + 23x77 + 85x86 + 42x87$

HARD OBJECTIVES:

$zh4 = 0.007x11 + 0.006x12 + 0.008x13 + 0.005x14 + 0.009x15 + 0.007x41 + 0.014x42 + 0.015x51 + 0.027x61 + 0.013x62 + 0.015x63 + 0.012x64 + 0.027x66 + 0.028x67 + 0.026x71 + 0.014x72 + 0.013x73 + 0.015x74 + 0.029x76 + 0.025x77 + 0.027x81 + 0.014x82 + 0.015x83 + 0.013x84 +$

$0.027x86 + 0.025x87$

$zh5 = 0.019x11 + 0.038x12 + 0.018x13 + 0.039x14 + 0.037x15 + 0.017x41 + 0.038x42 + 0.040x51 + 0.10x61 + 0.009x62 + 0.008x63 + 0.011x64 + 0.019x66 + 0.018x67 + 0.10x71 + 0.007x72 + 0.012x73 + 0.009x74 + 0.019x76 + 0.021x77 + 0.10x81 + 0.008x82 + 0.012x83 + 0.009x84 + 0.018x86 + 0.019x87$

$zh6 = 0.009x11 + 0.010x12 + 0.008x13 + 0.011x14 + 0.007x15 + 0.009x41 + 0.010x42 + 0.009x51 + 0.019x61 + 0.038x62 + 0.035x63 + 0.037x64 + 0.019x66 + 0.018x67 + 0.019x71 + 0.038x72 + 0.037x73 + 0.039x74 + 0.019x76 + 0.020x77 + 0.019x81 + 0.038x82 + 0.039x83 + 0.037x84 + 0.020x86 + 0.019x87$

$zh7 = 0.060x11 + 0.110x12 + 0.062x13 + 0.108x14 + 0.112x15 + 0.059x41 + 0.111x42 + 0.109x51 + 0.027x61 + 0.026x62 + 0.028x63 + 0.029x64 + 0.058x66 + 0.061x67 + 0.027x71 + 0.026x72 + 0.025x73 + 0.029x74 + 0.060x76 + 0.063x77 + 0.027x81 + 0.024x82 + 0.028x83 + 0.025x84 + 0.057x86 + 0.060x87$

SOFT CONSTRAINTS:

$s1$: $21.05x11 + 36.5x12 + 30.86x13 + 30.66x14 + 30.66x15 + 8.34x41 + 10.18x42 + 13.49x51 + 12.39x61 + 12.5x62 + 12.5x63 + 12.5x64 + 10.14x66 + 11.49x67 + 12.39x71 + 12.5x72 + 12.5x73 + 12.5x74 + 10.14x76 + 11.49x77 + 12.39x81 + 12.5x82 + 12.5x83 + 12.5x84 + 10.14x86 + 11.49x87 \leq 25000$

$s2$: $x66 \leq 45$
$s3$: $x67 \leq 5$
$s4$: $x76 \leq 22$
$s5$: $x77 \leq 2$
$s6$: $x86 \leq 21$
$s7$: $x87 \leq 2$

HARD CONSTRAINTS:

$h1$: $x11 + x12 + x13 + x14 + x15 \leq 667$
$h2$: $x12 + x13 \geq 138$
$h3$: $x41 + x42 \leq 162$
$h4$: $x51 \leq 104$
$h5$: $x61 + x62 + x63 + x64 \leq 456$
$h6$: $x61 \geq 46$
$h7$: $x64 \geq 46$
$h8$: $x63 + x64 \geq 182$
$h9$: $x71 + x72 + x73 + x74 \leq 220$
$h10$: $x71 \geq 22$

h11: x74 ≥ 22
h12: x73 + x74 ≥ 88
h13: x81 + x82 + x83 + x84 ≤ 217
h14: x81 ≥ 22
h15: x84 ≥ 22
h16: x83 + x84 ≥ 87

all x ≥ 0

Optimizing the Yield of an Extrusion Process in the Aluminum Industry

K. Masri and A. Warburton

Faculty of Business Administration, Simon Fraser University, Burnaby, B.C., Canada, V5A 1S6

1 Introduction

In this paper we outline a computer system for analyzing and improving the efficiency of aluminum extrusion operations. The system is capable of examining and optimizing all aspects of an extrusion operation. A primitive ancestor of the system was successfully applied at Alcan Vancouver Works, a division of Alcan Aluminum Ltd., where it led, after a capital outlay of approximately $1.2 million for plant modifications, to average annual savings of about $1.0 million. The pay back period for the investment was estimated to have been eight months.

In order to understand the computer system, it is necessary to understand certain factors related to extrusion operations. We begin by describing a typical extrusion process and discussing the variables that determine process efficiency. We then describe the computer system. Finally, we summarize how the Alcan improvements were achieved. Complete technical details regarding the system can be found in [2].

2 The Aluminum Extrusion Process

Aluminum extrusion is a metallurgical process used to produce elongated components of variable lengths and uniform cross-section. The most common application of extrusion is in the production of aluminum siding. Due to the competitive nature of the industry as well as the variable nature of the price of aluminum, it is important to a manufacturer to maintain an efficient extrusion operation. The efficiency of an extrusion operation can be measured by its *recovery rate*; that is, by the ratio of the weight of the metal delivered to the customer as a final product to the weight of "raw" metal used during production.

In essence, the extrusion process consists of forcing a block of material, commonly referred to as a *billet*, through a suitably shaped aperture in a die. The result is a product with a uniform and smaller cross-sectional area than the original block. The work material in the extrusion process may be solid metal (e.g., aluminum, copper, magnesium, steel) or plastic. The main advantage of extrusion over other forming processes (e.g., forging or casting) is the ability to achieve precise dimensional accuracy and shape of the final product in a single deforming operation. Also, the extrusion process provides the final product with excellent surface finish as well as superior mechanical properties [1]. Figure 1 shows a typical extrusion press setup.

The steps involved in extruding a billet can be summarized as follows:

- The billet is preheated and loaded into the extrusion press.
- A ram is used to push the billet into the container and through the die. For reasons to be discussed later, it is not possible to force the entire billet through the die, and a thin disk of material, commonly called the *butt*, is left in the container after the billet is extruded.
- As the billet is forced through the die, the resulting extruded section is "pulled" onto the runout table where it will subsequently be cut into shorter lengths. The end of the extruded section is deformed by the pulling operation and is discarded as *end scrap*.
- The container is separated from the die, the extruded section, and the butt.
- The butt is sheared from the face of the die using a "butt shear" and is discarded as *butt scrap*.
- The shear, container, and ram are returned to their initial position.

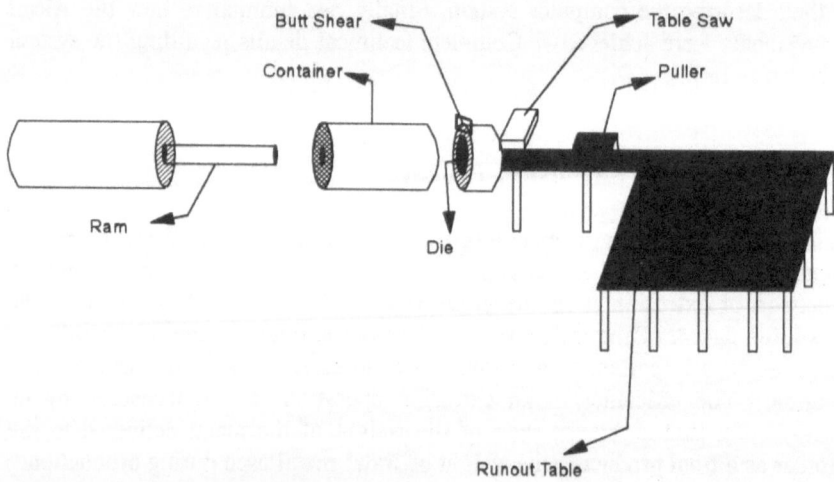

Figure 1. A schematic of a typical press setup.

2.1 Filling Customer Orders Because of long lead times and minimum order quantities, customer orders must be extruded from billets already held in inventory, and the extruder maintains an inventory of billets in several different lengths. A *single customer order* is defined by the *shape* of the extrusion required by the customer, the *amount of metal* ordered, and the desired *cut length* (the length of the final pieces ordered by the customer). If a customer requires the same shape with two different cut lengths, then the order is considered to be two distinct orders. For a number of reasons related to metal processing considerations, a single customer order is filled using billets of a single fixed length (chosen from the inventory of billets). Once a billet length has been chosen to fill an order, it is customary for the extruder to round the amount of metal ordered up or down so that an integral number billets of that length will be required to fill the order. Each billet is extruded, butt scrap and end scrap are discarded, and the remaining extruded section is cut into pieces of the length ordered by the customer. Thus, each billet used to fill the order yields the same amount of final product. It follows that the recovery rate for a single customer order is the same as the recovery rate for each individual billet used to fill it.

3 Factors Affecting Recovery Rate

Several controllable variables affect the recovery rate for a single customer order. We now consider these variables in more detail.

3.1 Minimum Butt Thickness During the extrusion process, the pressure applied by the ram to force the metal through the die is a function of the remaining billet length. Due to material flow principles, once the billet reaches a critical length inside the container, the pressure required to extrude the remaining metal will increase so much that the press will not be able to supply the necessary force to continue the extrusion of the billet. Therefore, a thin disk of material (the butt scrap) remains un-extruded, and is disposed of as scrap metal. It is important to note that the minimum butt scrap is *not* dependent on the original length of the billet and that a certain minimum buttscrap will remain whether a short or a long billet is used for extruding. For example, the current press setup at Vancouver Works requires a minimum butt thickness of 25 mm. So, for a 650 mm billet, a 25 mm minimum butt scrap would imply a minimum loss of at least $100(25/650)\% = 3.85\%$ in recovery rate.

3.2 Minimum and Maximum Extrudable Billet Lengths Due to material flow principles, the physical characteristics of the press, and the maximum pressure that can be applied by the press, there are certain minimum and maximum billet lengths that the press is able to extrude. At Vancouver Works, the press can handle minimum and maximum billet lengths of 400 mm and 850 mm, respectively.

3.3 End Scrap A further loss in recovery rate can be attributed to the end scrap generated during the extrusion operation. Prior to cutting the extruded metal into the pre-required customer-specified order lengths, each extrusion requires straightening. The straightening process involves "stretching" the metal. The stretching operation deforms the ends of the extrusion to such an extent that the end pieces become unsalvagable and must be scrapped (thus the term "end scrap"). The length of the end scrap depends only on the type of jaws used to grasp the metal during the stretching operation. At Alcan Vancouver Works, the length of the end scrap per extrusion is currently *constant* at approximately 2 m.

3.4 Runout Table Length The runout table length also affects recovery rate. If the runout table length is sufficiently long, billets of any extrudable length can be processed in a single "push" without exceeding the length of the runout table. Thus a long runout table affords the extruder flexibility when choosing a billet length to fill an order. For reasons to be discussed in 3.6 below, it is very undesirable to use a billet that is too long to be processed in a single push.

3.5 Billet Lengths A *press setup* is defined by the minimum and maximum extrudable lengths, the minimum butt thickness, the end scrap length and the runout table length. Ideally, *given a press setup*, a single customer order would be filled using billets of precisely the length that would yield, upon full extrusion, minimum butt thickness, the end scrap, and the maximum possible integral number of cuts that will fit on the runout table. This easily calculated length, called the *ideal billet length for the order*, yields the maximum possible recovery rate for the order given the press setup. This rate is called the *ideal recovery rate for the order*.

Given a press setup, the ideal recovery rates for each order in a set of single customer orders can be aggregated by order weight to determine the *ideal recovery rate for the entire set of orders*. In order to achieve the ideal recovery rate for an order set, it would be necessary to hold in inventory the ideal billet length for each individual order. Unfortunately, because only a few billet lengths are held in inventory, each order must be filled by selecting from the few lengths held in inventory the length that will yield the highest recovery rate for that order. Thus, the ideal recovery rate for a set of orders provides a *de facto* upper bound on the actual recovery rate that an extrusion facility can achieve. For several years, the ideal recovery rate at Alcan Vancouver Works has fluctuated minimally around 91%.

3.6 Multiple Pushes From time to time, an order will be received whose ideal billet length is less the minimum billet length that can be accepted by the extrusion press. In such a case, the extrusion operation must be halted temporarily when the extruded section reaches the end of the runout table. Because orders requiring multiple pushes slow production, they are considered to be very undesirable. An important secondary objective when modifying an extrusion operation is to try to minimize the frequency of multiple pushes. See [2] for a discussion of this issue, together with details on how the calculation of the ideal billet length changes when a multiple push is unavoidable.

4. The Billet Selection System

We are now in a position to describe our billet selection system. It is convenient to view the system in terms of its inputs and outputs.

Billet Selection System

Inputs
- Press setup parameters (controllable)
  minimum butt length
  endscrap length
  minimum extrudable billet length
  maximum extrudable billet length
  runout table length

- Inventory parameter (controllable)
  number of billets held in inventory

- Set of orders to be filled (not controllable)

Outputs
- Recovery maximizing set of billet lengths and amounts of each length

To obtain its outputs, the billet selection system must solve the *billet selection problem,* defined as follows: Suppose that we are given a press setup and a set of orders to fill. Further, suppose that we wish to fill these orders from a set of k distinct billet lengths held in inventory. What should these k lengths be and how much of each length should be stocked in order to maximize the recovery rate for the set of orders?

The system solves the billet selection problem by formulating and solving it as a combinatorial optimization problem. In effect, billets can be ordered in 1 mm length increments, so if LMAX and LMIN are, respectively, the maximum and minimum allowable billet lengths associated with the press set up, the billet selection problem can theoretically be solved by examining each and every possible combination of k billet lengths chosen from the LMAX-LMIN+1 lengths available. This "brute force" approach rapidly becomes impossible to implement as k and the number of orders increase. Therefore the system uses a variety of local search and related heuristics to obtain suboptimal solutions to the billet selection problem.

The performance of our heuristics has been checked for various values of k and order sets by comparing the heuristically obtained suboptimal solutions with the brute force solutions obtained by running a 80486 PC for many days. In every case, the heuristics produced optimal or nearly optimal solutions. If k is large, many days on the PC (or years) is not sufficient time to produce an optimal solution by brute force, and more sophisticated methods are required to check the accuracy of the heuristics. For these problems, we are currently developing a subgradient algorithm to solve the LP relaxation of the large scale integer programming problem that results from formulating the billet selection problem as an uncapacitated plant location problem. Solving the LP will provide an good upper bound on the achievable recovery rate. (See [3] for a discussion of the plant location problem). We note also that the billet selection problem can be modeled as a nonconvex optimization problem with k continuous variables, one for each billet length. Because of the nonconvexities involved, and the presence of many local optima, we have not actively pursued nonlinear programming approaches to the problem.

5. Using the Billet Selection System at Alcan Vancouver Works

We will use Alcan Vancouver Works to illustrate some of the ways in which the billet selection system has been and could be used as a tool for analyzing and improving the recovery rate of aluminum extrusion operations. The Alcan plant processes an average of about 3000 customer orders per year, and there are approximately 3500 different shapes that can be extruded using dies at the plant. In 1989, prior to the recovery improvement project, the plant used a butt shear that could provide a minimum butt thickness of 50 mm. The jaws used for pulling the extrusion deformed a total of 2.9 m per extruded section, independently of the total length of the extrusion. The runout table was short and there existed considerable potential for lengthening it. Finally, the company stocked three billet lengths for filling orders, with the actual lengths being chosen on the basis of "experience". With these press setup and inventory parameters, the plant's recovery rate was deemed to be unacceptably low. In fact, management considered an improvement of the recovery rate to be an essential condition for the survival of the extrusion operation at the Vancouver Works.

A primitive "brute force" and "coarse approximating" ancestor of the present billet selection system was used to generate information that led directly to a 12% increase in the recovery rate at the plant. To see how this improvement was achieved, first note that, as mentioned previously, long lead times and minimum order quantities make it necessary to fill orders using billets already in inventory at the times the orders are received. Therefore, the system had to estimate optimal billet lengths for present and future orders by using a set of historical order data. During the summer of 1989, using the previous six months of order data, various press setup configurations were tested with three billet lengths held in inventory. The runs were repeated with four lengths. Of the four billet lengths, three were free to vary, and the fourth was set at the maximum allowable billet length of 850 mm. At the time, it was deemed essential -- though later found to be unnecessary -- to stock the longest billet length in order to ensure that shapes having high weight per unit of length could be accommodated.

Associated with each run was a predicted recovery rate. Each percent increase in recovery rate led to an estimated annual saving of between $50k and $125k per year, depending on the volume of future orders and the price of aluminum. On the other hand, there were very substantial costs associated with changing the press setup. However, changes to inventory policy, including increasing the number of lengths held from three to four, could be accommodated at negligible cost.

After reviewing the outputs from the run, the press setup and the inventory policy were modified as shown below. The estimated recovery rate improvements attributable to each of the four changes are shown in square brackets.

- The runout table was lengthened by 12m (to the property line -- any further lengthening would have been prohibitively expensive). [2.3%].
- The butt shear was replaced with a sharp shear that reduced the minimum butt thickness by 50% to 25 mm. [4.4%].
- New jaw grips were added that reduced the buttscrap length by 0.9 m. [2.3%]
- The number of billets kept in inventory was increased to four. The lengths stocked were those chosen by the billet selection system. [3.6%].

Each of the modifications made a significant contribution to the improvement of the recovery rate. In total , the model predicted that the above changes would improve the recovery rate by 12.6% The payback period for the $1.2M required to modify the plant was predicted to be 12 months. The actual increase in recovery rate was 12%, and the actual pay back period for the modification expenses was only 8 months. After the initial modifications, the model was rerun every six months or so using the new press setup and the previous six month of order data. The inventory lengths held in stock remained quite stable and the improved recovery was generally maintained.

5.1 Parametric Analysis of Recovery Rate Many extrusion plants maintain more than four billet lengths in inventory, and, given the modified plant setup at Alcan, we decided to use the billet selection system to investigate the recovery improvements possible by keeping more billet lengths in stock. If the cost savings resulting from improved recovery were deemed to be more than the costs and operational complications associated with stocking more lengths, then the inventory policy could be altered accordingly. We also decided to investigate at the same time how well historical order data could be used to estimate optimal billet lengths for future order sets. These investigations required the full computational power of the current billet selection system and could not have undertaken using its 1989 ancestor.

We used Alcan Vancouver Works' production order data from July 1992 to May 1993. The data was divided into two sets. The first set, called the *92 data set*, includes the order data from July 1992 to December 1992. The *93 data set* contains the data from January 1993 to May 1992. The following computations were performed using as inputs the current Alcan plant setup and from 1 to 10 different billet lengths in inventory.

1. The 92 data set was initially used to determine the recovery maximizing billet lengths for the 92 data set. The suggested billet lengths and the ensuing maximum recovery rate, 92 OPT, were noted.
2. The billet lengths suggested for the 92 data set were used as the *actual* billet lengths to fill the orders in the 93 data set. The resulting suboptimal recovery rate for the 93 data set will be referred to as 93 SUB. Note that 93 SUB is the recovery rate that would actually be achieved (exclusive of waste caused by uncontrolled factors such as operator error) if the 93 orders were filled using billets of the length suggested by the optimization over the 92 data set.
3. The billet selection system used the 93 data set to determine the calculate the maximum recovery rate, 93 OPT, for the 93 data set.

The results of the above computations appear in Table 1.

Table 1. Recovery Rates as a Function of Number of Billets in Inventory.

	Number of Billets									
	1	2	3	4	5	6	7	8	9	10
92 OPT	80.73	84.46	85.92	86.81	87.45	87.97	88.29	88.54	88.74	88.96
93 OPT	80.83	84.97	86.40	87.24	87.84	88.29	88.58	88.80	88.99	89.16
93 SUB	80.59	84.97	86.40	87.20	87.71	88.23	88.58	88.71	88.99	89.12

As noted earlier, the ideal recovery rate at Alcan under the current press setup has fluctuated minimally through time around 91 percent. Looking at the 92 OPT and 93 OPT rows of the table we see how the maximum recovery rate increases towards the ideal rate --- rapidly at first, then more slowly as the number of billets increases beyond 7 or 8. We also see that the maximum recovery rates are similar in both time periods.

Comparing the 93 OPT and 93 SUB rows, we see that using historical data to predict optimal billet lengths was remarkably effective for the two time periods considered. Order data is no longer available to compare earlier periods in this way, However, historically, actual recovery rates, with approximately 6 month intervals between billet length optimizations, have remained quite stable.

We also performed experiments similar to those just described but using shorter time periods for the optimizations. The ensuing recovery rates were inferior to those obtained using 6 months of data for optimization, suggesting that better yields are obtainable using longer term order data than by trying to respond to month to month fluctuations in orders.

5.2 Postscript In spite of the recovery improvements already achieved, Alcan Vancouver Works has recently closed permanently. The closure came after a long strike at the plant, and appears to be part of a much larger global divestiture by Alcan Aluminum Ltd. of "non-core" operations such as extrusion. At the time of the closure, the desirability of increasing the number of billets lengths from the current four to six or seven was under investigation. Such a change would have led to predicted increase in recovery rate of about 2%.

References

[1] Blazynski, T.Z. (1986). *Design of Tools for Deformation Processes*, Elsevier Applied Science Publishers, New York.
[2] Masri, K. (1994). *Improving Extrusion Efficiency Alcan Aluminium Ltd. Vancouver Works*, MBA Thesis, Faculty of Business Administration, Simon Fraser University, Burnaby, Canada .
[3] Nemhauser, G. and Wolsey, L. (1988). *Integer Programming and Combinatorial Optimization*, Wiley Interscience Series in Discrete Mathematics and Optimization.

Projective and Symbolic Degeneracy-Reducing Techniques for Multiple Objective Linear Programming

Jean-Michel Thizy[1,2]

[1] Faculty of Administration, University of Ottawa, 136 J.-J. Lussier, Ottawa K1N6N5 Canada

Abstract. Multiple objective linear programming (MOLP) is almost universally implemented within a simplex algorithm and relies on some of its properties, in particular the use of bases. Alternative linear programming methods such as the projective method need not use bases. Yet, their most common approach to post-optimal analysis has been conservative, e.g. reconstructing a basis in order to use the classical framework. A second, less investigated approach, is to adapt the scope of MOLP to the new methods. The simplex and the new methods are affected differently by degeneracy that causes numerical difficulties both in the data representation and the solution process. Symbolic solvers can mitigate such difficulties. Formulating MOLP as disjunctive optimization related to logic programming, the constraint logic program CLP($\Re$) is selected for its symbolic treatment of algebraic constraints seamlessly embedded in a Prolog syntax.

Keywords. Multiple objective programming, interior point methods, logic programming

1 Introduction

Virtually all methods of multiple objective linear programming (MOLP) rely on the simplex algorithm (Dantzig, 1963). Recently, alternative methods of solving linear programs have become very popular, such as interior (Karmarkar, 1984), proximal point methods (Rockafellar, 1976; Wright, 1989), overrelaxation (Mangasarian, 1981), neural networks (Adler, 1994). Since the simplex algorithm has been the quasi-universal method of solving linear programs for several decades, its approach has influenced MOLP deeply and analysts have come to expect its framework. In fact, the simplex algorithm presents many advantages over its competing solution techniques, for example the calculation of extreme points and bases. One strategy to introduce new approaches is to equip them with the classical mechanisms provided by the simplex algorithm, such as post-optimal analysis. This article adopts a direct approach, i.e. surveys what aspects of MOLP can be readily answered by the new linear programming methods. The focus is placed on the exploration of the efficient region of an MOLP, as symbolized by the program:

[2] Supported by NSERC Grant OGP0042197 and AUCC Going Global Program.

eff {Cx}
s.t.
$Ax = b$
$x \geq 0$

where x is a vector of n variables, A is an m·n matrix, b is a vector of m coefficients, and C is a k·n matrix representing the criteria that define an efficient region.

Section 2 briefly motivates the study of large-scale MOLP. In Section 3, the absence of bases, an apparent liability of interior point methods, is first compensated by asymptotic characterizations via penalties, which are compared to those of goal programming. The lack of bases is then exploited to alleviate the effects of primal degeneracy which encumbers simplex-based MOLP. Degeneracy is viewed as a discrepancy between the symbolic representation of MOLP by constraints and its numerical solution. In response, Section 5 represents an MOLP solution as a system of constraints, using a constraint programming language that removes some of the preceding difficulties.

2 Large-Scale MOLP

Given the reputed difficulty of reporting the solution of a traditional MOLP, one may wonder whether large-scale MOLP is a relevant subject of investigation for practical decision-making. For example, the efficient region of an MOLP often consists of a number of faces that increases exponentially with the number of variables in the program. Thus, there exists no known polynomial procedure to describe the efficient region of every MOLP, which seems to preclude large-scale MOLP. However, there are many subclasses of large MOLP for which the efficient region can be described succinctly, such as those that can be characterized as:

eff {$Cx + Dy$}
s.t.
$Ax + Ey = b$
$x, y \geq 0$

where y is a vector of p variables (typically p is much greater than n), E is an m·p matrix, D is a k·p matrix formed by k identical rows. Although it is a large scale MOLP, the number of its efficient faces is related to the relatively small number n of variables x.

3 Interior Point Methods for Linear Programming

Preceded by some important theoretical analyses (Dikin, 1967; Khachiyan, 1979), the projective method of linear programming (Karmarkar, 1984) has generated widespread attention because it has provided the base for fast solution of very large linear programs on standard uniprocessors, and it is amenable to parallelization. The method differs from the simplex method that computes vertices of the feasible region. In contrast, the new method can produce a sequence of points interior to the feasible region. Karmarkar's polynomial-time, interior point algorithm and its ensuing affine scaling implementations (Barnes, 1986; Vanderbei, 1986) have revived interest in many applications of linear programming beyond the reach of the simplex method.

3.1 The Challenging Representation of Interior Points

The application of interior point methods forces a review of many aspects of linear programming and related disciplines such as MOLP. For example, interior point methods do not seek optimal extreme points. As an illustration, consider the linear program:

max $2 x_1 + 2 x_2$
s.t.
$x_1 + x_2 = 1$
$x_2 + x_3 = 1$
$x_1 , x_2 , x_3 \geq 0$.

A simplex algorithm yields the optima (1 0 1) or (0 1 0), whereas an interior point method may yield the solution (.5 .5 .5) . In the close interplay between MOLP and the simplex algorithm, the extreme solutions found by the simplex algorithm have provided a decisive help, while generating extreme points of an MOLP has itself come to be accepted as a norm. On the contrary, an interior point method can produce (.5 .5 .5) as an efficient solution of the following MOLP, e.g. by applying the standard MOLP method of point estimate weighted sum that can yield the preceding linear program.

eff $\{ x_1 - x_3 , x_1 + 3 x_2 + 2 x_3 \}$
s.t.
$x_1 + x_2 = 1$
$x_2 + x_3 = 1$
$x_1 , x_2 , x_3 \geq 0$.

Typical MOLP enumeration starts with one of the extreme optima and explores adjacent efficient vertices (Geoffrion, 1972; Zionts, 1976). Therefore it is attractive to devise procedures that can process an interior solution (.5 .5 .5) and *purify* it to yield an extreme optimum, thus fitting well-known procedures for MOLP. Such an approach has been adopted by commercial interior point software for sensitivity analysis of linear programs, i.e. retrieving a basis in order to report the requested information under the classical format. Although some polynomial algorithms have been devised to extract a basis from an optimal linear programming solution (Megiddo, 1991), the procedure may be slow in practice.

A direct interior point implementation avoiding basis retrieval could be patterned after parametric analysis of linear programs which is designed to produce neighboring vertices (Adler, 1992; Mehrotra 1992). However, even with an interior point method, most parametric approaches construct bases because direct reoptimization is prone to inefficiency, for reasons explained in the following section.

3.2 An Opportunity: Asymptotic Interior Efficiency and Goal Programming

Interior point methods have been increasingly treated as asymptotic logarithmic barrier methods, an equivalence first analyzed in (Gill, 1986):

$$\lim_{\mu \to 0} \min cx - \mu \log \mathbf{1}x$$
s.t.
$$Ax = b$$
$$x \geq 0$$

where μ is the barrier parameter and $\mathbf{1}$ is a vector of coefficients 1 of dimension conformable with x. The calculation of an improving step close to the optimum, i.e. when the barrier parameter μ becomes small, may encounter numerical difficulties (Megiddo, 1989). Thus, on the negative side, interior points are not well suited to perform exact reoptimization in a search for a neighboring vertex, unlike classical post-optimal analysis. On the positive side, interior point methods can be adapted to MOLP by allowing for tolerances, such as a search for a solution of the approximate program:

$$\min_{\mu \geq \varepsilon} \min cx - \mu \log \mathbf{1}x$$
s.t.
$$Ax = b$$
$$x \geq 0$$

where ε denotes a prespecified tolerance value. This approach resembles loosely a non-linear goal programming method (Charnes, 1961):

$$\min_{\mu \geq 0} \min cx + f(\mu,x)$$
s.t.
$$Ax = b$$
$$x \geq 0$$

The resemblance is further enhanced by new variants of interior point methods (Adler, 1991) that tailor the barrier parameters to each nonnegativity constraint, yielding the approximation:

$$\min_{\mu \geq \varepsilon} \min cx - \log \mu x$$
s.t.
$$Ax = b$$
$$x \geq 0$$

in which μ and ε are vectors of parameters. In summary, interior point methods lead naturally to approximate MOLP analysis or goal programming. As a fine difference, whereas goal programming penalizes departures from prespecified objective goals, the barrier parameters μ estimate duality gaps, which can offer some advantages such as scalability and adaptability. Barrier methods are therefore related both to the multiple dual of hierarchical goal programming (Ignizio, 1993) and approximate sensitivity analysis (Hansen, 1989).

3.3 Relieving Degeneracy with an Interior Point

For all its success in linear programming, the use of bases is hindered by degeneracy (Gal, 1990). Consider the following primal degenerate program:

eff $\{ x_1 , x_2 \}$
s.t.
$$x_1 + x_2 \leq 1$$
$$x_1 + 2x_2 \leq 2$$
$$x_1 , x_2 \geq 0$$

and suppose that an efficient point is calculated, using a set of criterion multipliers (.5 .5). The resulting simplex tableau is:

1	1	1	0	1
-1	0	-2	1	0
1	0	0	0	0
-1	0	-1	0	-1

where the last two rows evaluate the two criteria. In standard algorithms for MOLP, replacing the fourth column by the third one in the basis does not yield an efficient adjacent point. Tableau-oriented procedures to avoid such degenerate pivots may still take an exponential number of pivots to reach an efficient adjacent point in the worst case.

Since every coordinate of any interior point is positive, its calculation is not affected by degeneracy in a strict sense, which favors interior characterizations of efficiency. However, if a neighboring vertex is degenerate, the numerical difficulties mentioned with the barrier method will increase. The approximate approach suggested in the preceding section should naturally alleviate such situations. In fact, the logarithmic barrier function (Gill, 1986) demonstrates the effectiveness of truncated searches for degenerate optima, pointing out that interior point methods yield good approximate solutions more reliably than the simplex algorithm.

Approximate reoptimization is of course foreclosed if the current efficient point lies exactly at a degenerate vertex. In this case, one can turn again to the postoptimal analysis of degenerate linear programming optima. Technically, degeneracy can be caused either by null variables, i.e., variables that remain zero in the entire feasible region (a case that can be remedied practically) or by several bases representing the same point, which simplex-based procedures may enumerate unwittingly. Interior point methods can obviate such difficulties either by replacing bases with partitions (Adler, 1992; Jansen, 1993), or by a direct implementation of sensitivity and parametric analyses (Mehrotra, 1992) that may require lengthier calculations.

4 Symbolic MOLP

In summary, degeneracy occurs when several systems of constraints of a linear program define the same solution. This multiple definition can be avoided by representations in which the solution *is* the defining system itself, such as symbolic linear programs (Wolfram, 1991, Complete Logic Systems, 1987; Konopasek, 1984). The description of the efficient region of an MOLP is more requiring than that of an optimal linear programming solution because the former does not constitute a convex region, but a disjunctive system of constraints. Constraint logic programming languages (Brown, 1989; Cohen, 1990), designed to treat combinations of logic and algebraic specifications, are well equipped to represent disjunctive systems symbolically.

The constraint logic programming language CLP($\Re$) (Jaffar, 1990) extends the semantics of Prolog (Colmeraurer, 1982), the best-known declarative logic programming language. It recognizes as privileged predicates the constraints formed by arithmetic expressions over the real numbers and relations such as equality and inequalities ($\leq$, $\neq$, $\geq$). An important property of CLP($\Re$) is that the constraints are treated uniformly in the sense that they are used to specify the input parameters to a program, they are the only primitives used in the execution of a program, and they are used to describe the output of a program. Although CLP($\Re$) can evaluate complex arithmetic expressions, it is illustrated here by simple examples. Consider the exchange of Japanese Yens (Y) for ECUs (X). The exchange is expressed by the predicate:

```
exchange_rate( X, Y ) :- X = 0.0082 * Y.
```

and thus, the goal

```
?- exchange_rate( X, 2 ).
```

queries the ECU value of 2 Yens. Such a program can be written in standard logic programs such as Prolog. However, the goal

```
?- exchange_rate( 0.0164, Y ).
```

succeeds in the CLP($\mathfrak{R}$) whereas it would fail in Prolog where the equality simply denotes the assignment of an arithmetic expression to a variable (performed from right to left), not a constraint. Many extensions of Prolog propose to handle equations, using *local propagation*, but are rarely able to solve systems of simultaneous equations. In contrast, CLP($\mathfrak{R}$) is able to respond correctly when queried about the following simultaneous equations:

```
?- 4*X + 2*Y = 12,  2*X + 4*Y = 12.
```

Given a set of constraints, the solver first determines the solvability of that set and, if solvable, computes its solution. CLP($\mathfrak{R}$) uses a modified simplex method (Shambin, 1974) to solve sets of linear constraints. The solver returns the solution:

```
X = 2, Y = 2.
```

In fact, CLP($\mathfrak{R}$) produces symbolic output: as constraints are defined syntactically, not only do they operate as tests, but they can define results. For example the following goal:

```
?- X1 >= 1, X1 >= 4, X1 =< 5.
```

returns the answer below:

```
X1 >= 4, X1 =< 5.
```

Answering a goal is defined in terms of satisfiability of the corresponding set of constraints and not in terms of finding a value. At each step of the process, the interpreter has to deal with only a single question: are the constraints solvable? Therefore we need to represent the efficient region of an MOLP as a system of constraints.

Property (Steuer, 1986, Theorem 9.5):
A solution x of the program: eff $\{Cx\}$, $Ax=b$, $x \geq 0$, is efficient if and only if there exists a solution ($u>0$, $v \geq 0$, y) of the system: $uC + vD + yA = 0$, where D is a diagonal matrix with components 1 for vanishing coordinates of x.

For example, we represent the degenerate program:

eff $\{ x_1 , x_2 \}$
s.t.
$$x_1 + 2x_2 \leq 2$$
$$2x_1 + x_2 \leq 2$$
$$3x_1 + 3x_2 \leq 4$$
$$x_1 , x_2 \geq 0$$

as the following system of constraints with additional slack variables. The CLP($\Re$) program is simplified for an easy interpretation of the preceding property by giving each component of the variables an explicit name. Unfortunately, this scheme yields a cumbersome definition of the matrix D.

```
/*              Definition of the diagonal matrix        */

d1(1,0).
d1(0,X):-X>0.
d2(1,0).
d2(0,X):-X>0.
d3(1,0).
d3(0,X):-X>0.
d4(1,0).
d4(0,X):-X>0.
d5(1,0).
d5(0,X):-X>0.


/*              Definition of the feasible region        */

feasible(X1,X2,X3,X4,X5):-   X1+ 2*X2+ X3              =  2,
                             2*X1+   X2      + X4      =  2,
                             3*X1+ 3*X2            + X5 =  4,
                             X1>=0, X2>=0, X3>=0, X4>=0, X5>=0.

/*              Definition of the efficient region        */

eff(X1,X2):- feasible(X1,X2,X3,X4,X5),
             U1+ D1*V1+    Y1+ 2*Y2+ 3*Y3 = 0,
             U2+ D2*V2+ 2*Y1+    Y2+ 3*Y3 = 0,
                 D3*V3+    Y1             = 0,
                 D4*V4        +    Y2      = 0,
                 D5*V5             +    Y3 = 0,
             U1>0, U2>0, V1>=0, V2>=0, V3>=0, V4>=0, V5>=0,
             d1(D1,X1),
             d2(D2,X2),
             d3(D3,X3),
             d4(D4,X4),
             d5(D5,X5).
```

A concise, but less explicit CLP($\Re$) program will index each component as in:

```
/*              Definition of the diagonal matrix         */
d([],[]).
d([0|D_end],[X|X_end]):- d(D_end,X_end), X>0.
d([1|D_end],[0|X_end]):- d(D_end,X_end).
```

The efficient region of the program is given by the disjunctive series of system of constraints produced by querying CLP($\Re$) with the goal: eff(X1,X2).

```
CLP(R) Version 1.1
(c) Copyright International Business Machines Corporation
1989 (1991) All Rights Reserved

1 ?- eff(X1,X2).
X2 = 1
X1 = 0

*** Retry?   yes
X2 = 0
X1 = 1

*** Retry?   yes
X2 = 0.666667
X1 = 0.666667

*** Retry?   yes
X1 = -2*X2 + 2
0.666667 < X2
X2 < 1

*** Retry?   yes
X1 = -0.5*X2 + 1
X2 < 0.666667
0 < X2

*** Retry?   yes
*** No
```

Note that the output offers a minimal representation of a partition of the efficient region as a normal disjunctive system of constraints.

5 Conclusion

To manage constraints over the real numbers, CLP($\Re$) needs to rely on numerical approximations. It is hampered by two factors:
- speed: symbolic languages, as CLP($\Re$), are slow and practically not competitive with numerical linear programming solvers,
- precision: the proper selection of constraints to represent the solution hinges on a sophisticated, but slow output interface and some judicious tolerances chosen for the symbolic solver, which will degrade with the size of the problem and the relative magnitude of its data.

Although constraint logic programs such as CLP($\Re$) seem to be effective only for prototyping MOLP systems, its proponents contend that they offer a sound base for much faster implementation of concurrent constraint logic programming. Even within the sizes of problems that can be practically solved by CLP($\Re$), an immediate challenge is the effective and succinct representation of efficient regions, for example as efficiency graphs (Gal, 1994). An advantage of constraint programming is that the declaration of the program is in itself a representation of the efficient region, that can amended, expanded or simply called by further procedures, e.g. to narrow the region by a decision-maker's utility function.

References

Adler, I., Monteiro, R.D.C.: Limiting behaviour of the affine scaling continuous trajectories for linear programming problems. Mathematical Programming **50** (1991) 29-51

Adler, I., Monteiro, R.D.C.: A Geometric View of Parametric Linear Programming. Algorithmica **8** (1992) 161-176

Adler, I., Verma, S.: Linear and Convex Optimization by Neural Networks. ORSA Computer Science Technical Section Conference, Williamsburg, Virginia, January 6, 1994

Barnes, E.R.: A Variation on Karmarkar's Algorithm for Solving Linear Programming Problems. Mathematical Programming **36** (1986) 174-182

Brown, R.G., Chinneck, J.W., Karam, G.M.: Optimization With Constraint Programming Systems. in Impact of Recent Computer Advances on Operations Research, R. Sharda ed., North Holland (1989) 463-473

Charnes, A., Cooper, W. : Management Models and Industrial Applications of Linear Programming, John Wiley & Sons, New York, 1961

Cohen, M.: Constraint Logic Programming Languages. Communications of the ACM (1990) 52-68

Colmerauer, A.: Prolog and Infinite Trees, Groupe Intelligence Artificielle, Faculté des Sciences de Luminy, Université Aix-Marseille II, 1982

Complete Logic Systems: Trilogy User's Guide, North Vancouver, British Columbia, Canada, 1987

Dantzig, G.B.: Linear Programming and Extensions, Princeton University Press, Princeton, N.J., 1963

Dikin, I.I.: Iterative Solution of Problems of Linear and Quadratic Programming. Soviet Mathematics Doklady. **8** 3 (1967) 674-675

Gal, T.: Degeneracy problems in mathematical programming and degeneracy graphs. ORION **6** 1 (1990) 3-36

Gal, T., ed.: Degeneracy in Optimization Problems. Annals of Operations Research, 1994

Geoffrion, A.M., Dyer, J.S., Feinberg, A.: An Interactive Approach for Multiple Optimization with Applications to the Operation of an Academic Department, Management Science **19** (1972) 357-368

Gill, P.E., Murray, W., Saunders, M.A., Tomlin, J.A., Wright, M.H.: On Projected Newton Barrier Methods for Linear Programming and an Equivalence to Karmarkar's Projective Method. Mathematical Programming **36** (1986) 183-209

Hansen, P., Labbé, M., Wendell, R. E.: Sensitivity analysis in multiple objective linear programming: The tolerance approach. European Jour. of Operational Research **38** 1 (1989) 63-69

Ignizio, J.P.: Goal Programming and Extensions, D.C. Heath, Lexington, Massachusetts, 1993

Jaffar, J., Michaylov, S., Stuckey, P.J., Yap, R.H.C.: The CLP($\Re$) Language and System. Technical Report 16292 (#72336), IBM Research Division, Yorktown Heights, 1990

Jansen, B., Roos, C., Terlaky, T.: Optimal Bases versus Optimal Partitions for Postoptimal Analysis in Linear Programming, Technical Report 92-21, Fac. of Techn. Math./Computer Science, Delft University of Technology, revised in May 1993

Karmarkar, N.: A New Polynomial-Time Algorithm for Linear Programming. Combinatorica **4** (1984) 373-395

Khachiyan, L.G.: A Polynomial Algorithm in Linear Programming. Soviet Mathematics Doklady **20** 1 (1979) 191-194

Konopasek, M., Jayaraman, S.: The TK!Solver Book, Osborne/McGraw-Hill, Berkeley, Ca. 1984

Mangasarian, O.L.: Iterative solution of linear programs. SIAM Journal on Numerical Analysis **18** (1981) 606-614

Megiddo, N., Shub, M.: Boundary Behavior of Interior Point Algorithms in Linear Programming. Mathematics of Operations Research 14 (1989) 97-146

Megiddo, N.: On Finding Primal- and Dual-Optimal Bases. ORSA Journal on Computing **3** 1 (1991) 63-65

Mehrotra S., Monteiro, R.D.C.: Parametric and Range Analysis for Interior Point Methods, Technical Report, Department of Industrial Engineering, University of Arizona, April 1992

Rockafellar, R.T.: Monotone Operators and the Proximal Point Algorithm. SIAM Journal on Control and Optimization **14** 5 (1976) 877-898

Shambin, J.E., Stevens Jr., G.T.: Operations Research: A Fundamental Approach, McGraw-Hill, New York, 1974

Steuer, R.E.: Multiple Criteria Optimization: Theory, Computation, and Application, John Wiley & Sons Inc., New York, 1986

Vanderbei, R.J., Meketon, M.S., Freedman, B.A.: A Modification of Karmarkar's Linear Programming Algorithm. Algorithmica **1** (1986) 395-407.

Wolfram, S.: Mathematica: a system for doing mathematics by computer, Addison-Wesley Pub. Co., Redwood City, Calif., 1991

Wright, S.J.: Implementing Proximal Point Methods for Linear Programming. Mathematics and Computer Science Division, Argonne National Laboratory, preprint MCS-P45-0189, 1989

Zionts, S., Wallenius, J.: An Interactive Programming Method for Solving the Multiple Criteria Problem. Management Science **22** (1976) 652-663

Interactive Multiple Criteria Optimization for Capital Budgeting in a Canadian Telecommunications Company

Jean-Michel Thizy[1], Savvas Pissarides[2], Surendra Rawat[1,3] & Daniel E. Lane[1]⋆

[1] Faculty of Administration, University of Ottawa, Ottawa ON K1N 6N5, Canada,
[2] Bell Canada, 160 Elgin St., Ottawa, Ontario, K1G 3J4 Canada,
research conducted while at the University of Ottawa,
[3] Stentor Research Centre Inc., 160 Elgin St., Ottawa, Ontario, K1G 3J4 Canada.

Abstract. Decision Support System for Optimal Resource Allocation (DSS ORA) is an interactive mathematical programming system for optimal resource allocation developed to support decisions of investment in capital intensive telecommunications projects. The system strives to maximize corporate goals while respecting financial constraints such as the availability of capital funds, institutional requirements and varied dependencies or synergies among projects. The Analytical Hierarchy Process (AHP) is used for quantification of qualitative managerial judgment in regard to the relative value of projects through a two stage process. An initial resource allocation is found by a linear program, the objective coefficients of which are determined by the AHP. Then, a modified simplex algorithm proposes some rates of funding substitution. Users choose the amount of substitution or can override the substitutions proposed. Thus, users can build gradual confidence in the constraint checking mechanism of DSS ORA and come to rely on its efficiency-seeking capabilities. DSS ORA has been tested by several groups of managers responsible for the management and the implementation of project portfolios with significantly consistent results. The flexibility, user friendliness and quick time response of DSS ORA make it an effective negotiation tool in a group setting.

Keywords. Interactive multiple criteria decision making, optimization, capital budgeting, telecommunications, mathematical programming

1   The Capital Budgeting Process

While capital budgeting constitutes a classical resource allocation problem, its exercise by corporate planners is regulated by institutional procedures to ensure accountability, respect organizational structures, safeguard minority interests, etc. For example, determining the total amount of capital to be allocated, what activities should be considered for funding, which criteria are compatible with the

⋆ Research supported by Bell Canada, NSERC Operating Grants OGP 0042197 and OGP 0043693, and AUCC Going Global Program

direction of the firm are all steps that require careful judgment and deliberation much before the decision model leading to a capital allocation is defined.

The adequate design of an analytical model to meet these requirements is challenging: capital budgeting models are often hindered by a poor economic measure of the costs and benefits of each project and their uncertainty. Systematic methods for capital budgeting have been actively investigated by public utilities such as telecommunications companies (Salo, 1989; Hoadley, 1993). The decision support system described here centers around a multiple objective program (MOP) for the annual capital budgeting exercise of a Canadian telecommunications company. It has been used by a functional group of the Company which is responsible for the management of a portfolio of programs[4]. These programs are presented to the Company's central capital budgeting committee allocating resources among all functional groups of the firm. Each program coordinates a coherent set of projects meeting a common set of objectives. Given the resources awarded to a particular program, the method proposed can in turn help allocate these resources to individual projects participating in the program. Hence, in agreement with most of the literature on capital budgeting, we will generally use the narrower term *project*, reserving the term *program* to the familiar construct of mathematical optimization.

Limited resources prevent the Committee from funding all the projects according to the requests of the functional groups. Therefore the projects are prioritized, using explicit criteria that reflect the Company's mission. Currently, the evaluation of each project with respect to each criterion involves human judgment based upon the knowledge, influence and experience of each member in the budgeting Committee. The decision-makers must also consider a number of organizational and operational constraints, such as project interdependencies. The Committee makes decisions that may be contested by some functional groups' managers. Legitimate concerns prompt the Committee to reconsider the allocation by modifying the rules used to define the Company's criteria, the contributions of the projects to these criteria and the nature of project dependencies. The appeal mechanism provides evidence of the complexity and the uncertainty existing in the actual budgeting decision process. Much of the difficulty comes from qualitative and subjective views of the Company's mission and the role of the projects to accomplish this mission.

Our decision support system aims at forming a more explicit, quantitative evaluation of the mission and goals of the Company and the suitability of alternative capital projects to achieve them. It can be used either by the Committee or by different unit managers to prepare their funding request or to respond quickly in the event that resources requested are not fully allocated.

---

[4] It is precisely such *program* planning that lead to the first use of linear *programming* (Dantzig, 1963), originally called Programming in a linear structure.

2 A Multiple Objective Programming System for Capital Budgeting

We now focus on a mathematical formulation of a multiple objective program for capital budgeting.

2.1 Review of Previous Models

Linear programming was applied to capital budgeting from its inception; there exists a vast number of analyses of capital budgeting by linear programming. We briefly review landmark analyses for MOP; for extensive bibliographies, readers should consult monographs such as (Bierman, 1988; Bromwich, 1979; Clark, 1984; Crum, 1981; Dean, 1951; Wilkes, 1983).

Charnes, Cooper and Miller (1959) were the first to formulate a linear program to solve a capital budgeting problem:

$$\max \quad \sum_{j=1}^{n} c_j x_j$$

subject to

$$\sum_{j=1}^{n} a_{ij} x_j \leq b_i \text{ for } i = 1, \ldots, m$$

$$0 \leq x_j \leq 1 \text{ for } j = 1, \ldots, n$$

where:
c_j is the net present value of Project j,
a_{ij} is the amount of Resource i required by Project j,
b_i is the total amount of Resource i available,
x_j is the fraction of Project j accepted,
n is the number of projects competing in the allocation,
m is the number of resources considered in the allocation.

Spronk (1981) extended the use of goal programming in capital budgeting by proposing an interactive multiple goal programming method based on a mutual and successive interplay between a decision-maker and an analyst. Spronk viewed his method, which does not require explicit representation of the decision-maker's preference function or trade-offs among competing objectives, as superior to conventional goal programming techniques.

Deckro, Spahr, and Herbert (1985) presented a non-linear goal program with three basic sets of goals: maximization of net present value, cash flow budgeting and control of risk.

2.2 The Multiple Objective Program of DSS ORA

Decision Support System for Optimal Resource Allocation (DSS ORA) is an interactive multiple objective programming system for efficient resource allocation designed to support collective capital budgeting decisions.

The allocation of resources to the projects is made subject to a number of inequalities, the most important of which is the capital funds constraint that represents the total availability of capital that may be allocated to all projects. Other constraints include overall short term financial impact of the portfolio such as the maximum level of acceptable depreciation expense for each project for the planning period. Constraints can also be used to represent dependencies or synergies among projects: for example, some projects cannot be implemented unless at least a certain portion of another project is implemented. Finally, the funding of every project is limited by upper and lower bounds. Our precise formulation is:

$$\text{Efficient } \{\text{Mission satisfaction}(x_1, \ldots, x_n)\} \tag{1}$$

subject to:

$$\text{Availability of Capital:} \quad \sum_{j=1}^{n} x_j \leq C \tag{2}$$

$$\text{Depreciation Limit:} \quad \sum_{j=1}^{n} d_j x_j \leq D \tag{3}$$

$$\text{Employees for implementation:} \quad \sum_{j=1}^{n} e_j x_j \leq E \tag{4}$$

$$\text{Software and other expenses:} \quad \sum_{j=1}^{n} s_j x_j \leq S \tag{5}$$

$$\text{Dependencies \& Synergies:} \quad \sum_{j=1}^{n} a_{ij} x_j \leq b_i \text{ for } i = 1, \ldots, r \tag{6}$$

$$\text{Bounds:} \quad l_j \leq x_j \leq u_j \text{ for } j = 1, \ldots, n \tag{7}$$

where x_j measures the level of funding of Project j. DSS ORA assesses satisfaction of the corporate mission via a multi-valued function F in (1) that can be improved by increasing corporate criteria such as revenue generation, revenue protection, savings and strategic importance, themselves implicitly known functions of the allocation. Inequality (2) represents the capital constraint, in which C denotes the total budget to be allocated among the n projects. Inequality (3) represents the depreciation constraint, where the coefficients d_j are the number of dollars depreciated per dollar allocated and D is the total amount of depreciation allowed. In a similar way, e_j represents the number of employees needed per dollar allocated and E is the total number of employees; s_j denotes the software and other expenses per dollar allocated to Project j and S is the total budget for software and other expenses. Constraints (6) represent dependencies and synergies between projects. Finally, in (7), u_j and l_j represent the upper and lower limits of allocation to Project j. For ease of presentation, the model is summarized as:

$$\text{Eff } \{F(x)\} \tag{8}$$

s.t.

$$Ax = b \tag{9}$$

$$x \geq 0 \tag{10}$$

where $x = (x_1, \ldots, x_p)^T$, A is a full row rank $m \times p$ matrix, b is an m-vector and F is a multi-valued function. The operator Eff seeks the efficient region specified by the program, although the interactive procedure described in Paragraph 2.4 is designed for more flexible exploration.

2.3 The Analytic Hierarchy Process

An initial allocation will first be obtained by calculating linear approximations of the multi-valued criterion function: $F = (F_h(x))_{h=1,\ldots,H}$, where F_h is a linear criterion. The Analytic Hierarchy Process (Saaty, 1980) is used to assess the value of each of its coefficients, called *priority*.

Step 1: Decision makers must choose a number of criteria important to the Company's mission.

Step 2: The relative importance of each pair of criteria is measured in order to obtain an overall scale of importance. When individuals in the evaluation group disagree, a compromise on a given comparison can be obtained either by discussion, vote or by taking the geometric mean of every member's comparison. This measurement resembles the familiar MOP method of point estimate weighted sums that defines nonnegative multipliers λ_h for $h = 1, \ldots, H$ such that: $\sum_{h=1}^{H} \lambda_h = 1$; these can be used to reduce the criterion function to the linear objective:

$$\max \sum_{h=1}^{H} \lambda_h F_h(x_1, x_2, \ldots, x_n)$$

(the similarity will be refined in Step 4 of the procedure.)

Step 3: Assess the value of each of the coefficients of F_h, i.e. the relative impact of each pair of projects on each criterion h, by a series of pairwise comparisons analogous to those of Step 2.

Step 4: Overall priorities for each project are obtained as the inner product of the vectors obtained in Step 2 and Step 3 (for this project).

Consider for example the allocation of resources to four projects, P1, P2, P3 and P4.

Step 1: Suppose that the capital budgeting decision group chose the criteria: revenue generation, revenue protection, savings and strategic importance (labeled A, B, C and D). The corresponding hierarchy is displayed in Fig. 1.

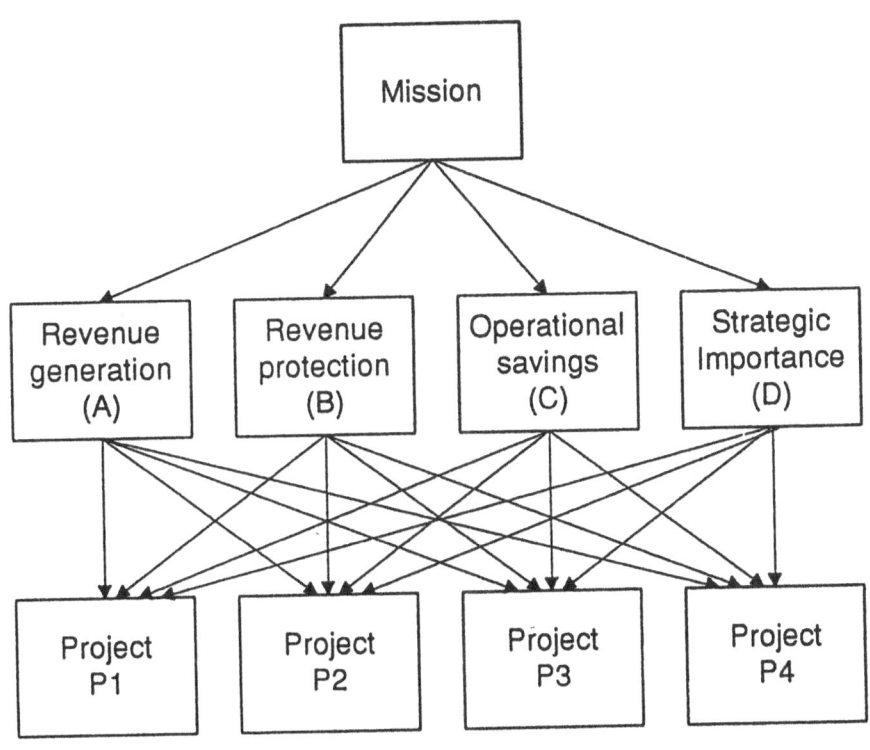

Fig. 1. AHP hierarchy for selection of telecommunications projects

Step 2: For the sake of simplicity, we assume that every comparison was made unanimously by the decision-making group. The following is the comparison matrix for the criteria (for instance, the second coefficient: 5 signifies that the criterion B is valued as more important than A):

$$
\begin{array}{c}
\begin{array}{cccc} A & B & C & D \end{array} \\
\begin{array}{c} A \\ B \\ C \\ D \end{array}
\begin{pmatrix}
1 & 5 & 3 & 5 \\
1/5 & 1 & 1/3 & 1 \\
1/3 & 3 & 1 & 3 \\
1/5 & 1 & 1/3 & 1
\end{pmatrix}
\end{array}
\quad \text{yielding } \lambda =
\begin{array}{c} A \\ B \\ C \\ D \end{array}
\begin{pmatrix}
0.558 \\
0.096 \\
0.249 \\
0.096
\end{pmatrix} .
$$

Step 3: Next are the matrices of comparisons of the projects under each criterion:

$$
\begin{array}{c|cccc}
A: & P1 & P2 & P3 & P4 \\
\hline
P1 & 1 & 1 & 7 & 5 \\
P2 & 1 & 1 & 7 & 5 \\
P3 & 1/7 & 1/7 & 1 & 1/3 \\
P4 & 1/5 & 1/5 & 3 & 1
\end{array}
\qquad
\begin{array}{c|cccc}
B: & P1 & P2 & P3 & P4 \\
\hline
P1 & 1 & 1/5 & 1 & 3 \\
P2 & 5 & 1 & 5 & 7 \\
P3 & 1 & 1/5 & 1 & 3 \\
P4 & 1/3 & 1/7 & 1/3 & 1
\end{array}
$$

$$
\begin{array}{c|cccc}
C: & P1 & P2 & P3 & P4 \\
\hline
P1 & 1 & 3 & 5 & 1/5 \\
P2 & 1/3 & 1 & 1 & 1/9 \\
P3 & 1/5 & 1 & 1 & 1/9 \\
P4 & 5 & 9 & 9 & 1
\end{array}
\qquad
\begin{array}{c|cccc}
D: & P1 & P2 & P3 & P4 \\
\hline
P1 & 1 & 5 & 7 & 1 \\
P2 & 1/5 & 1 & 3 & 1/3 \\
P3 & 1/7 & 1/3 & 1 & 1/7 \\
P4 & 1 & 3 & 7 & 1
\end{array}
$$

Following the method of Step 2, for each criterion, comparisons between projects yield one column of the following matrix:

$$
\begin{array}{c|cccc}
 & A & B & C & D \\
\hline
P1 & 0.424 & 0.153 & 0.199 & 0.440 \\
P2 & 0.424 & 0.632 & 0.066 & 0.121 \\
P3 & 0.050 & 0.153 & 0.058 & 0.052 \\
P4 & 0.102 & 0.062 & 0.677 & 0.387
\end{array}
$$

Step 4: By multiplying this matrix by the vector of multipliers λ found in Step 2, one gets the overall priorities

$$
\begin{array}{c}
P1 \\
P2 \\
P3 \\
P4
\end{array}
\begin{pmatrix}
0.34 \\
0.33 \\
0.06 \\
0.27
\end{pmatrix}
$$

The priorities are then used as objective coefficients of a linear program with constraints (2)-(7) to produce an initial allocation of resources:

$$
\max 0.34x_1 + 0.33x_2 + 0.06x_3 + 0.27x_4
$$

The constraints of the example include only Capital Availability (2) with an overall allocation of \$180 million, and a dependency constraint (6):

$$
0.5x_1 + x_3 \geq 80.
$$

Given individual bounds (7) displayed in Table 1, the capital allocation obtained by linear programming is contained in its last column:

Table 1. Allocation constraints (the rightmost three columns display $ million)

Project	Priorities	Dependency	Lower Limit	Upper Limit	Initial Allocation
P1	0.34	0.5	20	50	50
P2	0.33	0	25	35	30
P3	0.06	1	35	65	55
P4	0.27	0	45	55	45

Capital available : $ 180 million
Dependency right hand side : ≥ 80

2.4   The Interactive Allocation

The allocation delineated previously may need further refinement or sensitivity analysis. To these ends, the decision system offers an interactive procedure that leaves a great leeway to decision-makers while enforcing the budget or other constraints.

In fact, it was found that not only the values of the priorities, but even the analytical form of each criterion could be elusive. Thus, the decision system does not resort to an interactive optimization in the space of criterion multipliers λ as classical MOP methods propose, but allows users to assess their preferences directly in the space of budget allocations. At each step, a modified simplex algorithm proposes some rates of funding substitution. Then, users decide interactively what amount of substitution is preferable, and can override the rates proposed. Therefore users can propose their own solutions, build gradual confidence in the constraint checking mechanism of the simplex method, and come to rely of the efficiency-seeking capabilities patterned after (Geoffrion, 1972; Zionts, 1976).

Consider the preceding formulation (8)–(10). Let the current solution x be partitioned as $x = (x_N, x_B)$, where x_N is the subvector of nonbasic variables and x_B is the subvector of basic variables. Correspondingly, A is partitioned as $A = [N|B]$, where N consists of the nonbasic columns of A and B consists of the basic columns. Hence:

$$Nx_N + Bx_B = b.$$

Multiplying both sides by B^{-1} yields:

$$B^{-1}Nx_N + Ix_B = B^{-1}b,$$

which is equivalent to:

$$x_B = B^{-1}b - B^{-1}Nx_N.$$

Define the matrix $Y = B^{-1}N$. Each of its components y_{ij} describes the amount of decrease in the basic variable x_{Bi} caused by a unit increment in the non-basic variable x_{Nj}. Hence, its columns $y_1, y_2, \ldots, y_{p-m}$ can be used to define trade-off vectors. At the t-th iteration of the following algorithm for interactive resource allocation, each of the preceding quantities receives a superscript t.

Step 0: Set the iteration counter t at 0. Select an initial solution (e.g. from the initial linear program): $x_B^0 = (B^0)^{-1}(b^0 - N^0 x_N^0)$.

Step 1: For every nonbasic variable x_{Nj}^t, $j \in [1, \ldots, p-m]$, a trade-off vector y_j^t is calculated and presented to users.

Step 2: Either select a proper amount of trade-off Δx_{Nj}^t for some $j \in [1, \ldots, p-m]$, using:

$$\min_{i:y_{ij}^t > 0}\left\{\frac{x_{B_i}^t}{y_{ij}^t}\right\} \geq \Delta x_{Nj}^t \geq \max\left\{-x_N^t;\ \max_{i:y_{ij}^t < 0}\left\{\frac{x_{B_i}^t}{y_{ik}^t}\right\}\right\},$$

and let:

$$x'_B = x_B^t - y_j^t \Delta x_{Nj}^t,$$
$$x'_{Nk} = x_{Nk}^t \text{ for } k \neq j,$$
$$x'_{Nj} = x_{Nj}^t + \Delta x_{Nj}^t,$$

or terminate with solution x^t.

Step 3: If, for some index $q \in [1, \ldots, m]$, $x'_{Bq} = 0$, then let
$$x^{t+1} = (e_1, \ldots, e_{k-1}, e_{p-m+q}, e_{k+1}, \ldots, e_{p-m+q-1}, e_k, e_{p-m+q+1}, \ldots, e_n)x',$$

$$v_k = (-y_{1k}/y_{qk}, -y_{2k}/y_{qk}, \ldots, 1/y_{qk}, \ldots, -y_{m-1}/y_{qk}, -y_m/y_{qk})$$
and $(B^{t+1})^{-1} = (e_1, \ldots, e_{k-1}, v_k, e_{k+1}, \ldots, e_n)(B^t)^{-1}$;

else let $x^{t+1} = x'$.

Set $t = t+1$ and go to Step 1.

For example, at the optimum of the linear program displayed in Table 1, part of the rows of the matrix $-Y$ corresponding to the original variables x_1, x_2, x_3, and x_4 is shown below:

$$
\begin{array}{c}
P1 \\ P2 \\ P3 \\ P4
\end{array}
\left(
\begin{array}{cccc}
-1. & 0.0 & 0.0 & 0.0 \\
0.5 & -1. & -1. & -1. \\
0.5 & 1.0 & 0.0 & 0.0 \\
0.0 & 0.0 & 0.0 & 1.0
\end{array}
\right). \tag{11}
$$

To illustrate the use of the preceding matrix, suppose users choose a trade-off characterized by the direction of the first column, in an amount $\Delta x_{N_1}^t = 5$. The allocation to Project P1 decreases by 5, the allocations to Projects P2 and P3 each increase by 2.5. The allocation to Project P4 stays the same. Therefore after decreasing the funding of Project P3 by 5, the allocation is given by (12).

$$\begin{pmatrix} 45 \\ 32.5 \\ 57.5 \\ 45 \end{pmatrix} \tag{12}$$

The third column of the matrix (12). indicates that it is feasible to simply decrease the allocation to Project 2. Unlikely at first brush, such a decision of purely reducing resources for a project may be useful both from a practical standpoint (it enables decision makers to reduce the resources allocated) and from a theoretical one: users can avoid the pervasive assumption that their utility function must be pseudoconcave, under which larger allocations are always preferred to smaller ones.

3 Computer Implementation

DSS ORA provides a decision support environment that comprises modules for financial information management and resource allocation. Each module is implemented as a small library of objects written for Microsoft Windows (Pissarides, 1992). Financial information modules are designed to monitor capital resource availabilities and project requirements, e.g., by storing information about capital requirements, allowable depreciation expense, software and other expenses, manpower needs and relationships between projects.

- a module for project value assessment implements the AHP methodology to calculate priorities for each project,
- an optimizing module uses them to determine an initial allocation by linear programming, and
- a module of budget reallocation allows users to explore alternative funding by proposing some substitutions, one of which to be selected by decision makers.

Fig. 2 represents the mutual relationships between the modules of DSS ORA which are described in more detail next.

3.1 Project Value Assessment

This module implements the AHP methodology on a hierarchy formed by a top level comprising criteria for the corporate mission and a bottom level for the projects, as described in Sect. 2.3 and Fig. 1.

Comparisons of mission criteria
Users are first asked to input at most five names of criteria characterizing the corporate mission. Although there is no theoretical limit on the number of criteria, it has been found that five kept the number of comparisons within tolerable

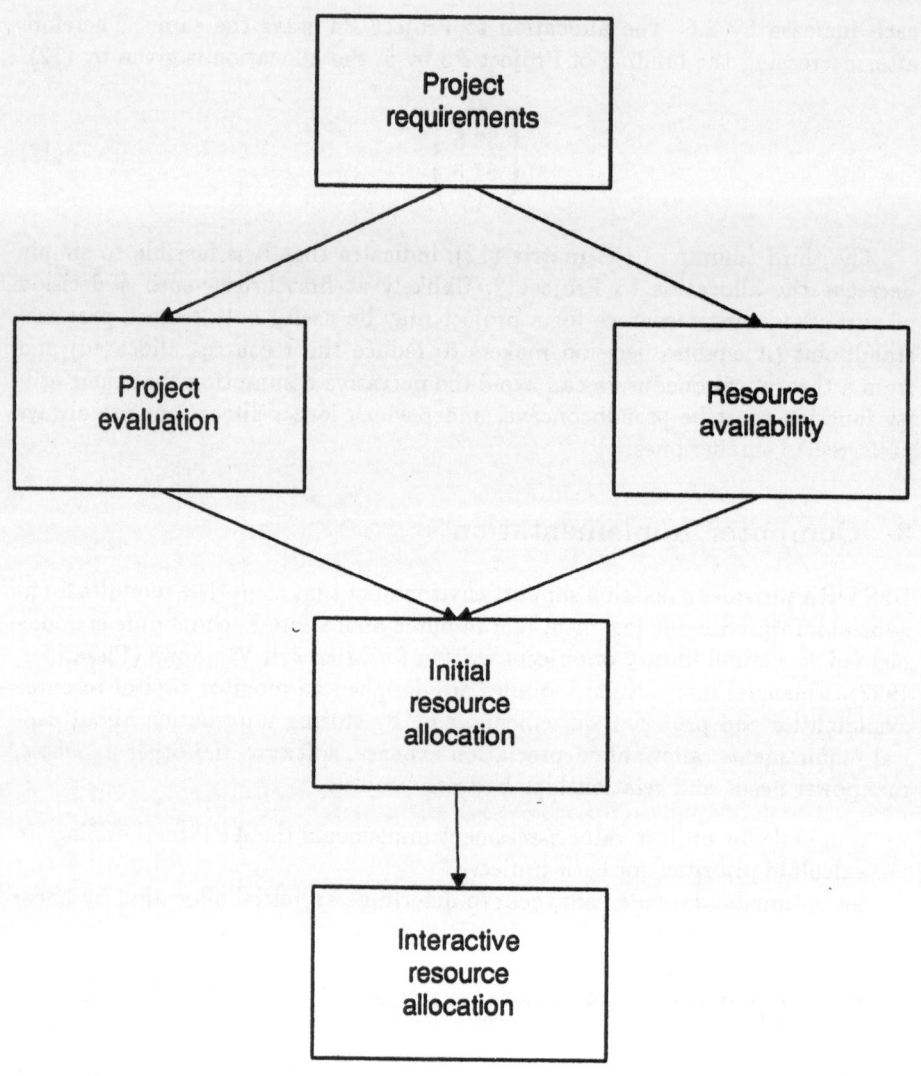

Fig. 2. DSS ORA module diagram

limits. Users must then compare each criterion with each of the other ones, choosing one of the following characterizations:

- Equal
- Moderately More Important
- Strongly More Important
- Very Strongly More Important
- Absolutely More Important

The interface used for the comparisons is shown in Fig. 3. Users are required to fill in the boxes with appropriate symbols. When all the comparisons are made, the system presents users with a vector of priorities for the criteria depicted in the rightmost column of Fig. 3.

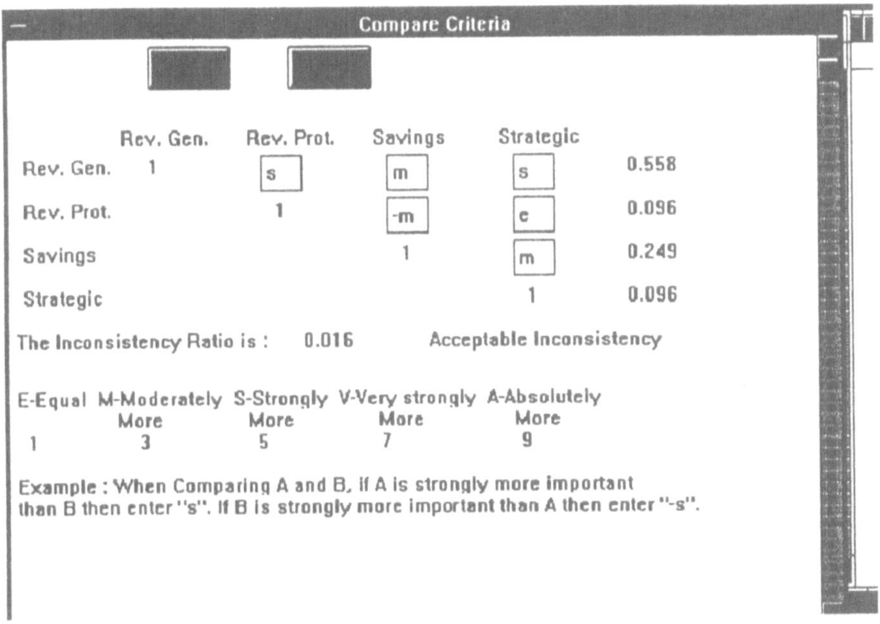

	Compare Criteria				
	Rev. Gen.	Rev. Prot.	Savings	Strategic	
Rev. Gen.	1	s	m	s	0.558
Rev. Prot.		1	-m	e	0.096
Savings			1	m	0.249
Strategic				1	0.096

The Inconsistency Ratio is : 0.016   Acceptable Inconsistency

E-Equal	M-Moderately More	S-Strongly More	V-Very strongly More	A-Absolutely More
1	3	5	7	9

Example : When Comparing A and B, if A is strongly more important than B then enter ''s''. If B is strongly more important than A then enter ''-s''.

Fig. 3. Project value assessment: mission criteria; projects.

Comparisons of projects

For the second level of the hierarchy, users must compare each pair of projects, as shown in Fig. 4. The projects are compared pairwise according to each criterion, following the same procedure as for criterion comparisons. At the end of the comparisons for a given criterion, the system presents users with the project priorities according to the current criterion together with an inconsistency ratio (Saaty, 1980). If this ratio is too high, users can go back and compare the pairs of projects again. When all the projects are compared according to all criteria, the system presents users with the overall project and criterion priorities together with the overall inconsistency ratio. Again, if the ratio is too high, users can revise the comparisons in order to obtain a better ratio.

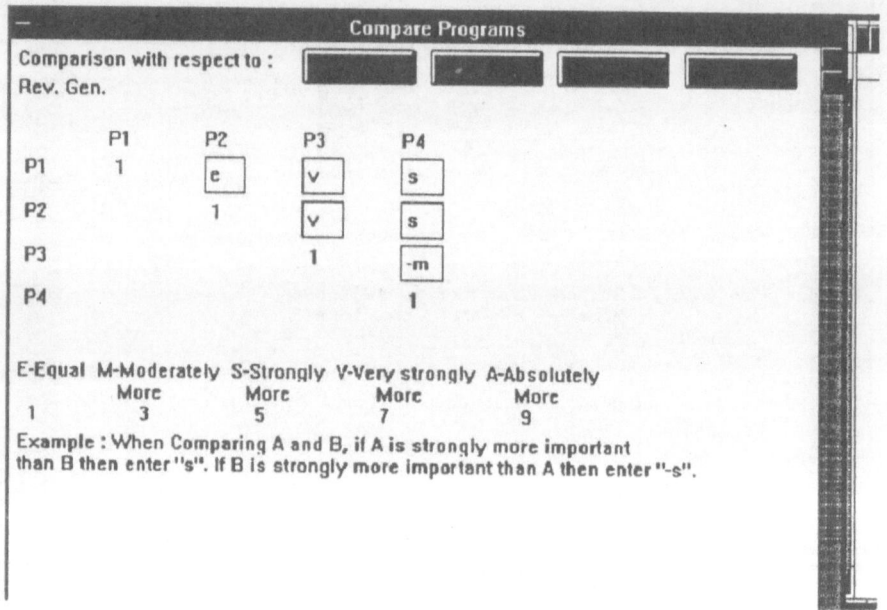

Fig. 4. Project value assessment: projects.

3.2   Initial Resource Allocation

The priorities are used as objective function coefficients of a linear program that determines an initial allocation of resources to projects. Designing a simplex algorithm was eased by implementing it as an object for an integrated manipulation of the simplex tableau, performing the following functions:

- keep track of all the elements in the tableau,
- keep track of the size of the tableau,
- perform pivot operations.

3.3 Interactive Resource Allocation

The module for interactive allocation displays the current allocation in an easy format. It allows users to change some of the inputs from the previous modules in order to obtain a more desirable allocation, or directly to swap some resources between projects without violating the constraints. For each project, the interactive analysis displays information on objective coefficients of the original linear program, its modified values and corresponding allocations, as shown in Fig. 5 and itemized below:

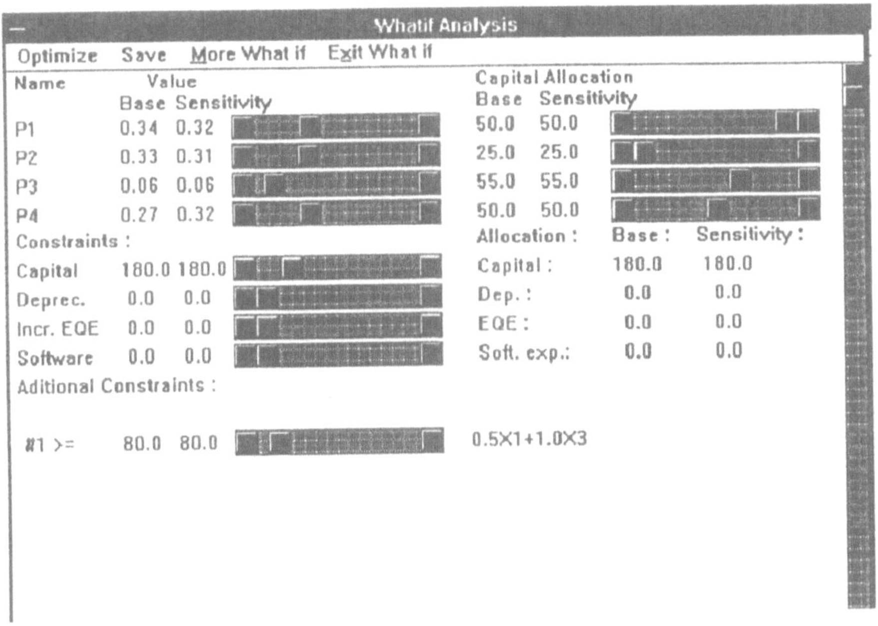

Fig. 5. User interface for interactive resource allocation by sensitivity analysis

- original objective coefficient determined by AHP analysis,
- current objective coefficient,
- a scroll bar to change the objective coefficient,
- original allocation determined by linear programming,
- current allocation,
- a scroll bar to change the allocation.

For sensitivity analysis of the resources, the module also displays the following information relative to each constraint:

- original right hand side value

- current right hand side value
- a scroll bar to allow the change of the right hand side value,
- original value of the left hand side determined by linear programming,
- current value of the left hand side.

The module offers four options to change the allocation of resources:

- change of objective function coefficients,
- change of amounts of resources (the right hand side of the constraints),
- computer-assisted change of funding allocation, and
- unassisted change of funding allocation.

In the first two options, the allocation is modified by selecting the menu item "Optimize" on the screen, which accepts the new data as input to the simplex algorithm contained in the module described previously and displays the new allocations both numerically and on scroll bars. Therefore users can perform sensitivity analysis on both the objective coefficients and the constraint right hand side values by dragging the scroll bar cursors in either direction. In Fig. 5, for a new objective function

$$\max 0.31x_1 + 0.31x_2 + 0.06x_3 + 0.32x_4,$$

the allocation becomes:

$$x_1 = 50, \ x_2 = 25, \ x_3 = 55, \ x_4 = 50.$$

The last two options, illustrated in Fig. 6, allow users to by-pass re-optimization by linear programming: in particular, they need not express the mission criteria as algebraic functions. In the last option, they can simply change the allocation of resources to the projects by changing the position of the cursors on the scroll bars, using the mouse. Each cursor on the scroll bars can move until one of the constraints is violated. For example, users can achieve a more desirable solution by decreasing the allocation to one project and increasing the allocation to another directly. Of course, the new solution may not be optimal with respect to the objective coefficients displayed by the interface.

In an unassisted change, the user can move the allocation scrollbars manually to decrease the allocation to Project P2 from 30 to 26 (for instance) and increase the allocation to Project P4 from 45 to 47. In this unassisted change, any arbitrary allocation will be possible within the constraints specified.

In an assisted change, users can select one allocation trade-off from a list generated by the system, shown in Fig. 6), that transposes the matrix of reduced coefficients (11).

Each trade-off indicates a rate of funding substitution. Users can decide to change the allocation by pressing a button on the screen. The changes are reflected by the position of cursors on the scroll bars and the corresponding display of values, both of which representing the allocation levels to each project. For example, choosing the first trade-off yields the new allocation given by (12).

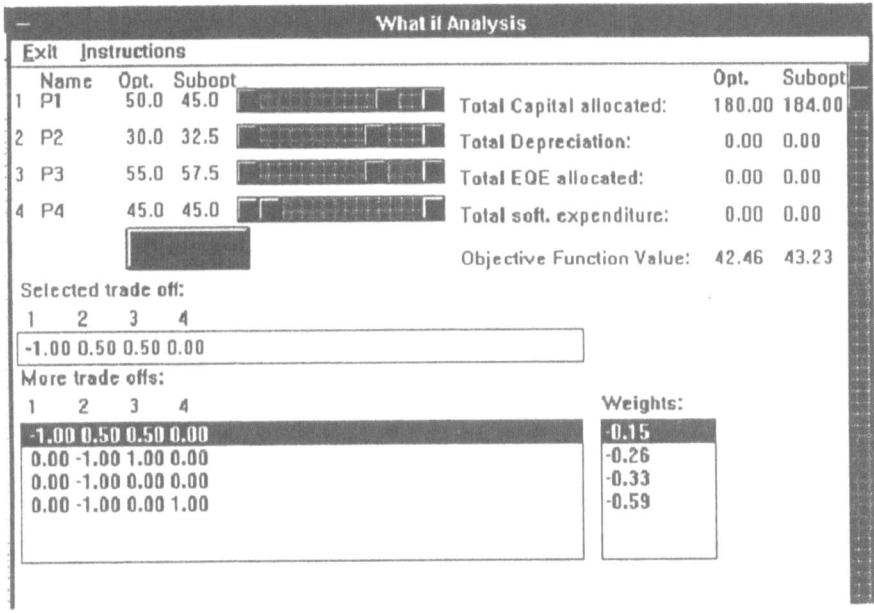

Fig. 6. User interface for interactive resource allocation: computer-assisted trade-offs

Users are allowed to switch to a different trade-off at any time. When one of the right hand side values becomes zero, the system performs a pivot and updates the list of trade-offs.

The algorithm is designed to help users get a better understanding of the problem at hand by exploring the region specified by the constraints. It does not require any analytic expression of the objective or utility function which in particular need not be pseudoconcave (Zionts, 1976), a qualification that would restrict the search to extreme points only.

4 Field Trial

A field trial was conducted to test whether DSS ORA could support group decisions as required by the Company's budgeting process. Two groups that worked with DSS ORA consisted of five managers. First, a functional group of staff managers responsible for the funding of a portfolio of five programs labeled P1, P2, P3, P4 and P5 (each consisting of numerous projects) had to develop tthe programs. The group held a managerial view of each project and its role in the portfolio, rather than a technical understanding of its functions and value in the corporate telecommunications network. Consequently, its members differed fairly substantially in timportance of each program toward each corporate mission criterion, as shown in Table 2. Yet, consensus was required of this request by the line managers responsible for implementation of these programs before the request could be forwarded to the control committee responsible for funding allocation.

Table 2. Project importance assessment results

Project	Based on inputs from Manager					Based on average inputs
	1	2	3	4	5	
P1	0.079	0.045	0.037	0.061	0.066	0.063
P2	0.322	0.170	0.466	0.351	0.251	0.311
P3	0.135	0.067	0.111	0.078	0.158	0.111
P4	0.239	0.467	0.169	0.125	0.086	0.206
P5	0.226	0.250	0.216	0.386	0.474	0.306

Saaty (1990) noted that the hierarchical representation and the pairwise comparisons have intuitive appeal, pinpointing internal inconsistencies of judgment. This was confirmed in our case, as a first evaluative session sparked a discussion aimed at developing a common understanding of the role of the projects in satisfying corporate mission criteria, and of critical underlying assumptions under which all projects had to compete for funding allocation. Attesting to the adequacy of the method, the outcome of the second session of pairwise evaluations was that the priorities of each of the managers for the programs were close enough to reach a unanimous consensus about the validity of the values based on the average input of the group, displayed in the last column of Table 2. The role each session of DSS ORA toward this progressive conciliation is represented in Fig. 7.

The second group using DSS ORA comprised line managers responsible for the implementation of the projects within the corporate telecommunications network. Consensus was reached in one session, and the resulting project evaluations were significantly close to the results obtained by the first group of staff managers. Given the consensus, in neither experiment did the managers feel that an interactive resource allocation was necessary.

5 Conclusion

DSS ORA has been tested with very consistent results by several groups of project managers. Beyond the interactive MOP methodology, the field trial proved that the system was an effective negotiation tool in a group setting. Applying DSS ORA to resource allocation within a functional unit limits the number of projects under consideration, with several ensuing benefits:

- familiarity of the managers with the projects,

145

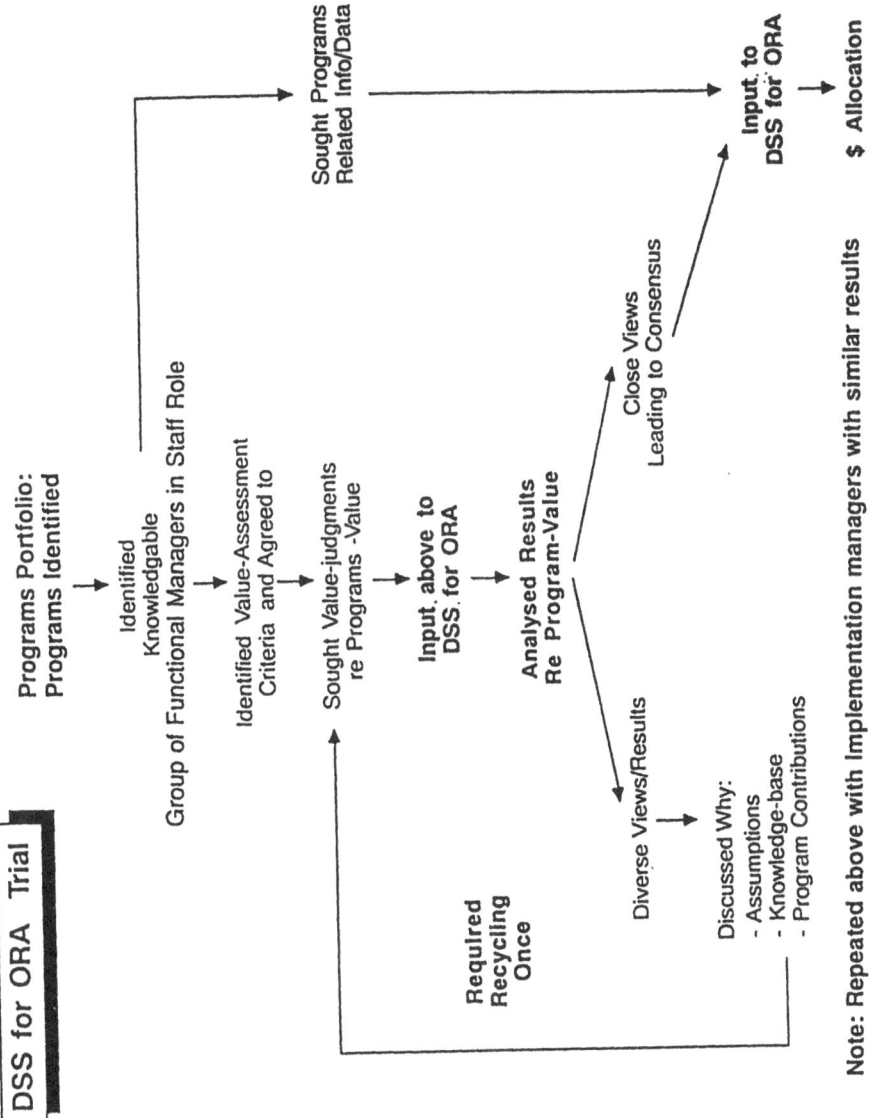

Figure 7: *Using DSS ORA for consensus building*

- relatively few pairwise comparisons,
- increased likelihood of consensus building,
- limited interference of external constraints with the consensual evaluations,
- possible acceptance of results as a base for implementation.

Capital budgeting in the Company spans many units and associated programs. To adapt DSS ORA to inter-unit budgeting, current research focuses on effective decomposition and aggregation techniques (Liang, 1994). A central issue is the formation of subportfolios of programs balancing two requirements:

- ease the comparisons of projects,
- circumscribe projects that share important technological or operational dependencies.

In this setting, resource allocation could proceed along several tiers: allocation to portfolios of programs, followed by intra-program allocation to the constituting projects.

References

Bierman, H. Jr.: Implementing Capital Budgeting Techniques. Ballinger Publishing Company, Cambridge, Massachusetts, 1988

Bromwich, M.: The Economic of Capital Budgeting. Pitman Publishing Limited, 1979

Charnes, A., Cooper, W., Miller, M. H.: An Application of Linear Programming to Financial Budgeting and the Cost of Funds. The Journal of Business (1959) 20–46

Clark, J. J., Hinderland, T. J., Pritchard, R. E.: Capital Budgeting. Prentice-Hall Inc., Englewood Cliffs, New Jersey, 1984

Crum, L. R., Derkinderen, F. G. J.: Capital Budgeting Under Conditions of Uncertainty. Martinus Nijhoff Publishing, 1981

Dantzig, G. B.: Linear Programming and Extensions. Princeton University Press, Princeton, New York, 1963

Dean, J.: Capital Budgeting. Columbia University Press, New York, 1951

Deckro, R. F., Spahr, R. W., Herbert, J. E.: Preference Trade-offs in Capital Budgeting Decisions. IIE Transactions 17 (1965) 332–337

Geoffrion, A. M., Dyer, J. S., Feinberg A.: An Interactive Approach for Multiple Optimization with Applications to the Operation of an Academic Department. Management Science 19 (1972) 357–368

Hoadley, B., Katz, P., Sadrian, A.: Improving the Utility of the Bellcore business. Interfaces 23 1 (1993) 27–43

Liang, Y.: Capital Budgeting Decision Making: Database, Aggregation and Disaggregation Methods for a Large Scale Problem. Master's thesis, University of Ottawa, 1994

Pissarides, S: Interactive Multiple Criteria Optimization for Capital Budgeting. Master's thesis, University of Ottawa, 1992

Saaty, T. L.: The Analytic Hierarchy Process. McGraw-Hill Inc., 1980.

Salo, A., Hamalainen, R. P.: A Modelling and Decisions Aid for Supporting Telecommunications Investments. Proceedings of the IIEE International Conference on Systems, Man and Cybernetics 1 (1989) 115–118

Spronk, J.: Interactive Multiple Goal Programming as an Aid for Capital Budgeting and Financial Planning with Multiple Goals, in: Capital Budgeting Under Conditions of Uncertainty. Martinus Nijhoff Publishings (1981) 188–212

Wilkes, F. M.: Capital Budgeting Techniques. Wiley, 1983

Zionts, S., Wallenius, J.: An Interactive Programming Method for Solving the Multiple Criteria Problem. Management Science 22 (1976) 652–663

The Ekeland's Principle and the Pareto ε-Efficiency

G. Isac

Department of Mathematics and Computer Science, Royal Military College of Canada, Kingston, Ontario, Canada, K7K 5L0

Abstract: We present in this paper a new variant of Ekeland's variational principle for vector valued functions with applications to the study of Pareto ε-efficiency. A new existence result for Pareto efficiency is also presented.

Key words: Ekeland's principle, Pareto efficiency , Pareto ε-efficiency, ordered vector spaces.

1. Introduction

Since its appearance in 1972 [9] the Ekeland's variational principle has been the subject of many papers, and it has found many applications in analysis and in applied mathematics [8], [10-14].

The Ekeland's principle is equivalent to the Caristi's Fixed Point Theorem, to the Drop Theorem and to the Petal Theorem and by these equivalencies it has interesting applications in nonlinear analysis and to the study of the geometry of Banach spaces [2], [4], [5-6], [15-17], [24], [27], [28], [30-31], [34]. Some extensions of this principle were also proved by several authors [1], [3], [7], [15], [25], [27].

The Ekeland's principle has also important applications to the Pareto optimization.

In 1983 we remarked that , an extension of Caristi's Fixed Point Theorem can be used to study the existence of critical points for generalized dynamical systems and hence, in particular to study the existence of Pareto efficient points [18]. Recently, a new variant of Ekeland's principle for Pareto Optimization was presented by P. Q. Khanh [21].

Our interest now, in this paper, is dedicated to the applications of the Ekeland's principle to the study of Pareto ε-efficiency.

The Pareto ε-efficiency has been recently considered by several authors [22], [25], [33], [35], [36], [37], [38] and it seems, this notion asks new variants of Ekeland's principle.

Our paper must be considered in this context and it is motivated by Tammer's recent papers [35], [36], [37].

In this sense, we present a new variant of Ekeland's principle for Pareto ε-efficiency. Our result is proved using simpler and more general assumptions as the assumptions used in [36].

2. Preliminaries

Let E be a vector space. We say that a subset $\mathbf{K} \subset E$ is a *convex cone* if $\mathbf{K} + \mathbf{K} \subseteq \mathbf{K}$, $\lambda \mathbf{K} \subseteq \mathbf{K}$ for all $\lambda \in \mathbf{R}_+$ and $\mathbf{K} \cap (-\mathbf{K}) = \{0\}$. If a convex cone $\mathbf{K} \subset E$ is defined, we associate to $\mathbf{K}$ the ordering "$\leq$" defined by, $x \leq y$, if and only if $y - x \in \mathbf{K}$, for all $x, y \in E$. Let $A \subset E$ be a non empty set. The *full hull* of A, denoted by $[A]_0$ is by definition $[A]_0 = \{x \in E \mid u \leq x \leq v, u, v \in A\}$. We say that a subset A of E is *full* or *0-convex* if and only if $A = [A]_0$. Let $E(\tau)$ be a topological vector space and $\mathbf{K} \subset E$ a convex cone.

We say that $\mathbf{K}$ is *normal* (with respect to the topology τ) if and only if , there is a neighborhood basis of zero for τ consisting of 0-convex sets. We recall the following normality tests.

1^0 *If* $\left(E, \| \ \| \right)$ *is a normed vector space, then the convex cone* $\mathbf{K} \subset E$ *is normal , if and only if, there is a constant* $\gamma > 0$ *such that* $0 \leq x \leq y$ *implies* $\gamma \|x\| \leq \|y\|$.

2^0 *If* $E(\tau)$ *is a locally convex space, then the convex cone* $\mathbf{K} \subset E$ *is normal, if and only if, there is a family* $\left\{p_\alpha\right\}_{\alpha \in A}$ *of seminorms generating the topology* τ *such that* $0 \leq x \leq y$ *implies* $p_\alpha(x) \leq p_\alpha(y)$, *for all* $\alpha \in A$

If $E(\tau)$ is a locally convex space and the convex cone $\mathbf{K} \subset E$ is normal, then for every $y \in \mathbf{K} \setminus \{0\}$ the set $[0, y]_0 = \{x \in E \mid 0 \leq x \leq y\}$ is τ-bounded.

Let E^* be the topological dual of E and let $\langle E, E^* \rangle$ be a duality .

If $\mathbf{K} \subset E$ is a convex cone the dual of $\mathbf{K}$, denoted by $\mathbf{K}^*$ is by definition $\mathbf{K}^* = \{y \in E^* \mid \langle y, x \rangle \geq 0, \text{ for all } x \in \mathbf{K}\}$. If $E(\tau)$ is locally convex space, and $\mathbf{K} \subset E$ is a normal convex cone, then we have $E^* = \mathbf{K}^* - \mathbf{K}^*$.

Our paper is dedicated to the study of a special problem concerning the ε-efficiency. An interesting notion of ε-efficiency has been recently studied by Chr. Tammer in [35], [36] and [37]. This concept is important since it contains as particular case the similar concepts defined and studied by P. Loridan [22], T. Staib [33] and I Vályi [38].

Let E be a linear topological space and D and B are proper subsets of E. Let $\varepsilon > 0$ and $k^0 \in E$ be such that $B + \alpha k^0 \subset B$ for all $\alpha > 0$ and $B_{\varepsilon k^0} = \varepsilon k^0 + (B \setminus \{0\})$.

As in [35], we say that an element $x_* \in D$ is an εk^0-efficient point of D with respect to B if there exists no point $x \in D$ such that $x_* \in x + B_{\varepsilon k^0}$.

We denote the εk^0-efficient point set of D with respect to B and k^0, by $Eff\left(D, B_{\varepsilon k^0}\right)$.

Also, we say that an element $x_* \in D$ is an efficient point of D with respect to B if there exists no element $x \in D$ with $x_* \in x + (B \setminus \{0\})$. We denote the efficient point set of D with respect to B by $Eff(D, B)$.

3. A Characterization of Normal Cones

The following characterization of normal cones is necessary for the principal result of this paper.

Proposition 1.   *Let $E(\tau)$ be a locally convex space. A convex cone $\mathbf{K} \subset E$ is normal, if and only if, every monotone increasing and order bounded sequence in E is weakly Cauchy.*

Proof. We suppose that $\mathbf{K}$ is normal and let $\{x_n\}_{n \in N} \subset E$ be a sequence such that

(1): $x_1 \leq x_2 \leq \cdots \leq x_n \leq \cdots \leq u$.

The normality of $\mathbf{K}$ implies that $\{x_n\}_{n \in N}$ is topologically bounded (since $\{x_n\}_{n \in N} \subset [x_1, u]_0$). For every $f \in \mathbf{K}^*$ we have that $\{f(x_n)\}_{n \in N}$ is convergent and because $E^* = \mathbf{K}^* - \mathbf{K}^*$, we deduce that, for every $h \in E^*$, the sequence $\{h(x_n)\}_{n \in N}$ is convergent, that is $\{x_n\}_{n \in N}$ is weakly convergent. In conclusion $\{x_n\}_{n \in N}$ is weakly Cauchy.

To prove the converse we suppose that $\mathbf{K}$ is not normal. Then, there is an element $v \in \mathbf{K}$ such that the order interval $[0, v]_0$ is not topologically bounded. Hence, there exists a continuos seminorm p and a sequence $\{x_n\}_{n \in N} \subset [0, v]_0$ such that

$p(x_n) \geq 7^n$, for every $n \in N$.

We consider the sequence $y_n = \dfrac{x_1}{2} + \dfrac{x_2}{2^2} + \cdots + \dfrac{x_n}{2^n}$, for every $n \in N$.

We have $y_1 \le y_2 \le \cdots \le y_n \le \cdots \le v$ and $\{y_n\}_{n \in N}$ is order bounded. Hence by assumption, $\{y_n\}_{n \in N}$ is weakly Cauchy and consequently is topologically bounded. On the other hand we have

$$p\left(y_n\right) \ge \frac{1}{2^n} p\left(x_n\right) - \frac{1}{2^{n-1}} p\left(x_{n-1}\right) - \cdots - \frac{1}{2} p\left(x_1\right) \ge \left(\frac{7}{2}\right)^n - \left(\frac{7}{2}\right)^{n-1} - \cdots$$

$$\cdots - \frac{7}{2} = \left(\frac{7}{2}\right)^n - \frac{\left(\frac{7}{2}\right)^n - \frac{7}{2}}{\frac{7}{2} - 1} = \frac{3}{5}\left(\frac{7}{2}\right)^n + \frac{7}{5}$$

for every $n \in N$ which implies that $p\left(y_n\right) \to +\infty$, that is, we have a contradiction. Hence, **K** must be a normal cone. ∎

4. A Variant of Ekeland's Principle

Let (X, d) be a complete metric space. We say that a mapping $\Gamma : X \to 2^X$ is a *dynamical system* if $\Gamma(x)$ is nonempty for any $x \in X$, and we say that an element $x_* \in X$ is a *critical point* of Γ if $\Gamma\left(x_*\right) = \left\{x_*\right\}$. Every critical point of Γ is a fixed point, but the converse is not true. To prove the variant of Ekeland's principle we will use the following result, which is a reformulation, for dynamical systems, of Dancs-Hegedus-Medvegyev's theorem [7].

Theorem 2. *Let (X,d) be a complete metric space and $\Gamma : X \to 2^X$ a dynamical system. If the following assumptions are satisfied:*

i_1) $\Gamma(x)$ *is closed for every* $x \in X$

i_2) $x \in \Gamma(x)$ *for every* $x \in X$

i_3) $x_2 \in \Gamma\left(x_1\right)$ *implies* $\Gamma\left(x_2\right) \subseteq \Gamma\left(x_1\right)$ *for all* $x_1, x_2 \in X$

i_4) *for every sequence* $\{x_n\}_{n \in N} \subset X$ *with the property* $x_{n+1} \in \Gamma\left(x_n\right)$,

for every $n \in N$, *we have* $\lim_{n \to \infty} d\left(x_n, x_{n+1}\right) = 0$,

then Γ has a critical point $x_* \in X$. Moreover, for every $\hat{x} \in X$ there is a critical point of Γ in $\Gamma(\hat{x})$. ∎

We denote by $(E(\tau), \mathbf{K})$ a locally convex space ordered by a closed convex cone $\mathbf{K} \subset E$. For the following we suppose given a such space.

We say that a mapping $\Phi . X \times X \to E$ is *a halfdistance* if the following properties are satisfied:

$i^{\circ})$ $\quad \Phi(x,x) = 0$, for all $x \in X$

$ii^{\circ})$ $\quad \Phi(x,y) \le \Phi(x,z) + \Phi(z,y)$, for all $x, y, z \in X$

The class of halfdistances is not empty, since for every $f : X \to E$, the mapping $\Phi(x,y) = f(y) - f(x)$, for all $x, y \in X$ is a halfdistance. Also, if F is an arbitrary vector space and $T : F \to E$ is a subadditive mapping such that $T(0) = 0$, then for every mapping $f : X \to F$ we can verify that $\Phi(x,y) = T(f(y) - f(x))$, for all $x, y \in X$ is a halfdistance.

The next theorem is the principal result of this paper and it is a variant of Ekeland's principle for vector valued mappings.

Theorem 3. *Let* (X,d) *be a complete metric space,* $(E(\tau), \mathbf{K})$ *an ordered locally convex space with* $\mathbf{K}$ *normal and* Φ *a halfdistance from* X *into* E. *If for an element* $k^0 \in \mathbf{K} \setminus \{0\}$ *the following assumptions are satisfied:*

$1^{\circ})$ *for every* $x \in X$ *the set* $\{y \in X | \Phi(x,y) + k^0 d(x,y) \in -\mathbf{K}\}$ *is closed,*

$2^{\circ})$ *there exist* $v_0 \in X$ *and* $w_0 \in E$ *such that* $\Phi(v_0, x) \ge w_0$, *for every* $x \in X$, *then, there exists* x_* *in* X *such that* $\Phi(x_*, x) + k^0 d(x_*, x) \notin -\mathbf{K}$, *for all* $x \in X \setminus \{x_*\}$.

Proof. First, we remark that if we consider the dynamical system $\Gamma_\phi(x) = \{y \in X | \Phi(x,y) + k^0 \, d(x,y) \in -\mathbf{K}\}$ for every $x \in X$, then the theorem will be proved if we show that Γ_ϕ has a critical point in X. To show this it is sufficient to verify the assumptions $i_1), i_2), i_3)$, and $i_4)$ of *Theorem 2*.

From assumption $1^{\circ})$ we have that $\Gamma_\phi(x)$ is closed for every $x \in X$, that is $i_1)$ is satisfied. Using the properties of d and Φ we have immediately that $x \in \Gamma_\phi(x)$, for every $x \in X$ which means that $i_2)$ is also satisfied. To verify $i_3)$ we consider two elements $x_1, x_2 \in X$ such that $x_2 \in \Gamma_\phi(x_1)$. We must show that $\Gamma_\phi(x_2) \subseteq \Gamma_\phi(x_1)$. Indeed since $x_2 \in \Gamma_\phi(x_1)$ we have

$(\alpha_1):$ $\quad \Phi(x_1, x_2) + k^0 \, d(x_1, x_2) \in -\mathbf{K}$

If $z \in \Gamma_\phi(x_2)$ is an arbitrary element, we have

$(\alpha_2):$ $\quad \Phi(x_2, z) + k^0 d(x_2, z) \in -\mathbf{K}$

We will have $z \in \Gamma_\phi(x_1)$ if we show that $\Phi(x_1, z) + k^0 d(x_1, z) \in -\mathbf{K}$.

From (α_1) there is an element $k_1 \in K$ such that, $\Phi(x_1, x_2) + k^0 d(x_1, x_2) = -k_1$ and from (α_2) there is an element $k_2 \in K$ such that $\Phi(x_2, z) + k^0 d(x_2, z) = -k_2$. Since Φ is a halfdistance there is an element $k_3 \in K$ such that $\Phi(x_1, z) = \Phi(x_1, x_2) + \Phi(x_2, z) - k_3$ and since $k^0 d(x_1, z) \leq k^0 d(x_1, x_2) + k^0 d(x_2, z)$ there is an element $k_4 \in K$ such that $k^0 d(x_1, z) = k^0 d(x_1, x_2) + k^0 d(x_2, z) - k_4$. Finally, we obtain that

$$\Phi(x_1, z) + k^0 d(x_1, z) = \Phi(x_1, x_2) + \Phi(x_2, z) - k_3 + k^0 d(x_1, z) =$$
$$= \Phi(x_1, x_2) + \Phi(x_2, z) - k_3 + k^0 d(x_1, x_2) + k^0 d(x_2, z) - k_4 =$$

$-k_1 - k_2 - k_3 - k_4 \in -K$, which implies that $\Gamma_\phi(x_2) \subseteq \Gamma_\phi(x_1)$, that is, i_3) is satisfied. To verify i_4) we consider a sequence $\{x_n\}_{n \in N} \subset X$ such that $x_{n+1} \in \Gamma_\phi(x_n)$ for every $n \in N$, with x_1 arbitrary in X. We have $\Phi(x_n, x_{n+1}) + k^0 d(x_n, x_{n+1}) \in -K$, which implies

(α_3): $k^0 d(x_n, x_{n+1}) = -\Phi(x_n, x_{n+1}) - k_5$, where $k_5 \in K$.

Since Φ is a halfdistance we have, $\Phi(v_0, x_{n+1}) \leq \Phi(v_0, x_n) + \Phi(x_n, x_{n+1})$ which implies, $-\Phi(x_n, x_{n+1}) \leq \Phi(v_0, x_n) - \Phi(v_0, x_{n+1})$ and hence, there is an element $k_6 \in K$ such that

(α_4): $-\Phi(x_n, x_{n+1}) = \Phi(v_0, x_n) - \Phi(v_0, x_{n+1}) - k_6$

From (α_3) and (α_4) we obtain $k^0 d(x_n, x_{n+1}) = \Phi(v_0, x_n) - \Phi(v_0, x_{n+1}) - k_5 - k_6'$ and if we denote $k = k_5 + k_6$ we have, $k^0 d(x_n, x_{n+1}) = \Phi(v_0, x_n) - \Phi(v_0, x_{n+1}) - k$, that is,

(α_5): $k^0 d(x_n, x_{n+1}) \leq \Phi(v_0, x_n) - \Phi(v_0, x_{n+1})$

If we consider now the series $\sum_{n=1}^{\infty} k^0 d(x_n, x_{n+1})$ and we denote

$S_m = \sum_{n=1}^{m} k^0 d(x_n, x_{n+1})$, we remark that $S_m \leq \Phi(v_0, x_1) - w_0$, for every $m \in N$, that is the sequence $\{S_m\}_{m \in N}$ is order bounded in E.

Since the sequence $\{S_m\}_{m \in N}$ is monotone increasing and K is normal we have, using **Proposition 1** that it is weakly Cauchy. Since $\{S_m\}_{m \in N}$ is weakly Cauchy on the line $L(k^0)$ generated in E by the element k^0, it is convergent in $L(k^0)$ and since $L(k^0)$ is homeomorphic to the real field by the homeomorphism $\varphi(tk^0) = t$ we obtain that

$\sum_{n=1}^{\infty} d\left(x_n, x_{n+1}\right)$ is a convergent series. Hence the sequence $\left\{d\left(x_n, x_{n+1}\right)\right\}_{n \in N}$ is convergent to zero and assumption i_4) is satisfied.

Applying finally **Theorem 2** we have that Γ_ϕ has a critical point and the theorem is proved. ∎

If we apply the second part of the conclusion of **Theorem 2**, we obtain the following corollary of **Theorem 3**.

Corollary. *If all the assumptions of Theorem 3 are satisfied, then there exists* $x_* \in \Gamma_\phi\left(v_0\right)$ *such that* $\Phi\left(x_*, x\right) + k^0 d\left(x_*, x\right) \notin -K$, *for all* $x \in X \backslash\left\{x_*\right\}$. *(The element* v_0 *is the same as in Theorem 3)*

The next result is the analogous for vector valued functions of the strong form of Ekeland's principle.

Theorem 4. *Let* (X, d) *be a complete metric space*, $(E(\tau), K)$ *an ordered locally convex space with* K *normal and* $f: X \rightarrow E$ *a mapping.*

If $\varepsilon > 0$ *is an arbitrary real number and there exist* $a \in X$ *and* $k^0 \in K \backslash\{0\}$ *such that:*

1). $f(a) \leq f(x) + \varepsilon k^0$, *for all* $x \in X$,

2). for every $x \in X$ *and every real number* $\alpha > 0$ *the set*

$$\left\{y \in X \mid f(y) - f(x) + \alpha d(x, y) k^0 \in -K\right\} \text{ is closed,}$$

then for every $\lambda > 0$ *there exists* $x_\lambda \in X$ *such that,*

i). $f\left(x_\lambda\right) \leq f(a)$

ii) $d\left(x_\lambda, a\right) \leq \lambda$

iii) $f(x) - f\left(x_\lambda\right) + \dfrac{\varepsilon}{\lambda} d\left(x_\lambda, x\right) k^0 \notin -K$, *for all* $x \in X \backslash\left\{x_\lambda\right\}$.

Proof. We consider the halfdistance $\Phi: X \times X \rightarrow E$ defined by $\Phi(x, y) = f(y) - f(x)$ and we remark that all the assumptions of **Theorem 3** are satisfied for the distance $\dfrac{\varepsilon}{\lambda} d$ if we consider $v_0 = a$ and $w_0 = -\varepsilon k^0$.

Then, by the Corollary of *Theorem 3* there exists

$$x_\lambda \in \Gamma_\Phi(a) = \left\{ y \in X \middle| f(y) - f(a) + \frac{\varepsilon}{\lambda} d(a,y)k^0 \in -K \right\} \text{ such that}$$

(β_1): $f(x) - f\left(x_\lambda\right) + \frac{\varepsilon}{\lambda} d\left(x_\lambda, x\right)k^0 \notin -K$ for all $x \in X \setminus \{x_\lambda\}$,

which is exactly (iii). Since $x_\lambda \in \Gamma_\Phi(a)$ we have that there is $k \in K$ such that

$\frac{\varepsilon}{\lambda} d\left(a, x_\lambda\right)k^0 = -k + f(a) - f\left(x_\lambda\right)$, that is, $\frac{\varepsilon}{\lambda} d\left(a, x_\lambda\right)k^0 + k = f(a) - f\left(x_\lambda\right)$ and

hence $\frac{\varepsilon}{\lambda} d\left(a, x_\lambda\right)k^0 \leq f(a) - f\left(x_\lambda\right) \leq \varepsilon k^0$, which implies $d\left(a, x_\lambda\right) \leq \lambda$ that is the

conclusion (ii) is satisfied.

Finally, since $f(a) - f\left(x_\lambda\right) - k = \frac{\varepsilon}{\lambda} d\left(a, x_\lambda\right)k^0$, we deduce,

$f(a) = f\left(x_\lambda\right) + k + \frac{\varepsilon}{\lambda} d\left(a, x_\lambda\right)k^0$ which implies $f\left(x_\lambda\right) \leq f(a)$, that is, (i) is

satisfied too, and the theorem is proved. ∎

5. Equivalencies

It is well known that the Ekeland's principle is equivalent to Caristi's fixed point, to the drop theorem and to the petal theorem [28], [5], [6], [15]. Recently , another theorem equivalent to Ekeland's principle was proved in [27].
By these equivalencies the Ekeland's principle has important applications [8], [10], [11], [13], [14], [5], [2], [16], [17], [30], [30], [31].
In this section we study for our variant of Ekeland's principle some similar equivalencies.
In this sense, we will show that *Theorem 3* is also equivalent to a new variant of Caristi's fixed point theorem and to a variant of the result established by Oettli and Théra in [27].
Theorem 5. *Let (X,d) be a complete metric space,* $\left(E(\tau), K\right)$ *an ordered locally convex space with* K *normal and* Φ *a halfdistance from X into E. Given*
$k^0 \in K \setminus \{0\}$, *we suppose the set* $\left\{ y \in X \middle| \Phi(x,y) + k^0 d(x,y) \in -K \right\}$ *to be closed*
for every $x \in X$. *For every* $v_0 \in X$ *such that, there exists* $w_0 \in E$ *with the property*

$\Phi(v_0, x) \geq w_0$ *for every* $x \in X$, *we denote* $\mathcal{B}_0 = \left\{ x \in X \middle| \Phi(v_0, x) + k^0 \, d(v_0, x) \in -\mathbf{K} \right\}$.

Then under the hypotheses indicated before , the following assertions are true:

Theorem A [Ekeland]. *There exists* $x_* \in \mathcal{B}_0$ *such that* $\Phi\left(x_*, x \right) + k^0 \, d\left(x_*, x \right) \notin -\mathbf{K}$,

for all $x \in X \setminus \left\{ x_* \right\}$

Theorem B.[Caristi-Kirk]. *If* $T{:}X \to X$ *is a multivalued mapping such that,*

$\left(\delta_1 \right){:} \left\{$ *for every* $\bar{x} \in \mathcal{B}_0$ *there exists* $x \in T(\bar{x})$ *such that* $\Phi(\bar{x}, x) + k^0 \, d(\bar{x}, x) \in -\mathbf{K}$

then, there exists $x_* \in \mathcal{B}_0$ *such that* $x_* \in T\left(x_* \right)$.

Theorem C. *If* $M \subset X$ *has the property*

$\left(\delta_2 \right){:} \left\{$ *for every* $\bar{x} \in \mathcal{B}_0 \setminus M$ *there exists* $x \in T(\bar{x})$ *such that* $x \neq \bar{x}$,

and $\Phi(\bar{x}, x) + k^0 \, d(\bar{x}, x) \in -\mathbf{K}$ *then there exists* $x_* \in \mathcal{B}_0 \cap M$.

Proof. **Theorem A** is true since **Theorem 3** and its Corollary. The theorem will be proved if we show that, **Theorem A**$\Rightarrow$**Theorem C**$\Rightarrow$**Theorem B**.

Theorem A$\Rightarrow$**Theorem C.** We know that **Theorem A** is true . Let all the assumptions of **Theorem c** be satisfied. From **Theorem A** we have an element $x_* \in \mathcal{B}_0$ such that

$$\Phi\left(x_*, x \right) + k^0 \, d\left(x_*, x \right) \notin -\mathbf{K}, \text{ for all } x \neq x_*. \text{ From condition } \left(\delta_2 \right) \text{ we have that}$$

$x_* \in M$ and hence $x_* \in \mathcal{B}_0 \cap M$, i.e. **Theorem C** is true.

Theorem C$\Rightarrow$**Theorem B.** We proved that under the assumptions of **Theorem 5**, **Theorem B** is true. We suppose satisfied all the assumptions of **Theorem B**.

We take $M := \left\{ \bar{x} \in X \middle| \bar{x} \in T(\bar{x}) \right\}$ and we remark that condition $\left(\delta_1 \right)$ implies condition $\left(\delta_2 \right)$. Hence, **Theorem C** furnishes an element $x_* \in \mathcal{B}_0 \cap M$ and from the definition of M we have certainly that $x_* \in T\left(x_* \right)$, i.e. the theorem is proved.∎

Theorem 6. *Under the assumptions of Theorem 5 the Theorems A, B, C are equivalent.*

Proof. Considering the assumption of *Theorem 5* we must show the implications: **Theorem B**$\Rightarrow$**Theorem C**$\Rightarrow$**Theorem A**. Indeed, **Theorem B**$\Rightarrow$**Theorem C**. We suppose true **Theorem B** and the hypothesis of **Theorem C**. Let $T{:}X \to X$ be the multivalued mapping defined by, $T(\bar{x}) := \left\{ x \in X \middle| x \neq \bar{x} \right\}$ for all $\bar{x} \in X$. If, for all $\bar{x} \in \mathcal{B}_0$ we have that $\bar{x} \notin M$ then condition $\left(\delta_1 \right)$ follows from condition $\left(\delta_2 \right)$ and by **Theorem B** there exists an element $x_* \in \mathcal{B}_0$ such that $x_* \in T\left(x_* \right)$, which is impossible since the definition of T.

Theorem C$\Rightarrow$**Theorem A..** Indeed, let **Theorem C** hold. For every $\bar{x} \in X$ we define $\Gamma(\bar{x}) = \left\{ x \in X \middle| x \neq \bar{x} \text{and } \Phi(\bar{x}, x) + k^0 d(\bar{x}, x) \in -\mathbf{K} \right\}$ and we consider the set $M := \left\{ \bar{x} \in X \middle| \Gamma(\bar{x}) = \phi \right\}$. If $\bar{x} \in \mathcal{B}_0 \setminus M$ then, from the definition of M there exists

$x \in \Gamma(\bar{x})$, such that $x \neq \bar{x}$ and $\Phi(\bar{x},x) + k^0 d(\bar{x},x) \in -K$, i.e. condition (δ_2) of *Theorem C* is satisfied and by this theorem there exists $x_* \in \mathcal{B}_0 \cap M$. We have $\Gamma(x_*) = \phi$, which means the conclusion of *Theorem A*.

Remark. Our variant of Caristi-Kirk's theorem (*Theorem B*) is different as the variant presented by Chr. Tammer in [36]. The hypothesis used in our variant are different and more simple as the hypothesis used in the variant presented in [36]. ∎

6. Application to ε-Efficiency

We give now an application of our main result to an important problem considered in the study of the ε-efficiency. Recently, Chr. Tammer proved the following result [36].

Theorem [Tammer] *Let (X,d) be a complete metric space, E a topological vector space, $K \subset E$ a convex cone with $k^0 \in \mathrm{int}\, K$ $B \subset E$ a cone with $clB + (K \setminus \{0\}) \subset \mathrm{int}\, B$ and $f : X \to E$ a mapping. If the following assumptions are satisfied:*

1) for each $r \in \mathbf{R}$ the set $M_r = \{x \in X | f(x) \in rk^0 - clB\}$ is closed

2) $f[X] \subset y + B$, for a certain $y \in E$,

then for every $\varepsilon > 0$ there exists some point $x_\varepsilon \in X$ such that:

$$(i^\circ): f_{\varepsilon k^0}(x_\varepsilon) \in \mathit{Eff}\left(f_{\varepsilon k^0}[X], K\right)$$

$$(ii^\circ): f_{\varepsilon k^0}(x_\varepsilon) \neq f_{\varepsilon k^0}(x), \text{ for all } x \neq x_\varepsilon \text{ where } f_{\varepsilon k^0}(x) := f(x) +.$$

$$k^0 d(x, x_\varepsilon)\sqrt{\varepsilon}$$

Remark In Tammer's theorem the set of efficient point is:

$$\mathit{Eff}\left(f_{\varepsilon k^0}[X], K\right) = \left\{ f_{\varepsilon k^0}(x_*) \Big| x_* \in X \text{ and } f_{\varepsilon k^0}[X] \cap f_{\varepsilon k^0}(x_*) . -(K \setminus \{0\}) = \phi \right\}$$

Using our variant of Ekeland's principle we obtain the following form of Tammer's theorem.

Theorem 7. *Let (X,d) be a complete metric space, $(E(\tau), K)$ a topological vector space ordered by the closed normal cone K and $f : X \to E$ a mapping. If the following assumptions are satisfied:*

1) for an element $k^0 \in K \setminus \{0\}$ the set $\{y \in X | f(y) - f(x) + k^0 d(x,y) \in -K\}$ is closed for every $x \in X$,

2) there exist $v_0 \in X$ and $w_0 \in E$ such that $f(x) - f(v_0) \geq w_0$, for all

$x \in X$ then for every $\varepsilon > 0$ there exists $x_\varepsilon \in X$ such that for $f_{\varepsilon k^0}(x) := f(x) + \sqrt{\varepsilon} d(x, x_\varepsilon) k^0$, we have:

$$(\theta_1): \begin{cases} f_{\varepsilon k^0}(x_\varepsilon) \in \mathit{Eff}\left(f_{\varepsilon k^0}[X], \mathbf{K}\right) \quad and \\ f_{\varepsilon k^0}(x_\varepsilon) \neq f_{\varepsilon k^0}(x), \quad for \quad all \quad x \neq x_\varepsilon \end{cases}.$$

Proof. We suppose that condition (θ_1) does not hold, that is, for a certain $\varepsilon > 0$ there exists no element x_ε with $f_{\varepsilon k^0}(x_\varepsilon) \in \mathit{Eff}\left(f_{\varepsilon k^0}[X], \mathbf{K}\right)$ and $f_{\varepsilon k^0}(x_\varepsilon) \neq f_{\varepsilon k^0}(x)$, for all $x \neq x_\varepsilon$.

This implies that for all $x \in X$ there exists an element $y \neq x$ with the property,

$(\theta_2): \{ f(y) + k^0 d_*(x, y) \in f(x) - \mathbf{K}, \quad where \quad d_* := \sqrt{\varepsilon}\, d.$

We define now the mapping $T: X \to X$ by $T(x) = y$, where y is given according to (θ_2).

Then obviously, we have $k^0 d_*(x, T(x)) + f(T(x)) - f(x) \in -\mathbf{K}$ and we remark that all the assumptions of **Theorem B (Caristi-Kirk)** are satisfied considering $\Phi(x, y) = f(y) - f(x)$ and $\mathcal{B}_0 = \left\{ x \in X \big| f(x) - f(v_0) + k^0 d_*(v_0, x) \in -\mathbf{K} \right\}$.

Hence, we obtain that T has a fixed point x_*, which is impossible, since $T(x) = y \neq x$ for all $x \in X$ and the theorem is proved. ∎

Comments. Our variant of the Ekeland's principle is very simple and **K** is not supposed to be with a nonempty interior or well based. When E is the real field **R** and $\mathbf{K} = \mathbf{R}_+$ we obtain exactly the classical Ekeland's principle considering $\Phi(x, y) = f(y) - f(x)$.

It is interesting to analyze if it is possible to obtain in the conclusion of **Theorem 7** that $f(x_\varepsilon) \in \mathit{Eff}\left(f[X], \mathbf{K}_{\varepsilon k^0}\right)$ and $d(x_\varepsilon, x^0) \leq \sqrt{\varepsilon}$, where

$f(x^0) \in \mathit{Eff}\left(f[X], \mathbf{K}_{\varepsilon k^0}\right)$ with $x^0 \in X$ given initially.

7. Application to Efficiency

Another possibility to apply our variant of Caristi-Kirk's fixed point theorem [**Theorem B**], to the Pareto optimization, is to use the idea developed in our paper

[18], that is, to apply this theorem to the dynamical system $\Gamma(x) = f[U] \cap (x + \mathbf{K})$, where $x \in f[U]$ and to study the existence of a critical point for Γ.

In this sense we consider the following variant of **Theorem B**, which is applicable to the study of critical points.

Theorem 8. *Let* (X, d) *be a complete metric space,* $(E(\tau), \mathbf{K})$ *an ordered locally convex space with* $\mathbf{K}$ *a closed convex normal cone and* Φ *a halfdistance from X into E. If the following assumptions are satisfied:*

$1°$) *there exists* $k^0 \in \mathbf{K} \setminus \{0\}$ *such that the set*

$$\left\{ y \in X \middle| \Phi(x, y) + k^0 d(x, y) \in -\mathbf{K} \right\} \text{ is closed for every } x \in X,$$

$2°$) *there exist* $v_0 \in X$ *and* $w_0 \in E$ *such that* $\Phi(v_0, x) \geq w_0$, *for every* $x \in X$,

then, for every multivalued mapping $T: X \to X$ *such that for all*

$$\bar{x} \in \mathcal{B}_0 = \left\{ x \in X \middle| \Phi(v_0, x) + k^0 d(v_0, x) \in -\mathbf{K} \right\} \text{ we have } T(\bar{x}) \neq \phi \text{ and}$$

$\Phi(\bar{x}, x) + k^0 d(\bar{x}, x) \in -\mathbf{K}$, *for any* $x \in T(\bar{x})$, *there exists* $x_* \in \mathcal{B}_0$ *such that*

$$T(x_*) = \{x_*\}$$

Proof. Supposing the assumptions satisfied, we have (applying **Theorem B**) that any choice map f on $\left\{ T(x) \middle| x \in \mathcal{B}_0 \right\}$ has a fixed point. Suppose T has no critical point. Considering a choice map f such that for any $x \in \mathcal{B}_0, x \in T(x)$ implies $f(x) \in T(x) \setminus \{x\}$, we have that a such map f can not have a fixed point, which leads a contradiction.■

Let U be a nonempty set and $(E, \| \; \|)$ a Banach space ordered by a closed normal convex cone $\mathbf{K} \subset E$ and $f: U \to E$ a mapping.

We say that $u_* \in U$ is a Pareto (maximal) point for f if

$$f(U) \cap \left[\mathbf{K} + f(u_*) \right] = \left\{ f(u_*) \right\}.$$

Theorem 9. *Let U be a nonempty set,* $(E, \| \; \|)$ *a Banach space ordered by a closed normal convex cone* $\mathbf{K}$ *and* $f: U \to E$ *a mapping.*

If there exists a complete subset $X \subseteq f(U)$ *and a halfdistance* Φ *from X into E such that:*

$1°$) *the multivalued mapping* $\Gamma(x) = f(U) \cap (\mathbf{K} + x)$, $\left(\text{with } x \in f(U) \right)$ *is such that* $\Gamma(X) \subseteq X$,

$2°$) *there exists* $k^0 \in \mathbf{K} \setminus \{0\}$ *such that the set* $\left\{ y \in X \middle| \Phi(x, y) + k^0 \| x - y \| \in -\mathbf{K} \right\}$ *is a closed set for every* $x \in X$.

$3°$) *there exist* $v_0 \in X$ *and* $w_0 \in E$ *such that* $\Phi(v_0, x) \geq w_0$, *for all* $x \in X$

$4°$) *for every* $\bar{x} \in \mathcal{B}_0 = \left\{ x \in X \middle| \Phi(v_0, x) + k^0 \| v_0 - x \| \in -\mathbf{K} \right\}$ *we have that*

$$\Phi(\bar{x}, x) + k^0 \|\bar{x} - x\| \in -\mathbf{K}, \text{ for all } x \in \Gamma(\bar{x}),$$

then f has a Pareto (maximal) point $\mathcal{U}_*$ in U, such that

$$\Phi\big(v_0, f(u_*)\big) + k^0 \|v_0 - f(u_*)\| \le 0,$$

Proof. The theorem is a consequence of **Theorem 8** where T is replaced by Γ. ∎

The next corollary is an extension to vector valued mappings of **Theorem 2** proved in the paper [18].

Corollary. *Let U be a nonempty set, $\big(E, \| \ \|\big)$ a Banach space ordered by a normal closed convex cone $\mathbf{K}$ and $f: U \rightarrow E$ a mapping.*

If there exists a complete subset $X \subseteq f(U)$ and a function $\varphi: X \rightarrow E$ such that :

1^0) the multivalued mapping $\Gamma(x) = f(U) \cap (\mathbf{K} + x), \big(\text{with } x \in f(U)\big)$ is such that Γ $(X) \subseteq X$,

2^0) there exists $k^0 \in \mathbf{K} \setminus \{0\}$ such that the set

$\big\{ y \in X \big| \varphi(y) - \varphi(x) + k^0 \|x - y\| \in -\mathbf{K} \big\}$ is closed for every $x \in X$,

3^0)there exist $w_* \in E$ such that $\varphi(x) \ge w_*$, for all $x \in X$.

4^0)there exists

$v_0 \in X$ such that for every $\bar{x} \in \mathcal{B}_0 = \Big\{ x \in X \big| \varphi(x) - \varphi(v_0) + k^0 \|x - v_0\| \in -\mathbf{K} \Big\}$ we

have that $\varphi(x) - \varphi(\bar{x}) + k^0 \|x - \bar{x}\| \in -\mathbf{K}$, for all $x \in \Gamma(\bar{x})$,

then f has a Pareto (maximal) point $u_* \in U$ such that $k^0 \|u_* - v_0\| + \varphi(u_*) \le \varphi(v_0)$.

Proof. We apply **Theorem 9** with $\Phi(x, y) = \varphi(y) - \varphi(x)$ and $w_0 = \varphi(v_0)$. ∎

Remark. Probably we can use the last Corollary to extend the concept of nuclear cone introduced in [18].

We consider now an existence result for efficient points using real valued functions.

Let $E(\tau)$ be a locally convex space ordered by a closed convex cone $\mathbf{K}$ and let $A \subset E$ be a nonempty set.

We say that $a \in A$ is an efficient (maximal) point of A if $A \cap (a + \mathbf{K}) = \{a\}$.

We denote the set of efficient points by $\text{Eff}(A, \mathbf{K})$.

If $A_0 \subset A$ is a nonempty subset we consider the dynamical system $\Gamma_0: A_0 \rightarrow 2^A$

defined by, $\Gamma_0(x) = A \cap (\mathbf{K} + x)$, for all $x \in A_0$. We denote,

$\mathbf{K}(A_0) = \big\{ v \in \mathbf{K} | v = v_1 - v_2 \text{ where } v_2 \in A_0 \text{ and } v_1 \in \Gamma_0(v_2) \cap A_0 \big\}$.

Theorem 10. *Let $E(\tau)$ be a locally convex space with the topology τ defined by the family of seminorms $\big\{ p_i \big\}_{i \in I}$. Let $A \subset E$ be a closed convex cone.*

If there is a nonempty subset A_0 of A such that:

1. A_0 is complete,

2. $\Gamma_0(A_0) \subseteq A_0$,

3. *for every $i \in I$ there is a continuos superadditive mapping $f_i : E \to R$ such that:*

$i^\circ). p_i(v) \leq f_i(v)$, *for every $v \in K(A_0)$,*

$ii^\circ). \sup\{ f_i(x) | x \in A_0 \} < +\infty$

then $Eff(A, K)$ is nonempty.

Proof. The proof is similar to the proof of *Theorem 3* proved in [19].∎

Corollary. *Let $E(\tau)$ be a locally convex space with the topology τ defined by the family of seminorms $\{p_i\}_{i \in I}$. Let $A \subset E$ a closed convex cone. If there is a nonempty subset A_0 of A such that:*

1. A_0 is complete,

2. $\Gamma_0(A_0) \subseteq A_0$,

3. for every $i \in I$ there exist $f_i \in E^*$ (the dual of E) and a continuous subadditive mapping $\varphi_i : E \to R$ such that,

$i^\circ) \ p_i(v) + \varphi_i(v) \leq f_i(v)$, *for every $v \in K(A_0)$*

$ii^\circ) \ \sup\{ f_i(x) | x \in A_0 \} < +\infty$

$iii^\circ) \ \inf\{ \varphi_i(x) | x \in A_0 \} > -\infty$

then $Eff(A, K)$ is nonempty. ∎

Open problem. It is interesting to extend *Theorem 10* to vector valued functions.

References

[1]. **Altman M.** : A generalization of Brézis-Browder principle on ordered sets, *Nonlin. Anal., Theory, Meth. and Appl. 6 Nr. 2 (1982), 157-165.*

[2]. **Banas J.** : On drop property and nearly uniformly smooth Banach spaces, *Nonlin. Anal., Theory, Meth. and Appl. 14 (1990), 927-933.*

[3]. **Brézis H. and Browder F. E.** : A general principle on ordered set in nonlinear functional analysis, *Adv. Math. 21 (1976), 777-787.*

[4]. **Caristi J.** : Fixed point theorems for mappings satisfying inwardness condition, *Trans. Amer. Math. Soc, 215(1976),241-251.*

[5]. **Danes J.** : A geometric theorem useful in nonlinear functional analysis, *Boll. Un. Mat. Ital. 6 (1972), 369-375.*

[6]. **Danes J.** : Equivalence of some geometric and related results of nonlinear functional analysis, *Comm. Math. Univ. Carolinae 26(1985), 445-454.*

[7]. **Dancs S., Hegedus M. and Medvegyev P. :** A general ordering and fixed-point principle in complete metric space, *Acta Sci. Math. (Szeged) 46(1983), 381-388.*

[8]. **De Figueiredo D. G. :** The Ekeland variational principle with applications and detours, *Tata Institute of Fundamental Research, Bombay (1989)*

[9]. **Ekeland I. :** Sur les problèmes variationnels, *C.R. Acad. Sci. Paris 275 (1972), A 1057-1059.*

[10]. **Ekeland I. :** On some variational principle, *J. Math. Anal. Appl. 47 (1974), 324-354.*

[11]. **Ekeland I. :** Nonconvex minimization problems, *Bull. Amer. Math. Soc. 1 (3), (1979), 443-474.*

[12]. **Ekeland I. :** Some lemmas about dynamical systems, *Math. Scand. 52 (1983), 262-268.*

[13]. **Ekeland I. :** The ε-variational principle revised, (Notes by S. Terracini), *Methods of nonconvex analysis (Ed. A. Cellina), Lecture Notes in Math. Springer-Verlag Nr. 1446 (1990), 1-15.*

[14]. **Elliot R. J. and Jarvis T. M. :** Prior play in a deterministic differential game, *J. Math. Anal. Appl. 86 (1982), 137-145.*

[15]. **Georgiev P. G. :** The strong Ekeland variational principle, the strong drop theorem and applications, *J. Math. Anal. Appl. 131 (1988), 1-21.*

[16]. **Giles J. R. and Kutzarova D. N. :** Characterization of drop and weak drop properties for closes bounded convex sets, *Bull. Austral. Math. Soc. 43 (1991), 337-385.*

[17]. **Giles J. R., Sims B. and Yorke A. C. :** On the drop and weak drop properties for a Banach space, *Bull. Austral. Math. Soc. 41 (1990), 503-507.*

[18]. **Isac G. :** Sur l'existence de l'optimum de Pareto, *Riv. Mat. Univ. Parma (4) 9 (1983), 303-335.*

[19]. **Isac G.:** Pareto optimization in ifninite dimensional spaces: the importance of nuclear cones, *J. Math. Anal. Appl. 182(1994), 393-404.*

[20]. **John J. :** Mathematical vector optimization in partially ordered linear spaces, *Peter Lang, Frankfurt (1986).*

[21]. **Khanh P. Q. :** On Caristi-Kirk's theorem and Ekeland's variational principle for Pareto extrema, *Preprint 357 Institute of Mathematics, Polish Academy of Sciences.*

[22]. **Loridan P. :** ε-solutions in vector minimization problems, *J. Optim. Theory Appl. 43 Nr. 2 (1984), 265-276.*

[23]. **Luc D. T. :** Theory of vector optimization, *Lecture Notes in Economics and Mathematical Systems, Springer-Verlag Nr. 319 (1989).*

[24]. **Montesinos V. :** Drop property equals reflexivity, *Studia Math. 87 (1987), 93-100.*

[25]. **Németh A. B. :** A Nonconvex vector minimization problem, *Nonlin. Anal. Theory, Methods and Appl. 10 (1986), 669-678.*

[26]. **Németh A. B. :** Between Pareto efficiency and Pareto ε-efficiency, *Optimization 20 Nr. 5 (1989), 615-637.*

[27]. **Oettli W. and Théra M. :** Equivalents of Ekeland's principle, *Bull. Austral. Math. Soc. 48 (1993), 385-392.*

[28]. **Penot J. P.** : The drop theorem, the petal theorem and Ekeland's variational principle, *Nonlin. Anal., Theory, Meth., Appl. 10 Nr. 9 (1986), 813-822.*

[29]. **Perressini A. L.** : Ordered topological vector spaces, *Harper and Row (1967), New York, Evanston and London.*

[30]. **Rolewicz S.** : On drop property, *Studia Math. 85 (1987), 27-35.*

[31]. **Rolewicz S.** : On Δ-uniform convexity and drop property, *Studia Math, 87 (1987), 181-191.*

[32]. **Schaefer H. H.** : Topological vector spaces *Mcmillan Company, New York , London (1966).*

[33]. **Staib T.:** On two generalizations of Pareto minimality, *J. Optim. Theory Appl. 59 Nr. 2 (1988), 289-306.*

[34]. **Takahashi W.:** Existence theorems generalizing fixed point theorems for multivalued mappings, Fixed point theory and applications *(ed. J. B. Baillon and M. Théra) Pitman Research Notes in Math., 252, Longman, Harlow (1991), 397-406.*

[35]. **Tammer Chr.:** A generalization of Ekeland's variational principle, *Optimization, 25 (1992), 129-141.*

[36]. **Tammer Chr.:** A variational principle and a fixed point theorem, *To appear (Proc. IFIP-Conf. Compiegne (1993)).*

[37]. **Tammer Chr.:** existence results and necessary conditions for ε-efficient elements, In: *B. Brosowski, J. Ester, S. Helding and R. Nehse (eds), Multicriteria Decision. Proc. 14 Th. Meeting of the German Working Group " Mehrkriterielle Entsheidung" Peter Larg, Frankfurt (1993), 97-110.*

[38]. **Valyi L:** Approximate saddle-point theorems in vector optimization, *J. Optim. Theory Appl. 55 Nr. 3 (1987), 435-448.*

Generation of Pareto Solutions by Entropy-Based Methods

A.M. Sultan[1] and A.B. Templeman[2]

Abstract

In recent years the Maximum Entropy Principle has been used to develop radically new approaches to various classes of optimization problems such as those of scalar non-linear constrained optimization, vector and minimax optimization. In this paper two new entropy-based approaches are developed, proved and applied to the problem of generating Pareto optimal solution sets for general multi-criteria optimization problems. The solution algorithms are applied to several test problems.

Keywords. Pareto set, Shannon Entropy, Multi-objective optimization

1 Introduction

Research into optimum engineering design now has a thirty-year history and continues to flourish. Over the past dozen years or so it has been recognized that real-world design is rarely adequately represented by a single criterion scalar optimization problem. Most design optimization problems require a multi-criteria formulation in order to represent the many conflicting goals which the designer is attempting to optimize simultaneously in the design process. Consequently, considerable attention has been given over the last decade to multi-criteria optimization methods.

A single criterion (or *scalar*) optimization problem may be stated as:

---

[1]P.O. Box 7361, Boulder, CO, 80306, U.S.A.

[2]Department of Civil Engineering, University of Liverpool, P.O. Box 147, Liverpool L69 3BX, U.K.

$$\text{Minimize:} \quad F(x)$$
$$x \in X$$

$$(1)$$

in which F is a single objective function of independent variables x_i, $i = 1, \ldots, N$, forming a variable vector x. X represents a constrained feasible region of x so problem (1) represents a standard constrained scalar optimization problem. A multi-criteria (or *vector*) optimization problem may be stated as:

$$\text{Minimize:} \quad F = (F_1(x), F_2(x) \ldots, F_C(x))$$
$$x \in X$$

$$(2)$$

in which F is now a vector of C different and independent objective functions, each a function of variable vector x and for each of which a reduction in value represents improvement. The constrained feasible region of x is again represented by X.

The interesting difference between scalar and vector optimization problems lies in the fact that scalar problem (1) usually has a unique solution, x^*, at least locally, whereas vector problem (2) usually has very many solutions; i.e., there is no unique vector x^* which simultaneously minimizes all the functions $F_1, F_2, \ldots, F_C$ which comprise F. It is therefore necessary to define very carefully what is meant by an 'optimum' solution of vector problem (2). One such definition is that of a Pareto solution.

A *Pareto solution* of problem (2) consists of any vector x such that when x is locally perturbed by a small amount Δx, none of the individual objectives F_j, $j = 1, \ldots, C$ in F decreases in value without at least one of the other objectives F_j increasing in value. Clearly, from this definition, if each of the individual objectives F_j in F is minimized independently of the others as a scalar optimization problem (1) then the resulting optimum variable vector x^* will be a Pareto solution of problem (2). Also, the scalar minimization of any linear combination of any of the F_j in F will also yield a Pareto solution. Figure 1 explains the idea of Pareto solutions graphically for the case in which problem (2) has two objective functions F_1 and F_2 which form the axes of the graph. Any feasible vector of variables x will yield particular values of the two objectives and will correspond to a unique point on the graph. The shaded region represents all possible feasible points. The curved boundary AB represents all Pareto solutions of the problem.

Engineering design may often be formulated as a vector optimization problem (2) in instances where the designer wishes to find some design in which several conflicting criteria must all be made as small in value as possible. In general, minimum cost and minimum failure probability are examples of conflicting criteria for structures; minimum cost and minimum surface roughness are similarly in conflict in machining operations. The set of Pareto solutions for such problems represents many possible solutions each of which is 'best' for a particular combination of weights or preferences which may be assigned to the individual objectives. Problem (2) must therefore be solved to generate the complete set of Pareto solutions. This is a difficult task and one which has been the subject of much research over the last decade. Having generated the Pareto set the designer may then select from among them that particular solution which best represents his own desired balance of preferences.

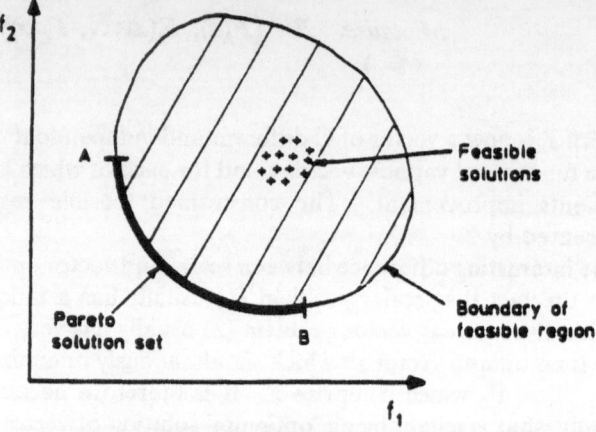

Fig. (1) Pareto Solutions

An alternative strategy to that of the last paragraph is for the designer first to articulate his own set of preferences of weights for each of the individual objectives in F. A sum of all the objectives, each weighted according to the desired preferences, may then be formed as a single objective function and the design problem solved as a scalar optimization problem (1). This will result in a unique solution which corresponds to one particular point in the Pareto set. This approach is much simpler to solve than the first but it requires the prior articulation of preferences which may not be known in the absence of any information about alternative good solutions such as is provided by the Pareto set.

This paper is concerned with the first approach above and presents two new methods for the generation of Pareto solution sets. Several methods already exist for generating Pareto solutions and a good survey is provided by Osyczka (1984). The methods described here are particularly useful in instances where C is large, i.e., the vector objective function has very many different conflicting objectives.

2 Preliminaries

In order to develop the Pareto set generation methods it is first necessary to introduce several established ideas, without further proof. The first of these is that a Pareto solution to problem (2) can be found by minimizing a scalar function consisting of a weighted sum of the individual objectives in **F**. If non-negative weights ω_j, $j = 1, \ldots, C$, which may be normalized without loss of generality, are associated with the objectives F_j in **F** then a Pareto solution to problem (2) may be found by solving the following scalar problem:

$$\text{Minimize: } \phi = \sum_{j=1}^{C} \omega_j F_j(x)$$
$$x \in X$$

(3)

with

$$\sum_{j=1}^{C} \omega_j = 1; \quad \omega_j \geq 0 \quad j = 1, \ldots, C$$

Problem (3) forms the basis of the *weighting objective functions method* for generating Pareto solution sets. This consists of solving problem (3) very many times, each time with a different set of weights ω, and so building up a set of Pareto solutions to problem (2). This is one of the most popular Pareto set generation methods because it is simple to use, but it has several disadvantages. Among these are the fact that different weighting sets may yield the same Pareto solution; the method cannot generate non-convex Pareto solutions, and it becomes very time-consuming to derive Pareto solutions for all combinations of all values of all the weighting coefficients when C is large.

A second prerequisite is a definition of *minimax optimization*. Minimax optimization is closely related to vector optimization as will be demonstrated later and can be stated as follows:

$$\text{Minimize Maximum: } < f_j(x) >$$
$$x \in X \quad j = 1, \ldots, C$$

(4)

Minimax optimization problem (4) requires that of the C individual objectives in vector optimization problem (2) the one with the currently largest value should be minimized over variables x. In any solution algorithm for problem (4) it is apparent that as x changes, the values of all the individual objectives will change. Consequently, the one with the largest current value will also change during the course of the solution process and this presents difficulties to any solution algorithm. A good example of a minimax optimization problem is that of the shape optimization of an engineering component which has been analysed by the finite element method. This will have calculated values of some stress resultant at C points in the component. The designer may then wish to alter the values of some design variables x of the component (such as thicknesses and boundary shape parameters) in such a way that the value of the maximum stress occurring at any of the C designated points in the component is made as small as possible. Problem (4) represents this design problem precisely.

A final prerequisite is the Shannon (1948) entropy function:

$$S = -k \sum_{j=1}^{C} p_j \ln p_j$$

$$(5)$$

where k is merely a positive constant depending on a suitable choice for the units of measure, and it is defined that $0 \ln 0 = 0$. It is a measure of the uncertainty in a discrete random process in which p_j is the probability associated with discrete event j. Jaynes (1957) used the Shannon entropy function to maximize S subject to the given information:

$$\sum_{j=1}^{C} p_j = 1$$

$$(6)$$

$$\sum_{j=1}^{C} p_j y_j(x) = E[y_j]$$

$$(7)$$

where E[•] is the expectation operator and it is axiomatic that $p_j \geq 0$. Equations (5) - (7) constitute a mathematical optimization problem and represent the mathematical form of Jaynes' Maximum Entropy Principle (MEP) which has an explicit solution:

$$p_j = \exp[\beta \cdot y_j(x)/k] / \left\{ \sum_{j=1}^{C} \exp[\beta \cdot y_j(x)/k] \right\}; \quad j = 1, \ldots, C$$

n which β is the Lagrange multiplier associated with the expected value constraint.

3 The Entropy-Based Weighted Method and its Aggregated Form

The first theorem shows that the solution of vector minimax problem (4) may be obtained by solving a scalar optimization problem.

3.1 Theorem 1:

The vector x^* which solves the vector minimax problem (4), where F is a vector of dimensionless objective functions, is generated by solving the scalar optimization problem, called the aggregated objective functions problem: (AOF)

$$\underset{x}{Min} \ \frac{1}{p} \cdot \ln \sum_{i=1}^{C} \exp[p \cdot F_i(x)]$$

$$(8)$$

where p is a positive parameter of increasing value towards infinity.

Proof:

This requires the use of Jensen's inequality (p-th norm inequality, Hardy, (1934)), which states that for any set of positive numbers U_i, $i=1,\ldots,C$ and $p \geq q \geq 1$,

$$\left\{ \sum_{i=1}^{C} U_i^p \right\}^{\frac{1}{p}} \leq \left\{ \sum_{i=1}^{C} U_i^q \right\}^{\frac{1}{q}}$$

$$(9)$$

Inequality (9) shows that the p-th norm of the set U decreases monotonically as its order, p, increases. An important property of the p-th norm is its limit as p tends towards infinity:

$$\lim_{p \to \infty} \left\{ \sum_{i=1}^{C} U_i^p \right\}^{\frac{1}{p}} = \underset{i \in C}{Max} <U_i>$$

(10)

Let

$$U_i = \exp[F_i(x)] : \quad i = 1,...,C$$

(11)

Since the objectives **F** are dimensionless, the U_i defined by equation (11) are all positive numbers and their substitution into result (10), gives:

$$\lim_{p \to \infty} \left(\sum_{i=1}^{C} \exp[p \cdot F_i(x)] \right)^{\frac{1}{p}} = \underset{i \in C}{Max} <\exp[F_i(x)]>$$

(12)

taking natural logarithms of both sides and noting that:

$$\ln \lim (\bullet) \equiv \lim \ln (\bullet)$$
$$\ln Max (\bullet) \equiv Max \ln (\bullet)$$

equation (12) becomes;

$$\lim_{p \to \infty} \left(\frac{1}{p} \right) \cdot \ln \sum_{i=1}^{C} \exp[p \cdot F_i(x)] = \underset{i \in C}{Max} <F_i(x)>$$

(13)

Equation (13) holds for any set of objectives **F(x)** including that set which results from minimizing both sides of equation (13) over $x \in X$. Thus, equation (13) may be extended to:

$$\underset{x \in X}{Min} \; \underset{i \in C}{Max} \; <F_i(x)> \; = \; \underset{x \in X}{Min} \; \left(\frac{1}{p}\right) \cdot \ln \sum_{i=1}^{C} \exp[p \cdot F_i(x)]$$

(14)

as p in the range $1 \le p \le \infty$ increases towards ∞. This completes the proof.

The next theorem shows how the scalar function (8), and hence, by virtue of theorem 1, the minimax optimization problem (4), is related to the general vector optimization problem (12). It is at this point that entropy is seen to be the link between the two problems (2) and (4).

3.2 Theorem 2:

For any value of P, a parameter with any non-zero positive or negative value:

$$\left(\frac{1}{P}\right) \cdot \ln \sum_{i=1}^{C} \exp[P \cdot F_i(x)] \ge \sum_{i=1}^{C} \lambda_i \cdot F_i(x) - \frac{1}{P} \sum_{i=1}^{C} \lambda_i \cdot \ln \lambda_i$$

(15)

with equality when the RHS is maximized over variables λ_i, i=1, ..., C, $\equiv$ λ. Such maximizing values of λ are:

$$\lambda_i = \exp[P \cdot F_i(x)] / \sum_{i=1}^{C} \exp[P \cdot F_i(x)]; \; i=1,...,C$$

(16)

Proof:

$$Let \; U_i = \exp[P \cdot F_i(x)] \; ; \; i=1,...,C$$

(17)

with dimensionless objectives **F**, as before. Thus the U defined by equation (17) will be positive numbers for any value of P. First, Cauchy's inequality (the arithmetic-geometric mean inequality, Hardy, (1934)) states that for U_i and λ_i; i=1, ..., C satisfying,

$$\sum_{i=1}^{C} \lambda_i = 1 \; and \; \lambda_i \ge 0$$

$$\sum_{i=1}^{C} U_i \geq \prod_{i=1}^{C} (U_i / \lambda_i)^{\lambda_i}$$

(18)

Taking natural logarithms of inequality (18) gives:

$$\ln\left(\sum_{i=1}^{C} U_i\right) \geq \sum_{i=1}^{C} \lambda_i \ln U_i - \sum_{i=1}^{C} \lambda_i \ln \lambda_i$$

(19)

substituting (17) into (19) yields:

$$\frac{1}{P} \ln \sum_{i=1}^{C} \exp[P \cdot F(x)] \geq \sum_{i=1}^{C} \lambda_i F(x) - \frac{1}{P} \sum_{i=1}^{C} \lambda_i \ln \lambda_i$$

(15)

Inequality (15) becomes an equality for any value of P when the RHS is maximized over variables λ subject to non-negativity and normality of the weights. Treating the RHS of (19) as a function to be maximized over λ subject to normality and non-negativity conditions, the Lagrangean is:

$$L(\lambda, \alpha) = \sum_{i=1}^{C} \lambda_i \ln U_i - \sum_{i=1}^{C} \lambda_i \ln \lambda_i + \alpha\left(\sum_{i=1}^{C} \lambda_i - 1\right)$$

(20)

There is no need to include the non-negativity conditions explicitly as the middle term of equation (20) imposes this indirectly. Stationarity of equation (20) with respect to λ_i ; i=1, ..., C and α gives:

$$\lambda_i = U_i / \sum_{i=1}^{C} U_i \; ; \; i = 1,...,C$$

(21)

Result (21) can be shown to correspond to a maximizing point of the RHS of inequality (19) by examining the second derivative matrix of the Lagrangean (20) which is negative definite.

Substituting result (21) into the RHS of inequality (19) gives, after some algebraic simplification:

$$\underset{\lambda}{Max}\left(\sum_{i=1}^{C} \lambda_i \cdot \ln U_i - \sum_{i=1}^{C} \lambda_i \ln \lambda_i\right) = \ln\left(\sum_{i=1}^{C} U_i\right)$$

Inequality (15), also, becomes an equality for any value of P when the RHS is maximized over variables λ , i.e., when the variables λ take values given by equation (21) on substitution of equation (17):

$$\lambda_i = \exp[P \cdot F_i(x)] / \sum_{i=1}^{C} \exp[P \cdot F_i(x)] \; ; \; i=1,...,C$$

(16)

and Theorem 2 is proved.

The RHS of inequality (15) consists of the vector optimization problem (2) in its scalar weighting form (3) but with an additional entropy term which is a function only of the multipliers (weights) λ and the parameter P. In its equality form with multipliers given by equation (16), relationship (15) shows that the entropy of the multipliers measures the difference between the scalar function (8) and the weighting objective function in problem (3).

It is clear that if both sides of relationship (15) in its equality form with multipliers given by equation (16) and minimized over variables $x \in X$ and with P becoming increasingly large and positive, the entropy term in (15) will tend towards zero. The LHS will generate a Pareto solution of the general vector optimization problem. This is expected, but it shows that the values of the multipliers ω in problem (3) which correspond to a minimax solution rather than just to any Pareto solution must be given by equation (16). Thus far we have shown that the scalar function (8) can be used to generate a vector minimax solution of problem (4) and that this solution is related through entropy to all Pareto solutions of the general vector optimization problem (2). The second main result of this section still remains to be proved and Theorem 3 formally states this.

3.3 Theorem 3:

The vector x^* which solves the scalar optimization problem (15) either in its aggregated form, the LHS, or its entropy-based weighted form, the RHS, for any non-zero value of P is a Pareto solution of the general vector optimization problem (2).

Proof:

From Theorem 2, relationship (15) is an equality for any P when λ is given by equation (16). This results holds for any set of objectives F including those evaluated at x^* which minimizes the RHS of equality (15).

Thus:

$$\frac{1}{P} \ln \sum_{i=1}^{C} \exp[P \cdot F_i(x^*)] = \sum_{i=1}^{C} \lambda^*{}_i F_i(x^*) - \frac{1}{P} \sum_{i=1}^{C} \lambda^*{}_i \ln \lambda^*{}_i$$

(22)

when λ^* is given by:

$$\lambda^*{}_i = \exp[P \cdot F_i(x^*)] / \sum_{i=1}^{C} \exp[P \cdot F_i(x^*)] \quad ; \quad i = 1, \ldots, C$$

(16)

x^* will be a Pareto solution of problem (2) if it is a solution of problem (3) with ω defined by (16), i.e., x^* must satisfy the necessary stationarity condition for problem (3) which is of the form (Kuhn-Tucker conditions for non-inferiority):

$$\sum_{i=1}^{C} u_i [\partial F_i(x^*) / \partial X_j] = 0 \quad ; \quad \forall j \in n$$

(23)

Name the LHS function and the RHS function in the equation (22) V_1 and V_2 respectively. The theorem will be proved if it can be shown that $(\partial V_1 / \partial x^*{}_j)$ and $(\partial V_2 / \partial x^*{}_j)$ are equal to (23).

Examining the first derivative of V_1 yields:

$$(\partial V_1 / \partial x^*{}_j) = \left(\frac{1}{P}\right)\left(\frac{\partial}{\partial x_j}\right)\left[\ln \sum_{i=1}^{C} \exp[P \cdot F(x^*)]\right] = \sum_{i=1}^{C} \lambda_i [\partial F_i(x^*) / \partial x_j] = 0$$

(24)

Examining the first derivative of V_2 yields the same result in equation (23).

3.4 Geometrical Interpretation

An important feature of Theorems 2 and 3 is that they impose no restrictions upon permissible values of P. Pareto solutions of the general vector optimization problem (2) may be generated by solving the scalar optimization problem (22) either in its aggregated form, the LHS, or its entropy-based weighted form, the RHS, for a range of different positive or negative values of P. This means that the scalar optimization problem, either side of (22), forms a very useful and efficient means of generating Pareto solutions. However, it is clear from Theorem 3 that values of P close to zero should not be used; both sides of (22) tend towards infinity as P tends towards zero.

In Theorem 1 a restriction is imposed upon P by the use of Jensen's inequality which is valid only in the range of $1 \geq P \geq \infty$. However, this theorem is used only to prove the vector minimax property of the solutions of either side of the problem (22) as P tends towards infinity. Consequently, the results of this section may be summarised in the following statement:

Pareto solution of the vector optimization problem (2) may be generated by solving the scalar optimization problem (22), on either side, for any values of P in the range $-\infty \leq \cdot P \leq \infty$ except those close to zero. For increasingly different non-zero values of P, the Pareto solutions generated will approach the solution of the vector minimax optimization problem (4).

It is clear that P plays an important role in generating individual solutions within the Pareto set. Any chosen value of P will generate a Pareto solution. Varying P will generate different Pareto solutions.

The Pareto solution generation properties of the scalar optimization problem (22), on either side, for any P have only recently become apparent. One big advantage that the method appears to have is that Pareto solution sets may be generated by specifying values for only one parameter, P. The currently popular weighting method (3) also involves scalar optimization but requires the specification of values for each of the objective weights ω_i; $i=1, \ldots, C$. Investigating the many different combinations of these weights values can be time-consuming, particularly for problems with many objectives.

However, the scalar optimization problem (22) can be interpreted geometrically. Consider the two criteria optimization problem presented in Fig. (2). In the space of objectives we can draw a line L with a slope $(-\lambda_1 / \lambda_2)$ or $[-\exp(P(F_1 - F_2))]$. The set L which represents this line is such that:

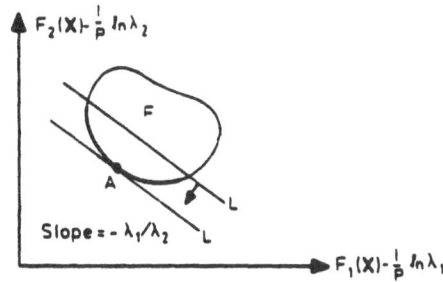

Fig. (2) Geometrical Interpretation of the EWOF Method
Based on (Osyczka, 1984)

$$\frac{1}{P}\ln\sum_{i=1}^{2}\exp[P\cdot F_i(X)] = \lambda_i\left[F_1 - \frac{1}{P}\ln\lambda_1\right] + \lambda_2\left[F_2 - \frac{1}{P}\ln\lambda_2\right] = \beta$$

where β is a constant. The minimization of (22), using either side, can be interpreted as moving in the line L with variable (λ_1 and λ_2) or [exp(PF$_1$) and exp(PF$_2$)] in a positive direction as close as possible to the origin, but keeping the intersection of the sets L and F. The point A for which L is tangent to F will be the minimum of (22). Note that for a non-convex problem, a great part of the set of non-inferior solutions may not be available, i.e., no values of λ_i or exp (PF$_i$) can locate the points in a certain region of the set F^{Pareto}. Consider the problem presented in Fig. (3). The line L$_1$ which is tangent at A with slope $(-\lambda_1 / \lambda_2)$ or $[-\exp (P(F_1 - F_2))]$ can be moved further in a positive direction until it is tangent at point B. Thus, the scalar optimization problem (23) with the values of λ_1 (x) or exp[PF$_i$(x)] will find point B but not A. Other values of λ_i (x) or exp[PF$_i$(x)] will find point B but not A. Other values of λ_i(x) or exp[PF$_i$(x)] will find point C. It is easy to see for this problem that the set of non-inferior solutions between D and E is not available.

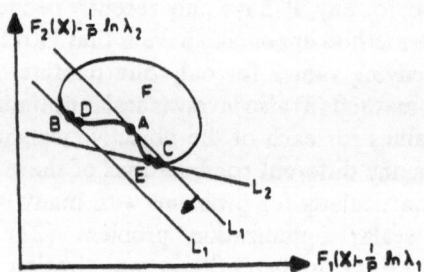

Fig. (3) The EWOF Method for a Non-Convex Problem
Based on (Osyczka, 1984)

In conclusion, the scalar optimization problem (22), which its LHS represents the aggregated objectives form and its RHS represents the entropy-based weighted objectives form, alone are inadequate since non-convex Pareto solutions can't be generated. Other entropy-based techniques which can generate non-convex Pareto solutions are needed so the two sets of methods can be used simultaneously to investigate gaps in the Pareto solution obtained by any of the sets alone and the next section deals with this.

4 The Entropy-based Constrained Method and its Compartmentalized Form

This section examines the development of the above mentioned method for the solution of vector optimization problems. Optimizing one objective while all of the others are constrained to some value is perhaps the most intuitively appealing generating technique (Cohon (1978)). Marglin (1967) appears to be the first to have suggested such an approach to vector optimization problems. The method follows directly, like the weighting method (3), from the necessary condition of non-inferiority (23). If x satisfies these conditions, then it is also an optimal solution to:

$$\underset{x \in X}{Min} \quad F(x, L) = \left\{ \begin{array}{l} F_h(x) \\ S.t.: F_i(x) \leq L_i; \; \forall i \neq h \end{array} \right\}$$

(25)

since

$$\nabla F(x,L) = u_h \cdot \nabla F_h(x) + \sum_{\substack{i=1 \\ i \neq h}}^{c} u_i \cdot \nabla F_i(x)$$

(26)

where the hth objective was arbitrarily chosen for minimization and L_i is preassigned upper bound on objective i. The RHS's in (25) seem to have appeared magically which, of course, they did. They are not at issue in the Kuhn-Tucker conditions (26), for which any RHS would do. They are, of course, important for feasibility.

4.1 Theorem 4:

The vector x^*_j which solves the vector minimax problem (4) is generated by solving the compartmentalized objective function problem: (COF)

$$\left. \begin{array}{l} \underset{x \in X}{Min} \dfrac{1}{p} \lambda_h(x) \cdot \ln \sum_{i=1}^{c} \exp\left[p \cdot F_i(x) \right] \\[4mm] S.t.: \dfrac{1}{p} \lambda_i(x) \cdot \ln \sum_{i=1}^{c} \exp\left[p \cdot F_i(x) \right] \leq 0 \; ; \; \forall i \neq h \in C \end{array} \right\}$$

(27)

with positive values of the parameter p increasing towards infinity.

Proof:

This requires recalling the entropy-based weighted problem:

$$\underset{x \in X}{Min} \sum_{i=1}^{C} \lambda_i F_i(x) - \frac{1}{p} \sum_{i=1}^{C} \lambda_i \ln \lambda_i$$

(28.a)

with

$$\lambda_i(x) = \exp[pF_i(x)] / \sum_{i=1}^{C} \exp[pF_i(x)]; \ \sum_{i=1}^{C} \lambda_i(x) = 1$$

which after substitution leads to the aggregated problem (8):

$$\underset{x \in X}{Min} \ \frac{1}{p} \cdot \sum_{i=1}^{C} \lambda_i(x) \cdot \ln \sum_{i=1}^{C} \exp[pF_i(x)]$$

(28.b)

Note that (28.b) is nothing else but the LHS of (15) since:

$$\sum_{i=1}^{C} \lambda_i(x) \cdot \ln \left(\sum_{i=1}^{C} (\cdot) \right) = \ln \left(\sum_{i=1}^{C} (\cdot) \right) \sum_{i=1}^{C} \lambda_i(x) = \ln \left(\sum_{i=1}^{C} (\cdot) \right)$$

Assuming that:

$$\sum_{i=1}^{C} \exp[p \cdot F_i(x)] = y(x \cdot p)$$

Then, (27) can be rewritten as

$$\underset{x \in X}{Min} \ \frac{1}{p} \exp[p \cdot F_h(x)] \cdot [\ln y(x,p) / y(x,p)]$$

$$S.t.: \frac{1}{p} \exp[p \cdot F_i(x)] \cdot [\ln y(x,p) / y(x,p)] \leq 0; \ \forall i \neq h \in C$$

Increasing p towards ∞ yields:

$$\left. \begin{array}{l} \underset{x \in X}{Min} \ \underset{h \in C}{Max} \ \ <F_h(x)> \\[2ex] S.t.: \ \underset{i \neq h \in C}{Max} \ \ <F_i(x) \leq 0 \end{array} \right\}$$

(29)

Equation (29) holds for any set of objectives F(x). Thus, equation (29) may be extended to:

$$
\left.
\begin{array}{l}
\underset{x \in X}{Min}\ \underset{h \in C}{Max} <F_h(x)> = \underset{x \in X}{Min}\ \frac{1}{p} \lambda_h(x) \cdot \ln \sum_{i=1}^{C} \exp[pF(x)] \\[2ex]
S.t.: \underset{i \neq h \in C}{Max} <F_i(x)> = \frac{1}{p} \lambda_i(x) \cdot \ln \sum_{i=1}^{C} \exp[p \cdot F_i(x)] \leq 0
\end{array}
\right\}
$$

$$(30)$$

as p in the range $1 \leq p \leq \infty$ increases towards ∞. This completes the proof. Theorem 4 does not directly involve entropy in any way. The next theorem shows how the compartmentalized problem (27), and hence, by virtue of theorem 4, the minimax optimization problem (29), is related to the general vector optimization problem (2). It is at this point that entropy is seen to be the link between the two problems (2) and (29).

4.2 Theorem 5:

For any value of P, a parameter with any non-zero positive or negative value:

$$
\left(\frac{1}{P}\right) \cdot \lambda_h(x) \cdot \ln \sum_{i=1}^{C} \exp[PF_i(x)] \geq \lambda_h(x) \cdot F_h(x) - \frac{1}{P} \lambda_h(x); \quad h \in C
$$

$$(31)$$

with equality when the RHS is maximized over variables λ. Such maximizing values of λ are:

$$
\lambda_h = \exp[P \cdot F_h(x)] / \sum_{i=1}^{C} \exp[P \cdot F_i(x)] ; \quad h \in C
$$

$$(32)$$

Proof:

Inequality (31) becomes an equality for any value of P when the RHS is maximized over variables λ subject to non-negativity and normality of the weights. The Lagrangean of (30) is:

$$
L(\lambda, \alpha) = \lambda_h(x) \cdot F_h(x) - \frac{1}{P} \lambda_h(x) \cdot \ln \lambda_h(x) +
$$

$$\sum_{i=1,\, i\neq h}^{C} \beta_i \left[\lambda_i F_i - \frac{1}{P} \lambda_i \ln \lambda_i \right] + \alpha \left[\sum_{i=1}^{C} \lambda_i - 1 \right]$$

$$(33)$$

There is no need to include the non-negativity conditions explicitly as the middle term of equation (33) imposes this indirectly. Stationarity of equation (33) with respect to λ_i ; $i = 1,...,C$ and α gives:

$$\lambda_h = \exp\left[P \cdot F_h(x)\right] / \sum_{i=1}^{C} \exp\left[P \cdot F_i(x)\right] \; ; \quad h \in C$$

$$(32)$$

substituting result (32) into the RHS of inequality (31) gives, after some algebraic simplification:

$$\underset{\lambda_h}{Max} \left[\lambda_h F_h(x) - (1/P) \lambda_h \cdot \ln \lambda_h \right] = (1/P) \lambda_h \cdot \ln \sum_{i=1}^{C} \exp\left[P \cdot F_i(x)\right]$$

and Theorem 5 is proved.

4.3 Theorem 6:

The vector $x^\bullet$ which solves the compartmentalized problem (27) or its entropy-based constrained form for any value of P is a Pareto solution of the general vector optimization problem (2).

Proof:

From Theorem 5, relationship (31) is an equality for any P when λ is given by equation (16). This result holds for any set of objectives F including those evaluated at $x^\bullet$ which minimizes (31), both sides. Thus:

$$
\left(
\begin{array}{l}
\displaystyle \mathop{Min}_{x \in X} \; \frac{1}{P} \lambda^*_{\,h} \cdot \ln \sum_{i=1}^{C} \exp\,[PF_i] \\[12pt]
\displaystyle S.t.: \frac{1}{P} \lambda^*_{\,i} \cdot \ln \sum_{i=1}^{C} \exp\,[PF_i] \leq 0
\end{array}
\right) \quad =
$$

$$
\left(
\begin{array}{l}
\displaystyle \mathop{Min}_{x \in X} \; \lambda^*_{\,h} F_h - \frac{1}{P} \lambda^*_{\,h} \cdot \ln \lambda^*_{\,h} \\[12pt]
\displaystyle S.t.: \lambda^*_{\,i} F_i - \frac{1}{P} \lambda^*_{\,i} \cdot \ln \lambda^*_{\,i} \leq 0
\end{array}
\right) \quad \forall \; i \neq h \in C
$$

$$(34)$$

when λ is given by (16).

x^* will be a Pareto solution of problem (2) if it is a solution of problem (25) with λ defined by (16), i.e., x^* must satisfy the necessary stationarity condition for problem (24).

5 Numerical Examples

Two numerical examples are presented. The Pareto set for each example is generated by different methods. The first example represents an optimal-shape design problem where the Pareto set performances and the associated designs, are presented as diagrams of the two parts of the beam. The form of these solutions goes through several distinct stages in progressing along the Pareto boundary of the feasible performance space. The form of these solutions and the trends they follow provide valuable design information for the designer or manufacturer. Even if none of them are chosen as the final design, the prescriptive advice that they offer on the likely form of good solutions in relation to different strategies performance can be used in the synthesis of that final design and is very difficult to obtain in any other way.

5.1 Example 1 (Osyczka, 1984)

This is a very well known example about beam design where two objectives must be satisfied:

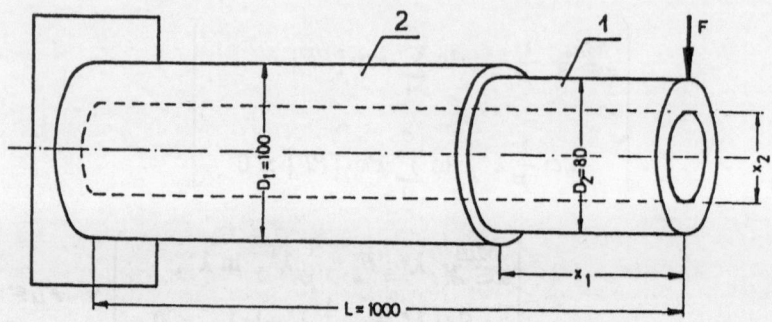

Fig. (4) Drawing of the Beam (Osyczka, 1984)

(1) the minimization of the volume of the beam, and
(2) the minimization of the static compliance of the beam.

Mathematically, the problem can be formulated as follows:

Optimize:

$$F_1(x) = 0.785[x_1 (6400.0 - x_2^2)$$

$$+ (1000.0 - x_1) (10^4 - x_2^2)] \, mm^3 \rightarrow Min$$

$$F_2(x) = 3.298 \times 10^{-5} \left\{ \left[\frac{1}{4.096 \times 10^7 - x_2^4} - \frac{1}{10^8 - x_2^4} \right] x_1^3 \right.$$

$$\left. + \frac{10^9}{10^8 - x_2^4} \right\} \, mm \, /N \rightarrow Min$$

S.t.:

$$g_1(x) = 180 - 9.78 \times 10^6 x_1 / (4.096 \times 10^7 - x_2^4) \geq 0$$

$$g_2(x) = 75.2 - x_2 \geq 0$$

$$g_3(x) = x_2 - 40 \geq 0$$

$$x_1, x_2 \geq 0$$

The results are summarized in Table (1) and Figs. (5-7). Referring to Table (1), the designer chooses one solution. This choice is based on his intuition and experience.

THE ENTROPY-BASED WEIGHTED OBJECTIVE FUNCTIONS METHOD	
No.　　$x = [x_1, x_2]$	$F(x) = [F_1(x), F_2(x)]$
1　　[165.29, 75.2]	[0.2943695e + 07, 0.49924662e - 03]
2　　183.58, 74.609]	[0.2961524e + 07, 0.495371176e - 03]
3　　[192.75, 74.307]	[0.2970895e + 07, 0.493597705e - 03]
4　　[202.39, 73.986]	[0.2981043e + 07, 0.491856365e - 03]
5　　[212.51, 73.644]	[0.2992045e + 07, 0.490157399e - 03]
6　　[219.88, 73.392]	[0.3000292e + 07, 0.489001861e - 03]
7　　[223.67, 73.262]	[0.3004618e + 07, 0.488433288e - 03]
8　　[152.71, 40.0]	[0.6162431e + 07, 0.340317842e - 03]
9　　[145.21, 40.0]	[0.6183632e + 07, 0.340058003e - 03]
10　　[137.95, 40.0]	[0.6204248e + 07, 0.33983076e - 03]
11　　[132.72, 40.0]	[0.6218924e + 07, 0.33968105e - 03]
12　　[131.05, 40.0]	[0.6223628e + 07, 0.33963611e - 03]
13　　[124.50, 40.0]	[0.634214e + 07, 0.339468941e - 03]
14　　[110.72, 40.0]	[0.6281057e + 07, 0.339171384e - 03]
15　　[101.02, 40.0]	[0.6308531e + 07, 0.339000951e - 03]
16　　[96.227, 40.0]	[0.6321995e + 07, 0.338929007e - 03]
17　　[74.569, 40.0]	[0.6385737e + 07, 0.33867266e - 03]
18　　[1.0, 40.0]	[0.6591174e + 07, 0.33846451e - 03]

THE ENTROPY-BASED CONSTRAINED OBJECTIVE FUNCTION METHOD			
1 [165.30, 75.199]	A		[0.2943734e + 07, 0.4992401e - 03]
2 [182.30, 74.651]			[0.2960245e + 07, 0.4956268e - 03]
3 [194.01, 74.265]			[0.2972194e + 07, 0.4933646e - 03]
4 [206.16, 73.859]			[0.2985099e + 07, 0.4912079e - 03]
5 [217.11, 73.487]			[0.2997181e + 07, 0.4894268e - 03]
6 [221.51, 73.336]			[0.3002174e + 07, 0.4887516e - 03]
7 [221.43, 71.549]			[0.3205612e + 07, 0.4663430e - 03]
8 [232.23, 71.191]			[0.3215188e + 07, 0.4652811e - 03]
9 [225.57, 67.174]			[0.3670394e + 07, 0.4277397e - 03]
10 [236.85, 66.252]			[0.3735082e + 07, 0.4232714e - 03]
11 [219.06, 66.058]			[0.3805428e + 07, 0.4189061e - 03]
12 [225.08, 65.402]			[0.3856105e + 07, 0.4156467e - 03]
13 [208.21, 64.240]	C		[0.4022139e + 07, 0.4063430e - 03]
14 [231.21, 62.351]			[0.4144831e + 07, 0.3994876e - 03]
15 [234.36, 62.070]			[0.4163319e + 07, 0.3985558e - 03]
16 [242.77, 61.416]			[0.4203013e + 07, 0.3966531e - 03]
17 [234.36, 58.967]			[0.4458198e + 07, 0.3850318e - 03]
18 [231.06, 58.955]			[0.4468607e + 07, 0.3845755e - 03]
19 [228.60, 58.361]			[0.4530231e + 07, 0.3820444e - 03]
20 [213.70, 57.052]			[0.4691012e + 07, 0.3758790e - 03]
21 [232.50, 55.351]			[0.4787910e + 07, 0.3725172e - 03]
22 [226.02, 54.971]			[0.4839180e + 07, 0.3707132e - 03]
23 [216.04, 54.552]			[0.4903343e + 07, 0.3685560e - 03]
24 [228.51, 53.746]			[0.4936624e + 07, 0.3675970e - 03]
25 [228.51, 52.481]			[0.5042167e + 07, 0.3644039e - 03]
26 [228.51, 51.345]			[0.5134723e + 07, 0.3617750e - 03]

27 [228.51, 50.906]		[0.5169965e + 07, 0.3608146e - 03]
28 [228.51, 50.241]		[0.5222748e + 07, 0.3594169e - 03]
29 [228.51, 49.383]		[0.5289871e + 07, 0.3577087e - 03]
30 [193.97, 48.748]		[0.5436395e + 07, 0.3538036e - 03]
31 [202.94, 47.265]		[0.5522785e + 07, 0.3518865e - 03]
32 [192.28, 46.483]		[0.5610471e + 07, 0.3499517e - 03]
33 [175.45, 46.399]		[0.5664181e + 07, 0.3488639e - 03]
34 [175.45, 44.620]		[0.5791270e + 07, 0.3463724e - 03]
35 [175.45, 42.870]		[0.5911458e + 07, 0.3442250e - 03]
36 [192.55, 40.0]		[0.6049859e + 07, 0.3421793e - 03]
37 [175.45, 40.0]		[0.6098175e + 07, 0.3412750e - 03]
38 [154.20, 40.0]		[0.6158251e + 07, 0.3403721e - 03]
39 [147.19, 40.0]		[0.6178043e + 07, 0.3401239e - 03]
40 [134.01, 40.0]		[0.6215292e + 07, 0.3397167e - 03]
41 [132.52, 40.0]		[0.6219488e + 07, 0.3396757e - 03]
42 [126.21, 40.0]		[0.6237336e + 07, 0.3395106e - 03]
43 [115.18, 40.0]		[0.6268511e + 07, 0.3392596e - 03]
44 109.69, 40.0]		[0.6284024e + 07, 0.3391511e - 03]
45 [103.31, 40.0]		[0.6302049e + 07, 0.3390382e - 03]
46 [92.308, 40.0]		[0.6333136e + 07, 0.3388738e - 03]
47 [85.273, 40.0]		[0.6353017e + 07, 0.3387872e - 03]
48 [77.439, 40.0]		[0.6375157e + 07, 0.3387062e - 03]
49 [51.448, 40.0]		[0.6448607e + 07, 0.3385353e - 03]
50 [48.834, 40.0]		[0.6455994e + 07, 0.3385250e - 03]
51 [44.642, 40.0]		[0.6467846e + 07, 0.3385106e - 03]
52 [42.974, 40.0]		[0.6472553e + 07, 0.3385057e - 03]
53 [0.0, 40.0]	B	[0.6593964e + 07, 0.3384645e - 03]

Table (1) Results for Calculations for Example (1)

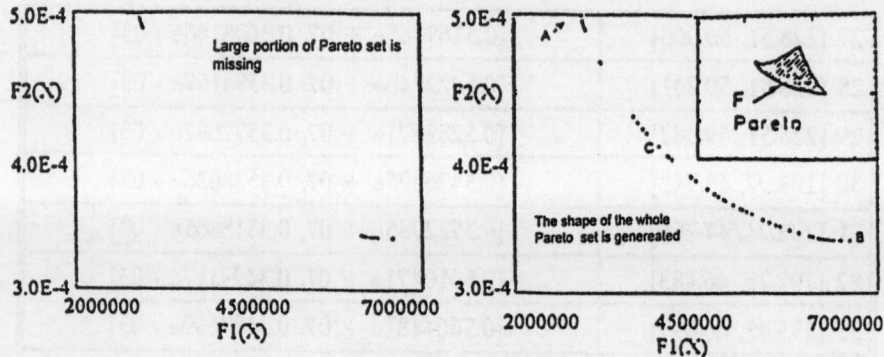

The Weighted Objectives Method   The Constrained Objective Method

Fig. (5) Pareto Sets Generated by the Entropy-Based Minimax Methods

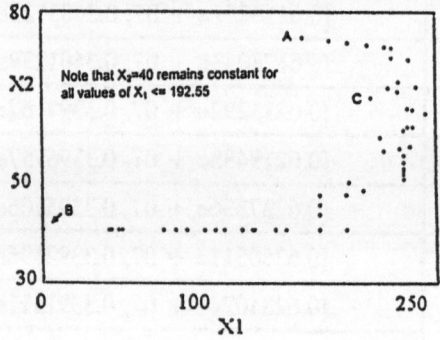

Fig. (6) The Design - Design Pareto Set Generated by the ECOF Method

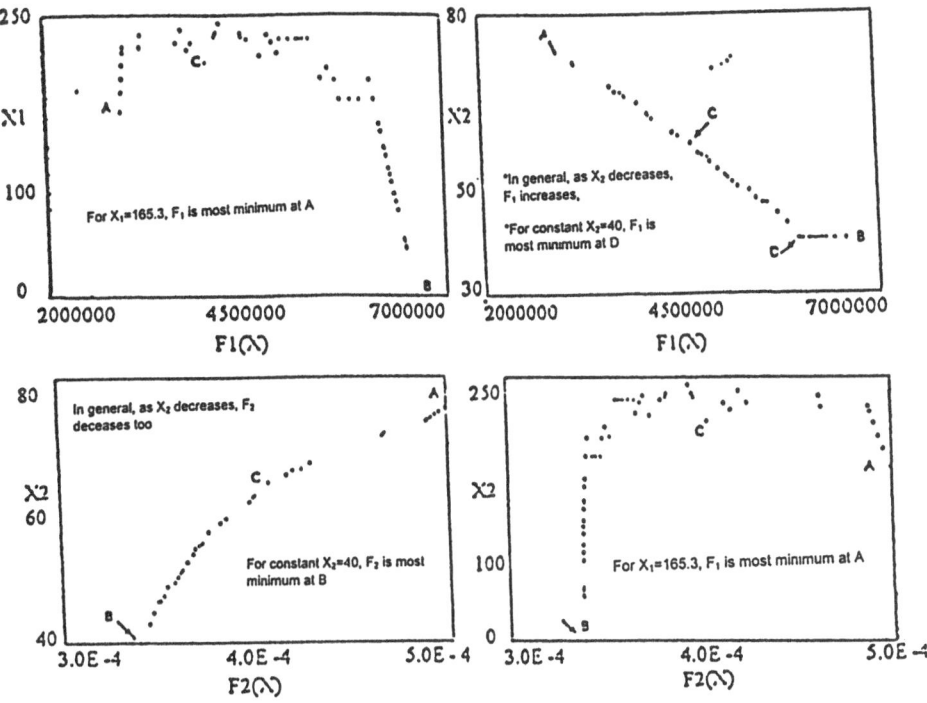

Fig. (7) The Design-Performance Pareto Set Generated by the ECOF Method

5.2 Example 2: (Singh, 1987)

The dressing depth of cut and the dressing lead are the two important dressing variables which significantly affect the surface roughness and specific grinding energy. Normally, to ensure good grinding performance, it is desirable to have, among other things, low surface roughness and, at the same time, low specific grinding energy. A finely dressed wheel gives low surface roughness but at the same time, a higher specific grinding energy, and vice versa. Therefore, both criteria of good grinding performance are in conflict and cannot be achieved simultaneously. Experiments were conducted to establish functional relationships between the dressing variables and the surface roughness as well as specific grinding energy under reciprocating plunge surface grinding conditions which may be cast as follows:

Optimize:

$$F_1(x) = -2.4798 + 0.55 x_1 + 0.22 x_2 \rightarrow Min$$
$$F_2(x) = 3.7366 - 0.56 x_1 - 0.27 x_2 \rightarrow Min$$

S.t.:

$$g_1(x) \equiv 3.18 - x_1 \geq 0$$
$$g_2(x) \equiv x_1 - 2.40 \geq 0$$
$$g_3(x) \equiv 2.70 - x_2 \geq 0$$
$$g_4(x) \equiv x_2 - 2.17 \geq 0$$
$$x_1, x_2 \geq 0$$

The results are summarized in Table (2).

THE ENTROPY-BASED WEIGHTED OBJECTIVE FUNCTIONS METHOD		
No.	$x = [x_1, x_2]$	$F(x) = [F_1(x), F_2(x)]$
1	[3.18, 2.7]	[-0.1368, 1.2268]
2	[2.4, 2.7]	[-0.5658, 1.6636]
3	[2.4, 2.5987]	[-0.5881, 1.691]
4	[2.40, 2.5159]	[-0.6063, 1.7133]
5	[2.40, 2.4594]	[-0.61873, 1.7286]
6	[2.40, 2.426]	[-0.6261, 1.7376]
7	[2.40, 2.373]	[-0.6377, 1.7519]

THE ENTROPY-BASED CONSTRAINED OBJECTIVE FUNCTION METHOD		
1	[3.18, 2.70]	[-0.1368, 1.2268]
2	[3.0560, 2.70]	[-0.2050, 1.2962]
3	[2.9648, 2.70]	[-0.2552, 1.3473]
4	[2.8292, 2.70]	[-0.3297, 1.4233]
5	[2.7395, 2.70]	[-0.3792, 1.4735]

6	[2.6024, 2.70]	[-0.4545, 1.5503]
7	[2.4794, 2.70]	[-0.5221, 1.6191]
8	[2.40, 2.2886]	[-0.6563, 1.7747]
9	[2.4, 2.17]	[-0.6824, 1.8067]

Table (2) Results of Calculations for Example (2)

6 Discussion

This paper has surveyed the new developments and use of informational entropy in the context of mathematical optimization processes. There are currently three areas in which entropy and the MEP appear to have considerable potential for methodological development. The first of these is as a means of measuring uncertainty, the second is as a means of making inferences in the presence of uncertain or incomplete information, and the third is as a means of solving a variety of optimization problems. The third of these has been studied here in some detail but the first two are certainly of equal importance and worthy of much more research.

The use of entropic inference in mathematical optimization processes is relatively new. The entropy-based weighted method or its aggregated form, and the entropy-based constrained method or its compartmentalized form developed by the authors for multi-criteria optimization have so far been the most successful. Computational work is concerned with assessing the efficiency of generating Pareto solution sets using the scalar minimization problem, either side of (22), and the constrained minimization problem, either side of (34), for varying P values in comparison with other conventional methods such as the weighting objective method (3), the constraint method (25). The objectives of developing new techniques to eliminate the difficulties described in the abstract are successfully achieved here. Because of that, the new methods are expected to replace the traditional methods for generating Pareto sets since considerable amounts of computational work and computer time can be saved. The two methods and their mean-entropy forms (2 & 4) are summarised in Table (3).

Finally, it is important to point out that two aspects have to be considered in computer programming:

1) In each of the minimax vector optimization methods, each feasible value of the entropy parameter P generates a Pareto point. A standard way of generating different values of P is by using the following formula:

$$P_i = P_{i-1} + \alpha$$

where α is a given step size between two consecutive values of P_i. Remember that the smaller step size we choose, the better the chance of having more Pareto solutions if there are any, $\alpha = 0.025$ has been used very often in this paper.

2) The entropy multipliers λ_1 can be written in FORTRAN as follows:

```
        DO 10 I=1,N
        A = 0.0
        DO 20 J=N, 1, -1
        AMBDA(I) = A+EXP(P*(F(J) - F(I))
        A = AMBDA(I)
   20   CONTINUE
        AMBDA(I) = 1.0/AMBDA(I)
   10   CONTINUE
```

where:

N = Number of objective functions.

Quadratic programming, dynamic programming or convex programming were developed originally for the solution of single criteria problems. However, they have been combined with the Pareto set generating techniques to solve multi-criteria optimization problems. In this paper, mathematical programming has been combined with the developed entropy-based techniques to solve multi-criteria optimization problems by generating its Pareto sets.

In the preceding section two examples were solved and their solutions were presented. Although it is difficult to select a preferred solution when the problem is purely numerical and has no physical or engineering interpretation, the only remaining purpose is showing the Pareto solutions. In such cases the shape of the Pareto optimal curve is the main significance to the decision maker and this can be reached by generating as many solutions as possible.

The first example is about design. The exploration of the relationships between the design variables (x_1 and x_2) and solution performances (F_1 and F_2) is possible only through multi-criteria optimization by generating the Pareto solutions. Three types of relationships can be recognised:

1) The performance-performance relationships, Fig. (5)
2) The design-design relationships, Fig. (6)
3) The design-performance relationships, Fig. (7)

The performance space and the design space relationships can directly be investigated by generating the Pareto set efficiently. Pareto solution A gives us the least volume while Pareto solution B gives us the least static compliance. These two solutions are extremes and the rest of the Pareto set solutions are in between, such as Solution C.

By the performance-performance relationships we mean the implications of choosing a certain level of performance in one criterion on the performances that are then attainable in other criteria. In terms of our beam design problem, the designer has to decide how much worsening of the beam volume, F_1, are we prepared to accept in order to lower static compliance. In this case, the Pareto set of performances follows the classic convex shape where improvement in volume has a steadily increasing rate of effect on static compliance. In the central portion, where C exists, the balance is even and even performances are likely to represent good compromise solutions.

By the design-design relationships we mean the implications of choosing or restricting values for one design variable on the values that need to be given to other design variables if acceptable performances are to be maintained. By studying these three types of relationships, the designer can easily select his/her preferred solution. The information we can get from the design-design space is that the interior diameter of the inner beam, x_2, stays constant for values of $x_1 \leq 154.2$. However, the relationship between x_1 and x_2 is not stable since design variables do not need to follow a stable relationship necessarily, because they are not objectives, but decisions vary in one way or another to satisfy optimality.

By the design-performance relationships, we mean the implications of choosing a certain level of one design on the performances that are then attainable in other criteria. The design-performance spaces, (F_1, x_2) and (F_2, x_2) follow stable shape while the design-performance spaces, (F_1, x_1) and (F_2, x_1), follow an unstable one for the same reasons mentioned before.

In the second example, we were able to obtain more Pareto solutions using the entropy-based methods than the ones obtained by non-entropy based methods. The shape of the Pareto optimal set defines a straight line which means the tradeoff between $F_1(x)$ and $F_2(x)$ will remain constant over the span of the Pareto optimal set. In addition, the same thing about the exploration of the relationships between the design variables (x_1 and x_2) and solution performances (F_1 and F_2) can be mentioned here.

By investigating the relative merits of the entropy-based weighted method and the entropy-based constrained method through solving several numerical examples, we have shown clearly that the entropy-based weighted method alone is inadequate to produce representative Pareto sets because of a large gap between two groups of solutions. There is no indication of whether these gaps are by reason of the set being concave in that region or there being no solutions in that region. The entropy-based constrained method provides more evenly distributed information and should be used in addition to the first method to investigate large gaps.

The numerical results obtained by the developed minimax methods make these methods very efficient computer-aided design tools. They are not only computationally efficient and easy to use, but they also generate very good representations of a complete Pareto set.

It is very important to note that, no matter how many criteria we consider, we always need to play with one parameter to generate the Pareto set. This property of the developed methods is very unique. In addition, it is notable that the size of the Pareto set generated by the developed methods is, in general, larger than the one generated by one of the known methods since generating Pareto sets can be done more quickly than before and more solutions are obtained.

In the preceding section, two examples were presented concisely on two-dimensional plots of the approximated Pareto optimal set in the performance space as shown before.

The results of the new entropic vector optimization techniques developed in this paper provide for the first time a quantity of information (objective values, design values, trade-offs) upon which designers can base their decisions at relatively low computational costs. This is a unique and important property, particularly when applying these techniques for problems with more than two objectives. The accuracy of the obtained results may be noted by comparing them with the corresponding graphical illustration figure or the reference taken from it. In all cases, previously published results obtained by other researchers were generated by the present methods to very high accuracy.

References

Cohon, J.L., *Multi Objective Programming and Planning*, New York: Academic Press, 1978.

Gere, J.S., *Optimization in Computer-Aided Design*, New York: North-Holand, 1983.

Hardy, G.H., Littlewood, J.E. and Palya, G., *Inequalities*, Cambridge: Cambridge University Press, 1934.

Jaynes, E.T., 'Information Theory and Statistical Mechanics', *The Physical Review*, Vol. 106, pp. 620-630 and Vol. 108, pp. 171-190, 1957.

Li, X.S., 'Entropy and Optimization', Ph.D. Thesis, Liverpool: Liverpool University, 1987.

Marglin, S., *Public Investment Criteria*, Cambridge, Mass.: MIT Press, 1967.

Osyczka, A., *Multicriterion Optimization in Engineering*, Chichester: Ellis Horwood, 1984.

Shannon, C.E., 'A Mathematical Theory of Communication', *Bell System Technical Journal*, Vol. 27, No. 3, pp. 379-428, 1948.

Singh, N. and Dutta, S.P., 'Some Multi-Objective Approaches to Diamond Dressing Optimization', *J. of Eng. Optimization*, Vol. 12, pp. 235-245, 1987.

Steuer, R.E., *Multiple Criteria Optimization: Theory, Computation and Application*, J. Wiley, 1986.

Sultan, A.M., 'Entropic Vector Optimization and Simulated Entropy: Theory & Applications', Ph.D. Thesis, Liverpool: Liverpool University, 1990.

Templeman, A.B. and Li, X.S., 'A maximum entropy approach to constrained nonlinear programming', *Eng. Optimization*, Vol. 12, No. 2, pp. 191-205, 1987.

MINIMAX VECTOR OPTIMIZATION METHODS

No	Name	Form
1.	The Entropy-Based Weighted Objective Functions Method, EWOF	$$\sum_{i=1}^{C} \lambda_i \cdot F_i(x) - \frac{1}{P} \cdot \sum_{i=1}^{C} \lambda_i \cdot \ln \lambda_i$$
2.	The Aggregated Objective Function Method AOF	$$\frac{1}{P} \cdot \sum_{i=1}^{C} \exp[P \cdot F_i(x)]$$
3.	The Entropy-Based Constrained Objective Function Method, ECOF	$$\lambda_h \cdot F_h(x) - \frac{1}{P} \cdot \lambda_i \cdot \ln \lambda_i$$ $$S.t.:\ \lambda_i \cdot F_i(x) - \frac{1}{P} \lambda_i \cdot \ln \lambda_i \le 0;\ i \ne h \in C$$
4.	The Compartmentalized Objectives Function Method, COF	$$\frac{1}{P} \cdot \lambda_h \sum_{i=1}^{C} \exp[P \cdot F_i(x)]$$ $$S.t.:\ \frac{1}{P} \cdot \lambda_i \cdot \sum_{i=1}^{C} \exp[P \cdot F_i(x)] \le 0;\ i \ne h \in C$$

$$\sum_{i=1}^{C} \lambda_i = 1; \; \lambda_i = \frac{e^{P \cdot F_i(x)}}{\underbrace{\sum_{i=1}^{C} e^{P \cdot F_i(x)}}} ; \; +\infty \geq P \geq -\infty$$

Table (3)

Part 2

Goal Programming

An Overview of Current Solution Methods and Modelling Practices in Goal Programming

M.Tamiz, D.F. Jones
University of Portsmouth, UK.

abstract

This paper presents an overview of the current state-of-the-art methods for modelling and solution of goal programming(GP) problems. Strategies for integrating these techniques into computerised software and thus specifications for producing an 'intelligent' GP solution and analysis package are suggested. Some recent criticisms of GP are detailed, together with how these perceived problems can be alleviated by means of such a package.

1   Introduction

Goal Programming(GP) is a multi-objective programming technique. The ethos of GP lies in the Simonan [22] concept of satisficing of objectives. That is, each objective under consideration (profit,safety,production level, etc.) is given a target or goal value to be achieved. The unwanted deviations from these objectives (under profit, under or over production, etc.) are then minimised , the distance function used for the minimisation being dependent on the type of GP used.

The roots of GP lie in a paper by Charnes, Cooper, and Ferguson in 1955 [4] in which they deal with executive compensation methods. A more explicit definition is given by Charnes and Cooper [5] in 1961 in which the term 'goal programming' is first used. Since 1961 the theory of GP has been formulated and many textbooks have been produced. The basic theory of GP and its extensions to the integer and non-linear cases is given by Ignizio [7]. The formulations of the dual of a GP and efficient solution algorithms are given by Ignizio [9, 8], many application areas are detailed by Lee [13] and Schniederjans [20], modelling methods for GP as well as a comprehensive GP bibliography are given by Romero [19] and a review of GP together with extensions into the field of artificial intelligence are given in a recent text by Ignizio and Cavalier [10].

The application areas of GP are. numerous, with over 350 references given by the bibliography in Romero [19]. Chronologically, a boom in GP application papers occurred in the late 1970's and early 1980's soon after the first major textbooks on GP appeared. Since then the level of applications has dropped to a steady, maintainable level [24].

GP models can be classified into two major subsets. In the first type the unwanted deviations are assigned weights according to their relative importance to the decision maker and minimised as an archimedian sum. This is known as weighted GP(WGP). The algebraic formulation of a WGP is given as

$$MIN \ a = \sum_{i=1}^{k}(w_{in}n_i + w_{ip}p_i)$$

Subject to,

$$f_i(\mathbf{x}) + n_i - p_i = b_i \ \ i = 1 \ldots k$$
$$\mathbf{x} \in C_s$$

Where $f_i(\mathbf{x})$ is a linear function(objective) of $\mathbf{x}$, and b_i is the target value for that objective. n_i and p_i represent the negative and positive deviations from this target value. w_{in} and w_{ip} are the respective weights attached to these deviations in the achievement function a. C_s is an optional set of hard constraints as found in linear programming.

In the other major strain of GP the deviational variables are assigned into a number of priority levels and minimised in a lexicographic sense. A lexicographic minimisation being defined as a sequential minimisation of each priority whilst maintaining the minimal values reached by all higher priority level minimisations. This is known as lexicographic GP(LGP). The algebraic representation of a LGP is given as:

$$Lex \ min \ \mathbf{a} = (g_1(\mathbf{n}, \mathbf{p}), g_2(\mathbf{n}, \mathbf{p}), \ldots \ldots, g_L(\mathbf{n}, \mathbf{p}))$$

subject to,

$$f_i(\mathbf{x}) + n_i - p_i = b_i \ \ \ i = 1, \ldots, k$$

This model has L priority levels, and k objectives. a is an ordered vector of these L priority levels. n_i and p_i are deviational variables which represent the under- and over- achievement of the i'th goal respectively. $\mathbf{x}$ is the set of decision variables to be determined. Any 'LP' style hard constraints are placed, by convention, in the first priority level. A standard 'g' (within priority level) function is given by:

$$g_l(\mathbf{n}, \mathbf{p}) = \alpha_{l_1}n_1 + \ldots + \alpha_{l_k}n_k + \beta_{l_1}p_1 + \ldots + \beta_{l_k}p_k$$

Where α and β represent inter-priority level weights, as in weighted GP, a zero weight is given to any deviational variable whose minimisation is unimportant.

The remainder of this paper is divided into four sections. Section two gives details of GP solution methods. Section three discusses GP modelling techniques. Section four details GP extensions. Finally, section five draws conclusions.

2 Solution Methods

2.1 Review of Solution Methods

WGP models are, in terms of simplex solution, equivalent to standard LP models and can hence be solved by any LP package although, for reasons discussed in Section 3, without specialised GP modifications this is not the most efficient means of solution either in terms of computation or of decision maker understanding.

Most of the work in development of GP solution algorithms has consequently been concerned with LGP solution. The first LGP solution method is detailed by Charnes and Cooper [5] and implemented as computer code by Jääskeläinen [11] in 1969, this code was limited to models with fifty or less variables. The next generation of LGP codes treated each priority level as a separate LP and added augmenting constraints at each priority level to safeguard the minimal value obtained at the previous level. This type of algorithm is known as sequential simplex and is described by Ignizio [9]. Further algorithms were produced by Authur and Ravindran [2], and Schniederjans and Kwak [21]. A comparison of these methods is given by Olson [18]. Ignizio [8] introduced a primal-dual method that allows dropping of dual constraints at each priority level and thus achieves a model size reduction at the end of each optimisation.

Algorithms are also available to handle integer lexicographic GP, papers by Lee and Luebbe [14] and Markland and Vickery [15] detail such codes.

2.2 Specialised Goal Programming Solution

More efficient Solution to both lexicographic and weighted GP's can be given by the employment as specialised GP speed-ups. This section presents a brief overview of these methods.

1. Feasible Basis Creation : An LGP, or the objective set in a WGP, does not require the addition of logical (slack,surplus,artificial) variables. The initial basic variable for an objective is given by the negative deviational variable if the target value is positive and by the positive deviational variable is the target value is negative. In the case of a zero target value, the deviational variable with least weight in the achievement function should be included in the starting basis.

   This method eliminates the need for 'phase 1' of the simplex method in LGP or WGP with no optional constraint set, the only 'phase 1' iterations required being concerned with the optional constraint set.

   To implement this method, it is clear that each deviational variable must appear in one, and only one, objective. This, as shown in the following section on modelling, does not reduce the flexibility of the GP model.

2. Variable fixing : As mentioned above, Ignizio [9] presents a method for LGP solution where variables are fixed on reaching optimality to a priority level in such a way as to ensure that the minimal value is maintained whilst minimising lower priority levels. This reduces the size of the model at each priority level and thus gives a faster solution time. This technique is reported as being able, on average, to solve a LGP with five priority levels in 1.5 times the time it takes to solve an LP of similar size. Further work by Ignizio produced a dual based method which eliminates rows at each priority level solution [9].

3. Restricted Pricing : Due to the nature of a GP, it is clear that both deviational variables cannot be in the basis simultaneously. This will never occur provided both variables have positive or zero weights in the achievement function. Thus when finding the simplex entering variable there is no need to consider a deviational variable if its corresponding deviational variable is already in the basis. For problems with a large number of objectives compared to decision variables, this leads to a significant reduction in the computational burden associated with determining the entering variable. In certain preference modelling situations detailed in the next section, basis restriction is essential in order to maintain the integrity of the model [25].

4. Deviational Variable Exchange : Techniques are available that exploit the fact that the column entries of a deviational variable of an objective in the simplex tableau have the opposite sign of the column entries for the other deviational variable in the same objective. Schniederjans and Kwak [21] present an algorithm which uses this fact to reduce matrix storage and iteration time. Tamiz and Jones [26] give an amended row-choosing procedure, NPSWAP, that allows swapping of deviational variables in the basis at each iteration.

These methods, when combined in a specialised GP solver, which can be produced by the adaption of an LP code (which - due to extensive LP research - tend to be efficient in terms of speed and size-capacity) produce significant speed-up as compared to solving a WGP by an LP solver or a LGP as a series of increasing size LP's by a macro-controlled LP solver.

3 Modelling Techniques

This section presents several of the most widespread criticisms and perceived shortcomings of GP, together with modelling methods for the alleviation of these criticisms and suggestions of how these methods can be incorporated in a large scale 'intelligent' GP package.

3.1 Incommensurability

Incommensurability occurs when deviations from objectives using different units of measurements are summed directly. This difference in scale corrupts the relative importance of the objective to the decision maker specified by their weights. The difference may become extreme (consider, for example, a cost goal in the order of 10^6 and a goal of .1 % variance in a sample batch) in which case, unbeknown to the decision maker, the larger magnitude goal may dominate the smaller goal even though approximate equality was specified by roughly similar weights.

The problem of incommensurability can be largely alleviated by the use of a normalisation procedure. The weight given to the deviational variable(s) of an objective is divided by some constant related to the magnitude of that objective thus bringing the formulation closer (although not exactly) to the original weights. All currently used normalisation methods, together with some theoretically good new methods are compared in a recent paper by Tamiz and Jones [27] and the most useful are given as:

1. Percentage Normalisation : Divisor is absolute value of right hand side of objective - this method should be used in the case when all deviations are required as percentage deviations from the target and hence is rather specialised. Care should be taken when comparing large (in percentage terms) positive and negative deviations (double the target value gives deviation = 100 % but halving target value gives deviation = 50 %). Clearly not suitable for zero target value.

2. Euclidean Normalisation : Divisor is the Euclidean sum of the coefficients of the decision variables in the objective. Probably the most general purpose normalisation method. Shown by Romero [19] to correspond to the Euclidean L_2 distance function.

3. Summation Normalisation : Divisor is the sum of the absolute values of the coefficients of the decision variables in the objectives. Provides a larger divisor that the Euclidean method and is useful in extreme cases of bad scaling of objectives and decision variables.

A normalisation method provides an approximate method of restoring original weights, but given the information that the problem has a degree of incommensurably the decision maker may like to adjust the weights manually to take account of the difference. In this case an incommensurability warning can be produced when the pre-analysis of the GP shows an objective dominating other objectives. This warning is designed only for decision maker information purposes.

A third option is a hybrid of the above two methods. In which incommensurability is measured before solution and if it reaches a certain threshold automatically triggers a normalisation method. This method can be extended to different normalisation methods being applied at different ranges of incommensurability. In the case of the incommensurability being below the lowest threshold the GP is solved without applying any normalisation.

3.2 Pareto Efficient Solutions

A major criticism of GP is its ability to produce Pareto Inefficient, sometimes called dominated, solutions [30]. Pareto efficiency in a GP is defined as a state in which no objective can be improved without the degradation of another objective. For an objective with both deviational variables penalised improvement is defined as movement towards the target value and degradation movement away from the target value. For an objective with a single deviational penalised improvement is defined as movement towards(penalised side) and away from(non penalised side) the target value and degradation is defined as movement away from(penalised side) and towards (non penalised side) the target value. The reason why movement away from the goal on the non-penalised value is classified as improvement is due to the fact that target values are not always , and in some case should not be [19], set at their ideal values. Pareto inefficiency in a GP occurs when some target values have been set too pessimistically by the decision maker.

Methods exist for the restoration of efficiency in an inefficient GP. Hannan [6] proposes a simple method for restoration. Romero [19] improves on this method by adding extra constraints that ensure that the efficient point found contains no objective at a lower value than that at the initial solution point.

Tamiz and Jones [28] give a method of detection of efficiency and inefficiency of both the model and each individual objective at the optimal solution point. They further suggest two methods of restoration of efficiency in the case where the solution at that point is found to be inefficient. The first method is based on the preferences given by the decision maker in the initial setting of weights and priority levels. The second method is based on an interactive process in which the decision maker specifies which of the inefficient objectives they would most like to see increased at each iteration.

These detection and restoration processes are computationally efficient and can be included in a specialised GP solver. The restoration stage also allows for detection of any pareto unbounded objectives and thus provides information concerning possible oversights on the part of the modeller.

3.3 Redundancy in Lexicographic GP

Another modelling error in GP is the setting of an excessive number of priority levels. In the variable fixing lexicographic solution algorithm variables whose entry into the basis would degregate the minimal value of that priority level are fixed. Thus the feasible region of any priority level is given by the set of alternative optima to all other higher priority levels. If at the end of a priority level all variables are classified as fixed there are no alternative optima and the feasible region for all subsequent priority levels is reduced to a point. In other words, no iterations are possible and any objectives associated with these priority levels play no part in the optimisation.

It is important in this case for the decision maker to re-adjust his priorities, weights and targets in order to eliminate this redundancy or to discard the redundant objectives.

The first requirement of an intelligent GP system is to check for the 'all variables fixed' condition at optimality for each priority level and if found to stop the analysis (no further iterations may take place in any case) and report all lower priority levels, and their associated objectives, as redundant. A further feature would be an analysis as to which unmet objectives have resulted in the fixing of decision variables as this points to unreasonably high target values. This process could then be connected to an interactive algorithm detailed in the next section.

3.4 Preference Modelling Techniques

The standard GP model allows only for a direct linear relationship between the penalty of a deviational variable and its relationship between the goal. The gradient in this relationship is given by the weight associated with that variable in the achievement function. This formulation is not sufficient to model all possible preference curves (a curve of penalty level against distance from the goal. Martel and Aouni [16] present the first generalised preference modelling method for GP based on the Prométhée method [3]. Tamiz and Jones [25] break all preference curves down to a combination of groups of four possible preference change types. These are presented below, together with discussion about their implementation in a GP package.

1. Increase in Preference : This occurs when a decision maker wants to increase the per unit penalty at some point distant from the goal. This would indicate problems caused by a large deviation from a goal target (e.g. low profit or a heavy amount of overtime worked). This is the well documented case of penalty functions as detailed by Romero [19].

   The most efficient means of modelling a penalty function in a specialised GP solver is to form another objective with the target value being equal to the value at which the penalty is to be increased. The deviational variable representing the region of increased per unit penalty is given a weight representing the increase in per unit penalty at that point. This formulation ensures that each deviational variable is placed in at most one objective and hence allows the use of the feasible basis creation technique given in section 2. A deviational variable swapping routine, again detailed in section 2, can be used to ensure a minimal amount of work associated with the extra objective.

2. Decrease in Preference : This occurs when the decision maker wishes to decrease the per unit penalty at some point distant from the goal. This would indicate a critical area around the goal followed by growing indifference as the benefits of being close to the goal are lost. This is the opposite of the first case and somewhat rarer in occurrence. It is modelled in the same way for similar reasons with the exception that the weight given to the new deviational variable is now negative to represent the decrease in penalty beyond the new goal value. A basis restriction technique, as outlined in section 2, is now necessary to ensure the integrity of the model.

3. Discontinuity in Preference : This case occurs when there is a sudden rise in penalty on crossing some threshold in an objectives value. This represents a sudden change in conditions (e.g. The need to rent another facility or a legal requirement being broken). This requires the use of a binary variable which is placed in the achievement function. Thus to include this preference type a GP solver must have some inbuilt binary variable handler and be able to optimise a mixed achievement function, where decision variable may be placed in the achievement function along with deviational variables. The first condition is not difficult if the GP solver has an underlying commercial LP solver, which have binary variable facilities. It will, however, produce a slow-down in computational speed if a large number of discontinuities are required. The second condition should be modelled in such a way as to allow the binary variables into the achievement function without affecting the GP analysis given by the other modelling techniques in this section. This can achieved by converting the sum of the binary variables in the achievement function to an objective with zero target value. This objective then requires special treatment as regards normalisation.

4. Non-Linear Preference Curve : This case is handled by means of a piece-wise linear approximation as detailed by Williams [29]. It is then reduced to a series of increases and decreases in preference.

4 Goal Programming Extensions

As well as the main areas of standard lexicographic and weighted GP and their modelling techniques as detailed in the previous section, a number of extensions and other varieties of GP exist. These formulations provide a valuable part of the GP paradigm and should not be neglected in a GP solver. If these techniques are readily accessible in a GP package it will, in the authors' view, increase both awareness and use of these methods. Consequently two main extensions of GP, together with suggestions for their implementation in a GP package and special rules for modelling and analysis are given in this section.

4.1 Fuzzy and Chebyshev GP

Fuzzy and Chebyshev GP formulations both work on an L_∞ metric rather than the L_1 metric used by weighted and lexicographic GP. That is, they seek to minimise the maximum deviation from the goals rather than a weighted sum of deviational variables. In traditional Chebyshev GP all deviations are taken as being from the best obtainable value for that objective. This value can be found by considering only that objective and optimising its value by means of solving an LP. Here we allow for the case in which the target value can be lower than the best obtainable value. This leads to a modified version of the standard form of a Chebyshev GP given as:

$$MIN \ z = \lambda$$

subject to,

$$f_i(x) - \lambda <= b_i \quad i = 1, \ldots, k_1$$

$$f_i(x) - \lambda <= b_i \quad i = k_1 + 1, \ldots, k_2$$

$$f_i(x) + \lambda >= b_i \quad i = k_1 + 1, \ldots, k_2$$

$$f_i(x) + \lambda >= b_i \quad i = k_2 + 1, \ldots, k$$

$$x \in C_s$$

Where the first k_1 objectives have only positive deviations penalised, the next $k_2 - k_1$ objectives have both deviational variables penalised, and the final $k - k_2$ objectives have only negative deviational variables penalised. Putting the Chebyshev GP in this form suggests it can be formed from a corresponding WGP by means of conglomeration of deviational variables into a single deviational variable λ with a weight of unity and creation of an extra constraint for each objective with both deviational variables penalised. This in turn suggests the fact that a decision maker can set an option in a GP solver for Chebyshev GP and their WGP model could be automatically converted into a Chebyshev GP and solved as an LP. The GP solution methods, such as feasible basis creation, deviational variable swapping, and basis restriction no longer apply to Chebyshev GP solution. Analysis should show the maximum deviation(s) and hence the λ value. Normalisation methods are still appropriate as incommensurability problems still exist and are of even more importance as the maximum deviation is now considered rather than a deviational variable sum. Also there are no weights to counteract the incommensurability.

The problems with incommensurability in Chebyshev GP lead on to the area of Fuzzy GP. Fuzzy GP is shown by Ignizio and Cavalier [10] to be similar to Chebyshev GP with all objectives scaled onto a zero-one range by dividing the deviation by the difference between the highest and lowest obtainable values for that objective (again found by optimising a series of LP's). The deviation divided by the range is known as the fuzzy membership function. This has the advantage of reducing all objectives onto a comparable scale but requires more computational time as 2k LP's, where k is the number of objectives, need to be solved before GP solution. It is also not usable in the case of unbounded objectives and is highly dependent on the accurate definition of the feasible region. Subject to these constraints on its use a fuzzy GP can be formed from a WGP in a similar manner as to a Chebyshev GP automatically by a GP solver.

4.2 Interactive Methods

In all GP types mentioned all parameters such as goal levels, weights, priority structures have been specified *a priori* of solution and the model is solved to a solution. Interactive methods allow progressive definition of the decision maker's weights during solution. At each iteration of the interactive procedure information concerning the objective levels at the current solution point is presented and the decision maker is asked to further articulate his preferences in some way. The GP is then re-optimised taking into account this new information. The method then proceeds until the decision maker is satisfied or no further changes are possible. Several interactive methods for WGP have been proposed. The most well known include Masud and Hwang's [17] algorithm, Lara and Romero's [12] adaption of the Zionts/Wallenius algorithm [31], and Spronks Interactive multiple goal approach [23]. For LGP, interactive methods are less common, although Ignizio [9] proposes an Augmented GP method.

In order for a interactive algorithm to be suitable for inclusion in an intelligent GP package it must fulfill the following design criteria.

1. It must be able to handle lexicographic GP - most of the application papers in GP have used a lexicographic structure [24]. If a method is only for WGP it may still be included but it is not, in the authors' view, sufficient as the only interactive method to be offered by the solver. An LGP interactive algorithm must be included.

2. It must be able to handle models of the same magnitude as current large-scale LP packages. The first consideration for such a model is that of computational time - the processing time should not be unreasonably large as to keep the decision maker waiting for long periods. The second consideration is that of information presented to the decision maker. The amount presented must be sufficient enough to allow the decision maker to make a rational choice but not so great as to overwhelm them. Thus, for example,

a method that presents k^2 pieces of information where k is the number of objectives will be difficult to implement for a large number of objectives.

3. The method should be compatible with the GP modelling techniques presented in the previous section. In particular it should be able to include pareto efficiency detection at each restoration if the interactive method does not guarantee pareto efficiency.

5 Conclusion

This paper has presented an overview of GP with reference to the incorporation of current GP solution and modelling techniques in an piece of intelligent GP software. The authors feel that the combination of these methods provide significant advances over the use of LP software for WGP solution or a series of calls to an LP solver for LGP solution. The benefits of the specialised GP solver can be divided into several categories

1. Speed : The gain associated with the use of the specialised GP speed-ups detailed in section two is of benefit when solving large-scale GP models. As a GP package can use an underlying LP solver for simplex iterations, the speed associated with LP solution technology is not lost.

2. Information : A GP solver can give specialised GP solution information such as percentage breakdown of the achievement function, objective and priority level output, information about pareto efficiency, etc. This information can be presented in a more readable manner than output from an LP solver.

3. Flexibility : With solution on an LP solver, normalisation can be expensive in terms of modelling time as it has to be applied manually. With a GP solver the effects of different normalisation can be measured simply by changing the input options to the GP solver. Likewise, some of the preference modelling techniques detailed in section 3 are more easily applied and are able to solve larger models than solution by an LP solver. Similarly conversion to Chebyshev or Fuzzy GP requires complete reformulation and input on an LP solver but can be set as an input option on a GP solver.

4. Learning Purposes : Techniques which may be available but not used by a decision maker due to time constraints, when presented as an option in a package, encourages use and hence expands the knowledge of, as well as use of, GP and its extensions.

5. Ease of Use : A specialised GP package is available to modellers without the need to know the underlying solution algorithms. To use GP options

in an LP solver requires some knowledge of the LP solution algorithm in order to adapt it for the solution of a GP with those options.

The suggestions made in this paper are incorporated in GPSYS [26] intelligent GP system currently under development at the University of Portsmouth.

Acknowledgments : The authors would like to thank the Engineering and Physical Sciences Research Council, UK. and the Numerical Algorithms Group, UK. for their sponsorship of this research

References

[1] AUTHUR, J.L. and RAVINDRAN, A. A Branch and Bound Algorithm with Constraint Partitioning for Integer Goal Programming Problems, *European Journal of Operational Research*, 4, 421-425. 1980.

[2] ARTHUR, J.A. and RAVINDRAN, A. An Efficient Goal Programming Algorithm Using Constraint Partitioning and Variable Elimination, *Management Science*, 24, 1109-1119. 1978.

[3] BRANS, J.P., VINCKE, P., and MARESCHAL, B. A Preference Ranking Organization Method, *Management Science*, 31, 647-656. 1985

[4] CHARNES, A., COOPER, W.W., and FERGUSON, R. Optimal Estimation of Executive Compensation by Linear Programming, *Management Science*, 1, 138-151. 1955.

[5] CHARNES, A. and COOPER, W.W. *Management Models and Industrial Applications of Linear Programming.* John Wiley and Sons, New York. 1961.

[6] HANNAN, E.L. An Assessment of Some of the Criticisms of Goal Programming, *Computers and Operations Research*, 12, 525-541. 1985.

[7] IGNIZIO, J.P. *Goal Programming and Extensions.* Lexington, Mass.: Heath (Lexington Books), 1976.

[8] IGNIZIO, J.P. An Algorithm for Solving the Linear Goal Programming Problem by Solving its Dual, *Journal of the Operational Research Society*, 36, 507-515. 1985.

[9] IGNIZIO, J.P. *Linear Programming in Single and Multiple Objective Systems.*, Prentice-Hall, Inc., Englewood Cliffs, New Jersey. 1982.

[10] IGNIZIO, J.P. and CAVALIER, T. Linear Programming, *Prentice Hall.* 1994.

[11] JÄÄSKELÄINEN, V. *Accounting and Mathematical Programming*, Contact Author, Helsinki. 1969.

[12] LARA, P. and ROMERO, C. An Interactive Multigoal Programming Model for Determining Livestock Rations: an Application to Dairy Cows in Andalusia, Spain. *Journal of the Operational Research Society*, **43**, 945-953. 1992.

[13] LEE, S.M. *Goal Programming for decision analysis.* Auerback, Philadelphia, 1972.

[14] LEE, S.M. and LUEBBE, R.L. A Zero–One Goal–Programming Algorithm Using Partitioning and Constraint Aggregation. *Journal of the Operational Research Society*, **38**, 633-640. 1987.

[15] MARKLAND, R.E. and VICKERY, S.K. The Efficient Computer Implementation of a Large–Scale Integer Goal Programming Model, *European Journal of Operational Research*, **26**, 341-354. 1986.

[16] MARTEL, J.M. and AOUNI, B. Incorporating the Decision-maker's Preferences in the Goal–programming Model, *Journal of the Operational Research Society,* **41**, 1121-1132, 1990.

[17] MASUD, A.S. and HWANG, C.L. Interactive Sequential Goal Programming, *Journal of the Operational Research Society*, **32**, 391-400. 1981.

[18] OLSON, D. A Comparison of Four Goal Programming Algorithms, *Journal of the Operational Research Society*, **35**, 347-354. 1984.

[19] ROMERO, C. *Handbook of Critical Issues in Goal Programming.* Pergamon Press, 1991.

[20] SCHNIEDERJANS, M.J. Linear Goal Programming, *Petrocelli Books* , 1984.

[21] SCHNIEDERJANS, M.J. and KWAK, N.K. An Alternative Solution Method for Goal Programming Problems : A Tutorial, *Journal of the Operational Research Society*, **33**, 247-251. 1982.

[22] SIMON, H.A. *Models of Man*, J. Wiley & Sons, New York. 1957.

[23] SPRONK, J. Interactive Multiple Goal Programming : Applications to Financial Planning, Martinus Nijhoff.

[24] TAMIZ, M., JONES, D.F., and EL-DARZI, E. A review of goal programming and its Applications, Presented at APMOD93, Budapest, Hungary. To be published in *The Annals of OR.* 1994.

[25] TAMIZ, M. and JONES, D.F. Expanding the Flexibility of Goal Programming via Preference Modelling Techniques, Technical Report, University of Portsmouth, UK. 1993.

[26] TAMIZ, M. and JONES, D.F. GPSYS : Preliminary User Guide, University of Portsmouth, UK. 1994.

[27] TAMIZ, M. and JONES, D.F. An Empirical Analysis of Normalisation Procedures Within Goal Programming, Internal Report, University of Portsmouth.

[28] TAMIZ, M. and JONES, D.F. Detection and restoration methods for pareto inefficient and unbounded solutions in goal programming , Internal Report, University of Portsmouth. 1994.

[29] WILLIAMS, H.P. *Model Building in Mathematical Programming*, J. Wiley & Sons. 1978.

[30] ZELENY, M. The pros and cons of Goal Programmng. *Computers and Operations Research*, 8, 357–359, 1982.

[31] ZIONTS S. and WALLENIUS J. An Interactive Programming Model for Solving the Multiple Criteria Problem, *Management Science*, 22, 652-663. 1976.

Goal Programming in Networks

Saul I. Gass[1]

[1] College of Business and Management, University of Maryland, College Park, MD 20742, USA

Abstract. The methodologies of network analysis and goal programming combine to form a powerful modeling framework that enlarges the application base that can be solved by the powerful computational procedures of networks. In this paper, we review and discuss goal programming network structures and some applications.

Keywords. goal programming, networks, manpower planning

1 Introduction

With the publication of Ford and Fulkerson's classic *Flows in Networks* in 1962, the mathematical foundations and the practical utility of network-based models were established. Over the past 30 years, a number of books on networks have been published: Hu (1969); Frank and Frisch (1971); Potts and Oliver (1972); Bazarra and Jarvis (1977); Jensen and Barnes (1980); Phillips and Garcia-Diaz (1981); Murty (1992); Glover, Klingman and Phillips (1992); Evans and Minieka (1992); and Ahuja, Magnanti and Orlin (1993). And, from Hillier and Liebman (1967) and Wagner (1969) on, we find that most Operations Research and Management Science texts treat network models as an important topic in its own right.

In contrast, goal programming, from the time it was first proposed by Charnes and Cooper (1961), took a bit longer to get established as a specialized theoretical and applied topic. The number of books dedicated to goal programming is not large and includes, among others, Ignizio (1976, 1982) and Schneiderjans (1984), Romero (1991), with chapters or sections in, for example, Zeleny (1982), Chankong and Haimes (1983), and Ringuest (1992). Just about all Operations Research/Management Science texts treat goal programming as an important modeling technique. We now find goal programming being included as an important tool for resolving multicriteria problems, Zeleny (1982), Romero (1991), Ringuest (1992). The general interest in goal programming and its extensive number of applications is, of course, due to its ability to resolve multicriteria problems. However, a researcher interested in determining whether and how the topics of networks and goal programming intersect would have to search far and wide before finding anything appropriate. In terms of books and texts (and we submit we may have missed some items), the only reference to goal programming considerations in network models is the book by Glover, Klingman and Phillips (1992). These authors have pioneered network applications, in general, and have shown how to incorporate goal programming procedures into networks so that the mathematical form of the network is maintained. There are a few journal papers that discuss goal programming and networks, in particular, Klingman and Phillips (1984), and Ignizio (1983a, 1983b).

In this paper, we review the basic structure of some network models, discuss how these structures can be extended to include goal programming considerations, and describe a few examples. We note that: (1) networks are powerful formulative tools and decision aids, (2) networks are computationally attractive in that very large-scale network problems can be solved readily and rather cheaply on serial or parallel computers, Barr and Hickman (1994), and (3) by using multicriteria conditions we can, via goal programming, incorporate more realism into network structures without any formulative or computational complications.

2 The Basic Minimum-Cost Network-Flow Transshipment Problem

A network consists of a set of nodes, with each node connected by an arc (edge) to at least one other node, such that goods (material, personnel) can be shipped from some subset of the nodes, called origins (sources), to some other subset of the nodes, called destinations (sinks). Such a network is usually designated as $G = (N, A)$, where N is the set of nodes and A is the set of arcs. A node that is not an origin or a destination is called an intermediate node. The shipments flow from the origin nodes, across connecting arcs and through the intermediate nodes until they arrive at the destinations nodes. The shipments are constrained by the supplies available at the origins and the demands at the destinations, with a shipment x_{ij} from a node i to a node j being constrained by a lower bound, l_{ij}, and an upper bound (capacity), u_{ij}, for that arc, i.e., $l_{ij} \leq x_{ij} \leq u_{ij}$. The lower bounds are often taken to be zero.

Each arc begins at some node i and ends at some node j, with the flow x_{ij} across the arc being directed from node i to node j. Such an arc represents a one-way link (street). However, a given pair of nodes (i,j) can have an arc with flow going from i to j and a second arc with an other flow going from j to i. These two arcs represent a two-way link between nodes i and j. An arc with its flow going from i to j is called directed and is denoted by the ordered number pair (i, j). Associated with each arc is a nonnegative cost, c_{ij}, of shipping one unit of the goods from i to j. These costs are assumed linear in that if x_{ij} is the amount shipped from i to j, the total cost is given by $c_{ij}x_{ij}$. Associated with each node i is a number b_i. If $b_i > 0$, then node i is a source node with a supply equal to b_i, if $b_i = 0$, then node i is an intermediate node, and if $b_i < 0$, then node i is a destination node with a demand equal to $-b_i$. The underlying basis of the problem is that we want to ship the goods from the origins to the destinations such that the total cost of the shipment is minimized. Under the important assumption of conservation of flow through a node, that is, what goes into a node must equal to what goes out of the node, the minimum-cost network-flow problem can be expressed mathematically as follows:

Minimize $\Sigma_i \Sigma_j c_{ij}x_{ij}$

subject to

$$\Sigma_j x_{ij} - \Sigma_j x_{ji} = b_i \quad \text{(for all i in N)} \quad (1)$$

$$0 \leq l_{ij} \leq x_{ij} \leq u_{ij}$$

The constraints (1) are the conservation of flow equations. There is one equation for each node and one variable for each arc. If the node i is a source node, then its corresponding conservation of flow equation has a positive right-hand-side; if the node is an intermediate node, then its right-hand-side is zero; and, if the node is a destination node, then its right-hand-side is negative.

Conservation of flow through a node n is illustrated by the following Figure 2.1:

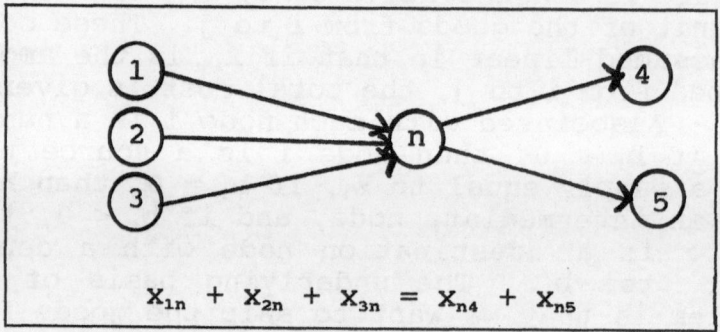

Fig. 2.1. Conservation of Flow Through a Node

$$X_{1n} + X_{2n} + X_{3n} = X_{n4} + X_{n5}$$

The network described by system (1) is referred to as a pure network problem. It has the property that if the b_i are integers, then an optimal solution in integers exists (assuming that the problem is feasible).

A typical directed arc, along with its parameters, is usually pictured as follows, Figure 2.1a:

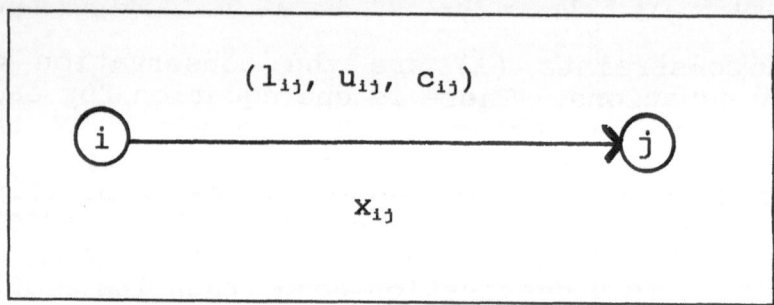

$$(l_{ij}, u_{ij}, c_{ij})$$

$$x_{ij}$$

Fig. 2.1a. Typical Directed Arc

An example of a minimum-cost problem is shown in Figure 2.2; the corresponding flow equations follow.

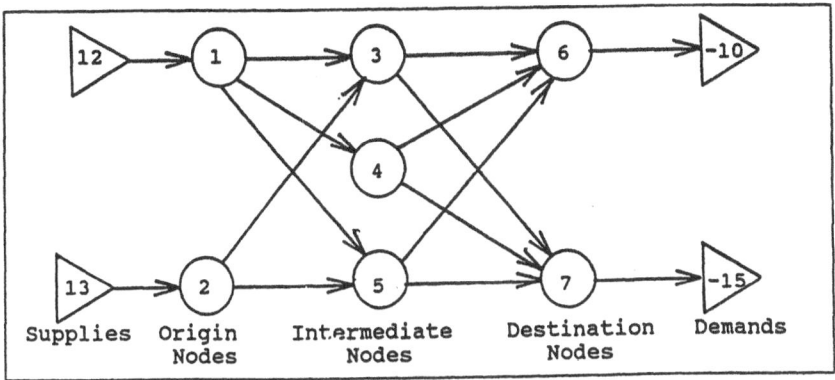

Fig. 2.2.   Typical Network Flow Structure

Mathematical Model of Figure 2.2 Network

$$\text{Minimize } \Sigma_i \Sigma_j c_{ij} x_{ij}$$

subject to

$$
\begin{array}{llllll}
x_{13} + x_{14} + x_{15} & & & & & = 12 \\
& x_{23} + x_{25} & & & & = 13 \\
- x_{13} & - x_{23} & + x_{36} + x_{37} & & & = 0 \\
- x_{14} & & & + x_{46} + x_{47} & & = 0 \\
- x_{15} & - x_{25} & & & + x_{56} + x_{57} & = 0 \\
& & - x_{36} & - x_{46} & - x_{56} & = -10 \\
& & - x_{37} & - x_{47} & - x_{57} & = -15 \\
& & & & l_{ij} \le x_{ij} \le u_{ij}
\end{array}
$$

3 Transportation and Assignment Problems

The simplest goal programming network problem
is that of a transportation (assignment) problem in which the total supply is less thank the
total demand.   This unbalanced transportation
problem is usually not presented as a goal
problem, but just as a variation of the basic
transportation problem.   For example, the
following network is for an assignment problem
in which employees in four different skill
categories (supplies) must be assigned to meet

requirements at for four different jobs (demands), with the total number of available employees (100) less that the total number required (120). Each skill class can be assigned to specific jobs at a cost c_{ij}, as depicted in the Table 3.1.

Table 3.1. Assignment Problem Supplies, Demands, and Costs

Employees by Skill	Supply	Jobs (j)			
		1	2	3	4
	1 (10)	10	7	-	-
	2 (20)	-	8	-	10
	3 (30)	-	9	10	-
	4 (40)	7	-	10	8
		(30)	(30)	(30)	(30)	Demand

Here, we treat the demands as goals, which can only be underachieved. The situation is depicted in Figure 3.1. There, the origin nodes, 1, 2, 3, 4, have been augmented by a shortage origin node 9 that has a pseudo-supply of 20 employees. A penalty of c_{9j} (j = 5, 6, 7, 8) is assigned to each the shortages x_{9j} in the usual linear fashion. The variables x_{9j} are the goal underachievement variables. This augmentation preserves the transportation network structure and the standard transportation simplex algorithm can be used to solve it.

If the shortage cost of not supplying a worker is $100 for each job type, then the optimal solution has a cost of $2,840, job 3 will be short 20 employees, with the optimal assignment of the 100 employees shown in Table 3.2.

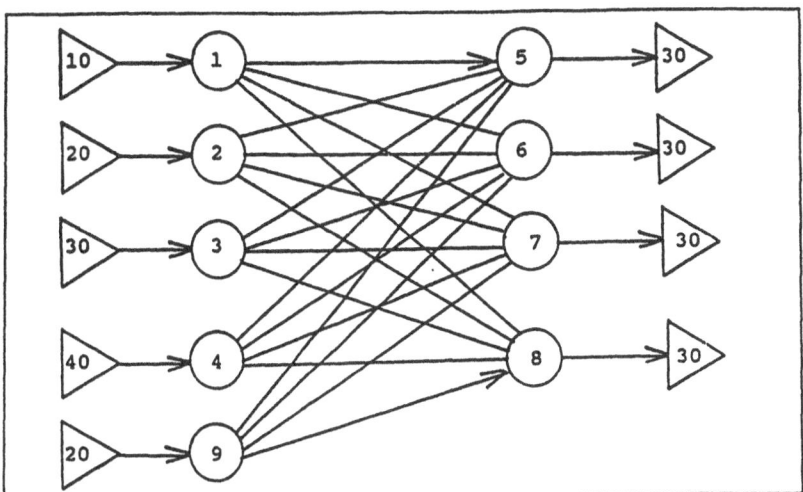

Fig. 3.1. Assignment Problem Network

Table 3.2. Solution to Assignment Problem

		Jobs (j)			
	Supply	1	2	3	4
	1 (10)	0	10	-	-
	2 (20)	-	0	-	20
Employees by Skill	3 (30)	-	20	10	-
	4 (40)	30	-	0	10
		(30)	(30)	(30)	(30) Demand

4 Manpower Planning Problems

As much of our work in networks and goal programming has been in the application area of
manpower models, we shall illustrate the goal-
programming aspects of networks using concepts
from manpower models, Gass (1984, 1988, 1991).
The basic multi-year manpower policy planning
problem can be stated as follows:

Given a work force at the beginning of the planning period, we want to determine the hiring, firing, promotion and training policies that should be used so that the ending work force best meets future work force requirements (forecasts).

The problem situation can be viewed as in Figure 4.1.

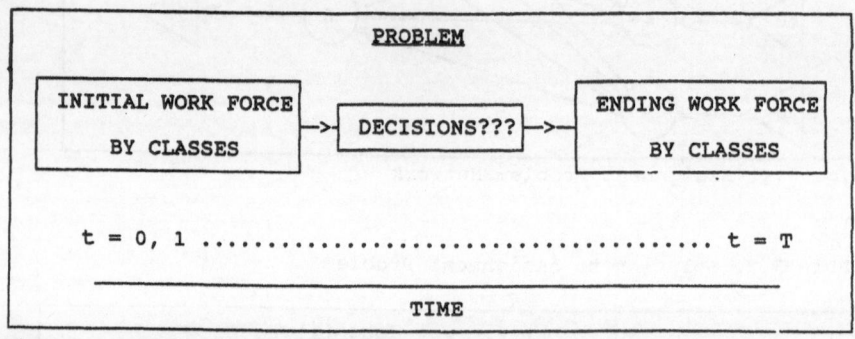

Fig. 4.1. The Manpower Planning Problem

Multi-year planning models can be often represented as standard linear programming or network problems. However, as they are, in most cases, infeasible, that is, the system can not meet the ending workforce conditions, goal programming procedures are key to determining acceptable, compromise solutions.

A person's job status changes over time. An individual's job profile consists of being hired, changing skills, changing jobs, being promoted or fired, quitting, retiring, dying. Individuals are described and classified by the time hired (seniority, years of service), skills (such as machinist, programmer), function (such as data processing or accounting), and job title (such as manager, rang, grade). Each person is a member of one and only one class, but a person can change classes based on transition assumptions. These distinctions are captured by the following notation:

$X(g,s,y,t)$ = the number of individuals in grade g, with skill s, with years of service y in planning period t.

A combination of (g, s, y, t) is called a state. An individual can be in only one state at time period t, with the initial state conditions given by $X(g,s,y,0)$. One feature of networks is that the characteristics of flows into a node can be different, but the flow out of the node can be identified only by the common characteristics. For example, let us assume for time period t that we have three similar personnel groups flowing into a node n, with the personnel groups having the same grade and skill, but differing in years of service. Then, the flow out of node n can be only characterized by the incoming common characteristics of grade and skill. This is illustrated in Figure 4.2.

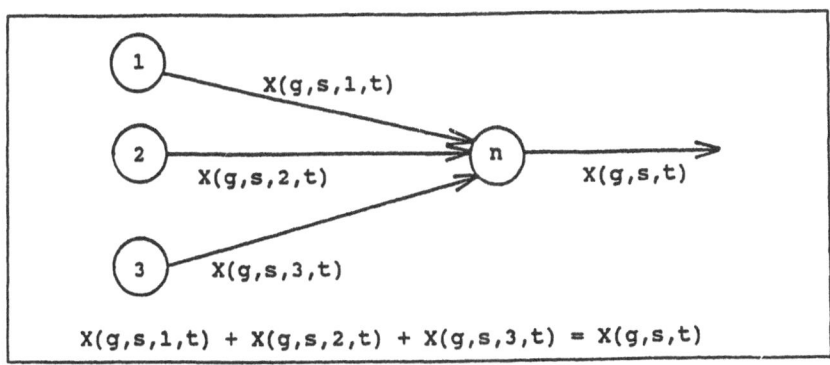

$$X(g,s,1,t) + X(g,s,2,t) + X(g,s,3,t) = X(g,s,t)$$

Fig. 4.2. Flow Identification

5 Goal Programming Conditions in Networks

5.1 Goal Conditions for Demand Nodes

We next illustrate some manpower goal programming conditions and how they may be expressed by standard network structures. We assume that the network optimization requires the objective function to be minimized and we are discussing a single planning year. At the end of the year, we would like the total personnel in each skill category to be equal to a given target goal $T(s)$, that is,

$$\Sigma_{g,y} X(g,s,y,1) = T(s) \quad \text{(for all s)} \quad (2)$$

Note that these conditions correspond to a set of demand nodes in the network representation. We assume that achieving equations (2) is not be possible due to hiring availabilities, attrition rates, and the general initial and final conditions of the planning model. That is, the problem is not feasible. We then need to consider equations (2) as soft goal constraints. In essence, we transform (2) to the goal constraints

$$\Sigma_{g,y} X(g,s,y,1) + n_i - p_i = T(s) \quad \text{(for all s)} \quad (2')$$

where the n_i are underachievement variables and the p_i are overachievement deviation variables. We shall consider conditions (2') in three different manpower goal programming settings, as follows.

We want to force the flow $\Sigma_{g,y} X(g,s,y,1)$ for skill s to be no more than $T(s)$, and, if $T(s)$ is attained, we then have the option of increasing the flow above $T(s)$ and having a surplus of personnel with skill s. From a network perspective, this can be accomplished by forcing the flow of all personnel with skill s across two final skill arcs joining nodes s1 and s2, which together act as surrogate nodes for a single demand node s. The lower and upper bounds and the penalty weight (cost) for the flow X1(s) across the first arc are, respectively, [0, T(s),P(-)], where P(-) represents an appropriate negative penalty. The lower and upper bounds and the weight for the flow X2(s) across the second arc are, respectively, [0, 00, P(+)], where the P(+) represents an appropriate positive penalty. This structure is illustrated in Figure 5.1.

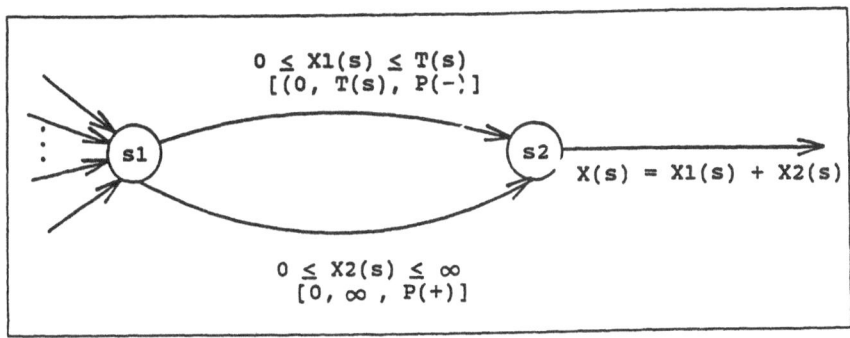

Fig. 5.1. Network Structure for Overachievement of Target Goal

Next, we assume the skill target structure as above, but with the additional condition that for each skill there is a high priority for achieving a minimum number of employees with that skill. This minimum amount, sometimes referred to as a fill goal, will be denoted by F(s), with F(s) < T(s). Thus, at the end of planning year, we would like the total personnel in each skill category to be at least equal to or above F(s), and, if possible, exactly equal to the target goal of T(s). The network is structured as in Figure 5.1, except there are now three arcs with the first arc carrying the flow up to the amount F(s), the

second arc carrying the flow that represents
the difference between [T(s) − G(s)], and the
third arc carrying any flow greater than T(s).
This is illustrated in Figure 5.2.

The network goal structures illustrated in
Figures 5.1 and 5.2 force the flow into demand
nodes to meet their target goal conditions from
below and then penalize any flows above the
goal. Next, we see how conditions (2′) can be
structured in network terms as standard goal
programming conditions. Here, for the demand
node, either underachievement or overachievement or exact meeting of the goal is allowed.
The appropriate network structure is shown in
Figure 5.3. Note that the top arc in Figure
5.3 measures the overachievement and the bottom
arc the underachievement. We show that the
bottom underachievement arc needs to be connected to a dummy supply node that can furnish
up to T(s). Both the overachievement and
underachievement flows have positive penalties
P(+).

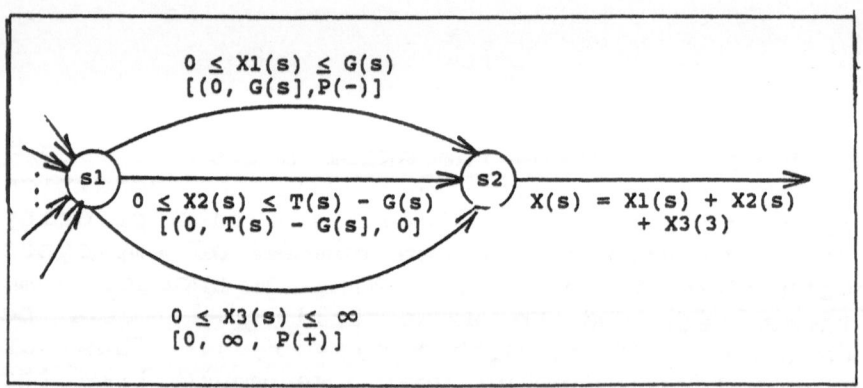

Fig. 5.2. Network Structure for Fill Goal

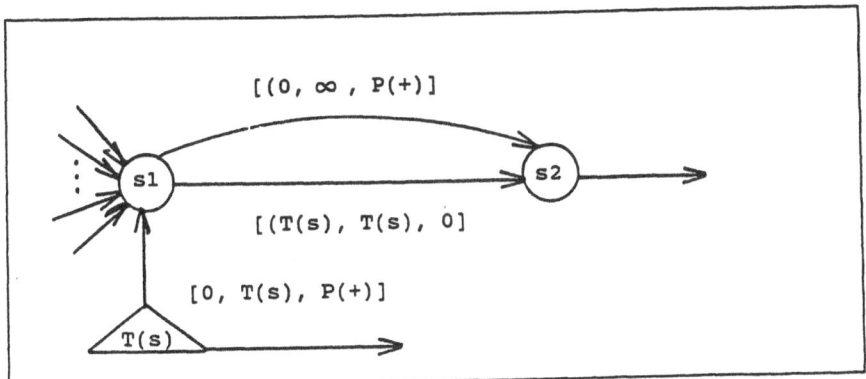

Fig. 5.3. Network Structure for Standard Goal Constraint

For some minimizing goal-programming problems, there are different costs associated with
the amounts overachieved or underachieved. For
example, let a demand node s have a goal of
100. For the first 10 units over or under, a
cost of $10 per unit is incurred; for the next
10 units over or under, a cost of $20 per is
incurred, and so on. Note that the costs are
decreasing economies of scale. Glover, Klingman and Phillips (1992) show how to handle
these conditions in a network structure, similar to that of Figure 5.3 by the addition of
overachievement and underachievement arcs
flowing out of or into the target node s1,
respectively. The costs on these arcs are the
$10, $20,..., with the lower and upper bounds
being (0, 20), respectively. Glover, Klingman
and Phillips (1992) illustrate goal programming network situations for a variety of production, manufacturing and manpower applications; they also show how some preemptive
priority conditions can be structured in network terms.

The three situations illustrated Figures
5.1, 5.2 and 5.3 are for target goals that are
associated with demand nodes; the flows through

the demand nodes do not enter other nodes. Figure 5.4 shows this situation for a general goal programming demand node i situated at output side of the network. Here we have added a dummy supply node (shortage pool) and a dummy demand node (excess pool).

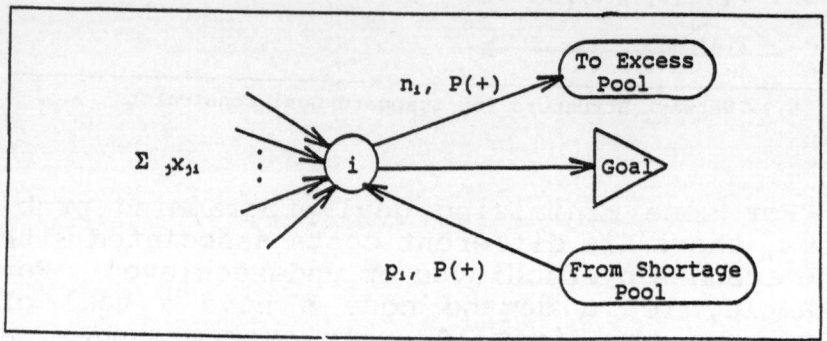

Fig. 5.4. General Goal Programming Network Demand Node

5.2 Goal Conditions for Intermediate Nodes

We next discuss the situation in which a flow between two intermediate nodes is subjected to a goal condition and this flow then connects to other intermediate or demand nodes. This is the standard goal-programming case in which we desire the total flow into a node i to be equal to or as close as possible to a specified target goal $T(i)$. That is, the constraint

$\sum_j x_{ji} = T(i)$ is treated as the goal constraint

$\sum_j x_{ji} + n_i - p_i = T(i)$. This situation arises

in manpower network models when, for example, we have an individual target goal for each skill, with the added condition that the sum of

all the skill flows must also meet a total
manpower goal. Here, the total manpower flow
is represented by nodes and arcs, as in Figure
5.1, while the individual skill goals are
represented by nodes and arcs which are connected to the total manpower node. We next
describe how the individual skill goals can be
transformed into goal-programming network
conditions. Note, that if the problem was
known to be feasible, the skill goal can be
forced exactly by connecting node i to a node k
by an arc (i,k) that has $l_{ik} = u_{ik} = T(i)$. We
illustrate the basic goal network structure in
Figures 5.5, 5.6 and 5.7 for the underachievement, overachievement and exact achievement of
the goal, respectively. Note that in each
figure that node i is connected to a node k,
with node i having a dummy demand of $T(i)$ and
node k having a dummy supply of $T(i)$.

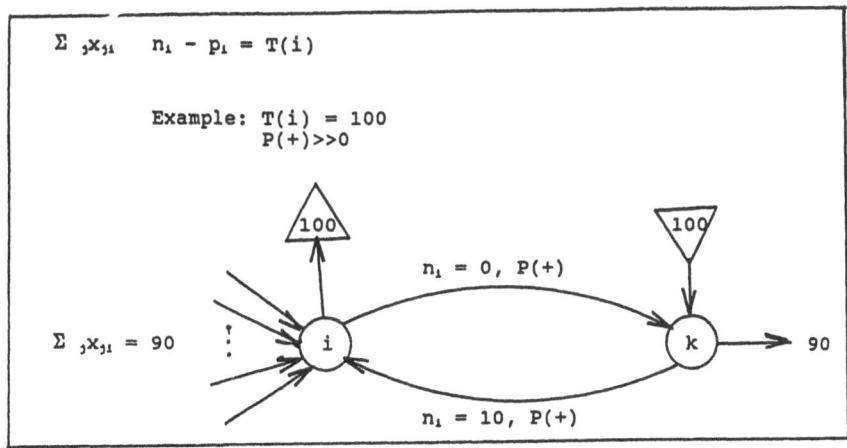

Fig. 5.5. Underachievement of Goal

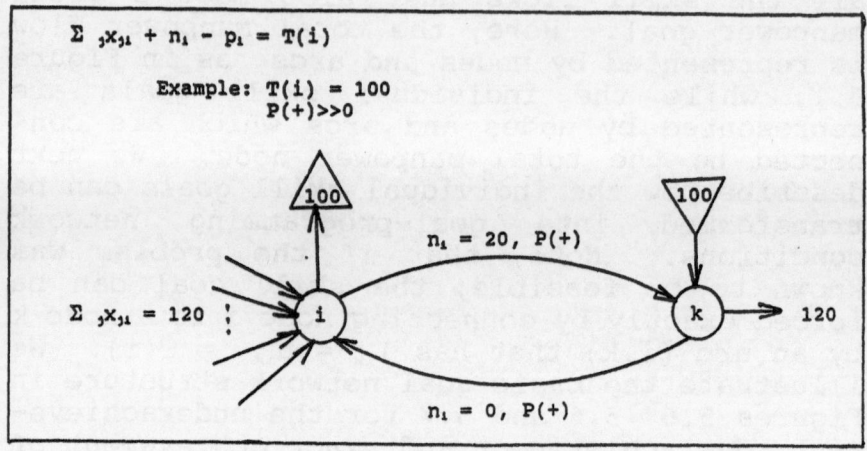

Fig. 5.6. Overachievement of Goal

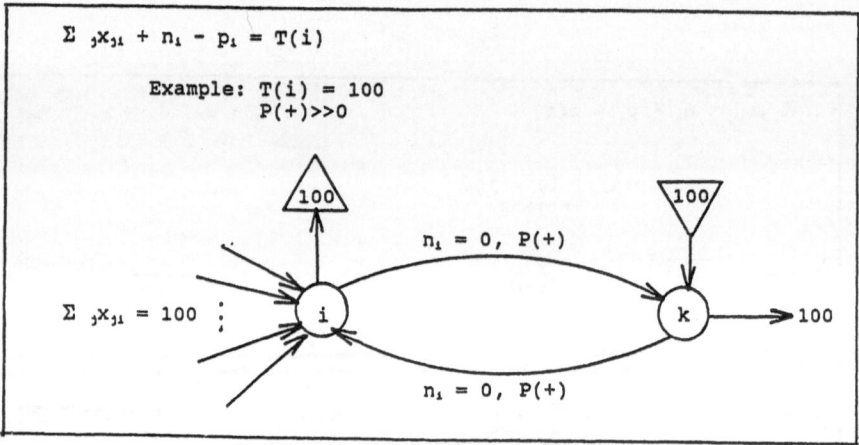

Fig. 5.7. Exact Achievement of Goal

To illustrate how the intermediate goal
constraint and the demand goal constraint can
be tied together in a network, we show, in
Figure 5.8, a flow for two grades and two
skills in which each grade total is subjected
to a goal constraint and, finally, the sum of
all personnel in both grades is subjected to a
total manpower goal.

Figure 5.9 illustrates a military manpower
application that involves goal programming

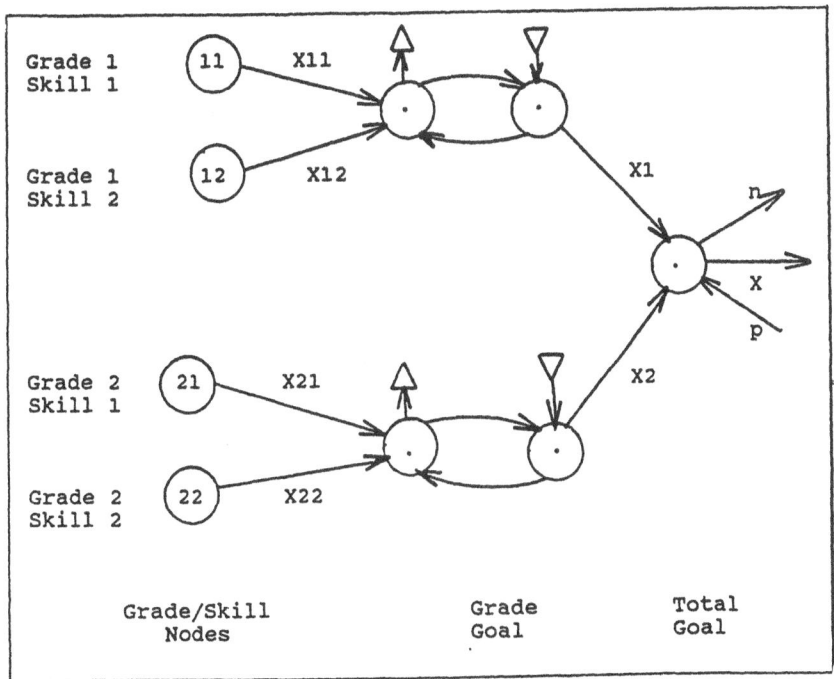

Fig. 5.8. Combined Goal Programming Network Structure

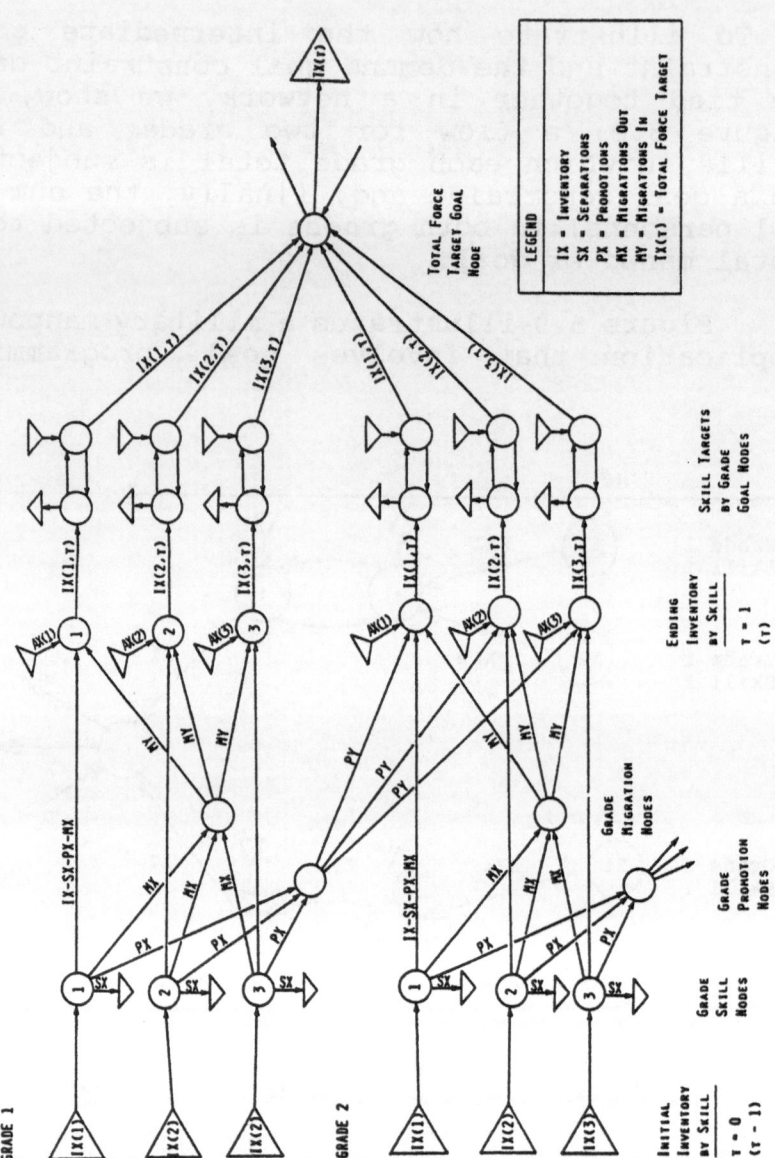

Fig. 5.9. Personnel Model with Goal Constraints

conditions that was used to explain a multiyear planning model, Gass et al. (1988). Here, initial inventories I(X) are indexed by grade and skill. This initial force is either separated from service, promoted to the next higher grade, or remains in the current grade, and can change skills either in grade or at a higher grade. We assume known or planned rates of separation or promotion; computed separations and promotions attempt to meet these target rates. Possible accessions, by grade and skill are given and added, as required, to form part of the next year's inventory. New inventories are formed to meet grade target requirements for the next year, as well as a yearly force target. Note that the illustrated network structure is not able to keep the flows indexed by grade and years-of-service as required to carry the flows from one year to the next. But an illustration like Figure 5.9 is a powerful means of describing and explaining complex situations without having to resort to the writing of the mathematical conditions. By invoking the conservation of flow assumption through a node, nontechnical persons can visualize a mental model of how the flows move and combine without resorting to pouring over complex mathematical equations.

As network-based models tend to be rather large in terms of constraints and variables, especially manpower planning models with goal constraints, a related problem is how to establish weights for the numerous overachievement and underachievement variables. A procedure for doing so that uses the Analytic Hierarchy Process (AHP) is described for large-scale linear programs in Gass (1986).

6 Summary

We have illustrated the main approaches for capturing goal programming conditions in min-imum-cost transshipment-networks. Being able to handle such conditions extends the power of network models, especially into the realm of long-range planning. Goal programming approaches for networks are due to the pioneering work of Fred Glover and Darwin Klingman, with whom the author had the pleasure of working with on a number of military manpower applications.

References

R. K. Ahuja, T. L. Magnanti, and J. B. Orlin (1993) *Network Flow,* Prentice-Hall, Englewood Cliffs, New Jersey.

R. S. Barr and B. L. Hickman (1994) "Parallel Simplex for Large Pure Network Problems: Computational Testing and Sources of Speed up," *Operations Research,* 42, 1, 65-80.

M. S. Bazaraa and J. J. Jarvis (1977) *Linear Programming and Network Flows,* John Wiley & Sons, NY, NY.

V. Chankong and Y. Y. Haimes (1983) *Multiobjec tive Decision Making,* Elsevier Science Publishing Co., NY, NY.

A. Charnes and W. W. Cooper (1961) *Management Models and Industrial Applications of Linear Programming,* Vol. I, John Wiley and Sons, NY, NY.

J. R. Evans and E. Minieka (1992) *Optimization Algorithms for Networks and Graphs,* Marcel Dekker, Inc., NY, NY

L. K. Ford, Jr. and D. K. Fulkerson (1962) *Flows in Networks,* Princeton University Press, Princeton, New Jersey.

H. Frank and I. T. Frisch (1971) *Communication, Transmission, and Transportation Networks,* Addison-Wesley, Reading, Massachusetts.

S. I. Gass (1984) "On the Development of Large-Scale Personnel Planning Models," *Systems*

Modelling and Optimization, (edited by P. Throft-Christensen, Springer, Berlin, 743-754.

S. I. Gass (1986) "A Process for Determining Priorities and Weights for Large-Scale Linear Goal Programmes," *J. Operational Research Society*, 37, 8, 779-785.

S. I. Gass, R. W. Collins, C. W. Meinhardt, D. M. Lemon, and M. D. Gillette (1988) "The Army Manpower Long-Range Planning Systems," *Operations Research*, 36, 1, 5-17.

S. I. Gass (1991) "Military Manpower Planning Models," *Computers and Operations Research*, 18, 1, 65-73.

F. Glover, D. Klingman, and N. C. Phillips (1992) *Network Models in Optimization and Their Applications in Practice*, John Wiley & Sons, NY. NY.

F. S. Hillier and G. J. Lieberman (1967) *Introduction to Operations Research*, Holden-Day, Inc. San Francisco, California.

T. C. Hu (1969) *Integer Programming and Network Flows*, Addison-Wesley Publishing Co., Reading Massachusetts.

J. P. Ignizio (1976) *Goal Programming and Extensions*, Heath Lexington Books, Lexington. Massachusetts.

J. P. Ignizio (1982) *Linear Programming in Single- & Multiple- Objective Systems*, Prentice-Hall, Englewood Cliffs, New Jersey.

J. P. Ignizio (1983a) "An Approach to the Modeling and Analysis of Multiobjective Generalized Networks," *Euro. J. of Operational Research*, 12, 357-361.

J. P. Ignizio (1983b) "GP-GN: An Approach to Certain Large Scale, Multiobjective Integer Programming Models," *Large Scale Systems*, 4, 177-188.

P. A. Jensen and J. Wesley Barnes (1980) *Network Flow Programming*, John Wiley & Sons, NY, NY.

D. Klingman and N. Phillips (1984) "Topological and Computational Aspects of Preemptive Multicriteria Military Personnel Assignment

Problems," *Management Science*, 30, 11, 1362–1375.

K. G. Murty (1992) *Network Programming*, Prentice-Hall, Englewood Cliffs, New Jersey.

D. T. Phillips and A. Garcia-Diaz (1981) *Fundamentals of Network Analysis*, Waveland Press, Inc., Prospect Heights, Illinois.

R. B. Potts and R. M. Oliver (1972) *Flows in Transportation Networks*, Academic Press, NY, NY.

J. L. Ringuest (1992) *Multiobjective Optimization: Behavioral and Computational Considerations*, Kluwer Academic Publishers, Norwell, Massachusetts.

C. Romero (1991) *Handbook on Critical Issues in Goal Programming*, Pergamon Press, Oxford, England.

M. J. Schniederjans (1984) *Linear Goal Programming*, Petrocelli Books, Inc., Princeton, New Jersey.

H. M. Wagner (1969) *Principles of Operations Research*, Prentice-Hall, Englewood Cliffs, New Jersey.

M. Zeleny (1982) *Multiple Criteria Decision Making*, McGraw-Hill Book Co., NY. NY.

SOLUTION OF NONLINEAR FIELD PROBLEMS BY GOAL PROGRAMMING

Kevin Y. K. Ng

Operations Research & Analysis
Department of National Defence
Ottawa, Ontario
Canada
&
Faculty of Administration
University of Ottawa
Canada

ABSTRACT   This survey paper reports on the application of goal programming techniques toward the solutions of nonlinear field problems.   The proposed approach involves approximating the solution by a set of trial functions containing unknown coefficients. The technique then minimizes in a weighted residual sense the absolute deviations of the system equations and/or the performance function residuals by nonlinear goal programming. Since the approximate solutions will in general not be able to satisfy all conditions, such as the boundary and initial conditions and at the same time minimize the residual errors, our approach involves reformulating the field problem as a preemptive goal programming model via the use of hard and soft constrains (or stiff and weak springs) in linear programming (or mechanics).   Examples from fluid dynamics and control theory will be employed to illustrate the methodolgy. In conclusion, the advantages, limitations and other related issues on the goal programming model in solving nonlinear problems will be discussed.

Key words: Goal programming, Method of Weighted Residuals, Fluid Dynamics, Optimal Control.

1 INTRODUCTION

The method of weighted residuals (MWR) has long been considered a cornerstone in the numerical solution of field equations (Finlayson, 1972; Minkowyez et al., 1988). In all of these problems, the unknown solution is approximated by a set of trial functions containing unknown coefficients. These coefficients are chosen by various error criteria to give the "best" approximation for the selected family. Obviously, there exists a wide choice in selecting the error criteria. The most commonly used methods are the collocation method, the least squares method, the Galerkin method, the subdomain method and the orthogonal collocation method.

Except for the collocation methods, the application of other MWR techniques to nonlinear problems often requires the evaluation of integrals of the trail function. Whereas for the collocation (or orthogonal collocation) method, it is only necessary to evaluate the residuals at the collocation points (or the roots of the orthogonal polynomials). The problem is thus reduced to the solution of a system of nonlinear algebraic equations. However, the solution of nonlinear equations can be extremely complicated, notwithstanding the system might be ill-conditioned. The nonlinear boundary conditions, such as those caused by thermal radiation (Salomatov and Puzyrez, 1974; Faghri and Sparrow, 1980; Yeh and Chung, 1984) can in addition increase the complexity in acquiring a solution. Boundary collocation points have to be chosen to ensure the satisfaction of the boundary conditions. Also, the collocation solution to (heat and mass) transport phenomena is further perplexed by integral boundary conditions. A typical example is the free convection problems due to a point or line heat source (Ng, 1978; Morton, 1966) with the measure of the strength of the heat source expressed as an integral boundary condition. (The integral expression results in that the convective heat flux must be constant for all values of the vertical coordinates because of continuity.)

This paper summarizes the work done by the author (Ng, 1981, 1987, 1989, 1991) in solving nonlinear field equations by means of goal programming. The aim of the goal programming approach in the solution of physical science problems are two fold. First, it provides a means to overcome some of the deficiencies of the collocation and orthogonal collocation methods, which retaining some of their better features, such as accuracy and simplicity of the method. Secondly, our approach illustrates that management science techniques, which have long been regarded as an inexact science, can be employed to obtain reasonably accurate solutions to problems in physical (or exact) science. This dispels the myth that goal programming, like all other management science techniques, is nothing more than a soft science technique.

The goal programming approach towards the solution of field equations, (be it the nonlinear boundary value problems or the optimal control problems), resembles the orthogonal collocation method, where the solution is approximated by sets of trail functions containing unknown coefficients. The heart of the idea is that approximate solutions will, in general, not be able to satisfy all conditions, such as boundary and initial conditions, at the same time as minimizing the error at the collocation points. The field problem is cast into the framework of preemptive goal programming via the use of hard and soft constraints in mechanics or linear programming (Kendall, 1975). The approach then lexicographically minimizes a prioritized set of absolute value deviations for the nonlinear boundary conditions (and/or the state or control variables constraints), the field (or dynamics) equations (and the performance function for the control problem).

The organization of this survey paper is as follows. The methodology of our approach, as well as the hard and soft constraints interpretations, is explained in the next section. The algorithm for solving the nonlinear field equation is described. Examples from fluid dynamics and optimal control theory are employed to illustrate the methodology. The advantages, limitations, as well as future research issues on the goal programming approach in solving nonlinear problems are discussed.

2 GOAL PROGRAMMING FORMULATIONS

2.1 Boundary Value Problems

Consider the boundary value problem:

$$F_k\left(x, u, u^{(1)}, \ldots, u^{(i)}, \ldots, u^{(n)}\right) = 0, \, x \varepsilon\left[x_o, x_f\right],$$

$$k = 1, 2, \ldots, k, \quad (1)$$

$$B_k\left(x, u, u^{(1)}, \ldots, u^{(m)}\right) = 0, \quad x = x_0$$

$$\text{and} \quad x = x_f, \text{respectively}, \quad (2)$$

$$IB\left(x, u, u^{(1)}, \ldots, u^{(p)}\right)$$

$$= \int_{x_o}^{x_f} \emptyset\left(x, u, \ldots, u^{(p)}\right) dx - C = 0, \quad (3)$$

where

$$u = (u_1, u_2, \ldots, u_L)^t, \; t = \text{transpose}, u^{(1)} = \mathrm{d}^i u / \mathrm{d}x^i,$$

F_k = nonlinear differential operator,

B_k = boundary operator,

IB = integral boundary condition and C = constant.

An assumed mode is taken in the form:

$$\hat{u} = \left(\sum_{i=1}^{l_1} C_{li} V_{li}, \sum_{i=1}^{l_2} C_{2i} V_{2i}, \ldots, \sum_{i=1}^{l_t} C_{Li} V_{Li} \right), \quad (4)$$

where C_{li}, $1=1,2,..,L$; $i=1,2,..,I_L$, are unknown parameters and V_{li} are monomials or orthogonal polynomials. The trial functions (4) are substituted into the differential system (1) and boundary conditions (2), (3) to form the residuals:

$$RF_k(c_{li}, x) = F_k(x, \hat{u}, \hat{u}^{(1)}, \ldots, \hat{u}^{(n)}), \quad (5)$$

$$RB_k(c_{li}) = B_k(x, \hat{u}, \hat{u}^{(1)}, \ldots, \hat{u}^{(m)}), \quad (6)$$

$$RIB(c_{li}) = IB(x, \hat{u}, \ldots, \hat{u}^{(p)})$$

$$= \int_{x_0}^{x_f} \varnothing(x, \hat{u}, \ldots, \hat{u}^{(p)}) dx - C. \quad (7)$$

If the trial functions were the exact solution, the residuals RF_k, RB_k, $k=1,2,..,K$ and RIB would be zero. In the approximation (4), the coefficients c_{li}, are chosen so that $\hat{u}$ satisfies the associated boundary conditions (2), (3) and $\hat{u}$ satisfies the differential system (1) in the weighted residual sense.

The initial step in solving the nonlinear boundary value problem via the goal programming technique is to perceive the system (1-3) in the light of hard and soft constraints in mechanics or linear programming (Kendall, 1975). A hard constraint or limit can be compared with a stiff spring; if a stiff spring (hard constraint) is to be stretched (violated) by a significant amount, a relatively great force (expense) is required. Similarly, a soft constraint can be compared with a weak spring; the stretching (violation) of such a spring (constraint) requires relatively little proportional force (expense).

The boundary conditions (2) represent physical limits, conditions or restrictions imposed on the solution governing the physical phenomena. They are rigid or hard constraints and thus must be closely satisfied. In the context of goal programming, the hard constraints (2) are goals that one cannot do without and are assigned absolute priority. In heat and mass transfer problems, integral boundary constraints normally represent average heat or mass fluxes. It is conceivable that the assumed mode (4) might never satisfy constraint (3) exactly. To ensure that the continuity of flow is not violated, it is desirable to have the assumed modes satisfy constraint (3) as closely as possible without violating the absolute restrictions of the solution, namely, hard constraint (2). In other words, equation (3) can be regarded as restriction that represents "wants" rather than "musts" to be satisfied as much as possible if the penalty for doing so is not too great. Thus integral boundary condition (3) can be interpreted as a "soft" constraint and is assigned the second highest priority level. The performance of the algorithm is then judged by how well the assumed mode û (eq.(4)) satisfies the differential system (1). This is achieved through the minimization of the deviations from the respective goal levels by either the MINSUM or MINMAX criterion.

We define the weighted integrals of the residuals as

$$\int \delta(x - x_j) RF_k(c_{lj}, x)dx = RF_k(c_{lj}, x_j), \quad (8)$$

where δ is the displayed Dirac delta function. We choose M collocation points, x_j, $j=1,2,..,M$, $M>N$, $N = \max\{I_1,I_2,...,I_L\}$. We then anticipate trial functions (4) to approximate the solution of systems (1-3) as closely as possible at the chosen M collocation points x_j. Here the boundary residual (6), the integral boundary residual (7) and the sum of the absolute value deviations of $RF_k(C_{lj},X_j)$ (eq.(8)) are minimized over the collocation points with (6), (7) and (8) ranked at priorities 1,2 and 3 respectively. In the goal programming formulation, the integral boundary condition (3) is regarded as "satisfied", whenever

$$-\varepsilon < RIB(C_{li}) < \varepsilon, \quad (9)$$

where ε is a predetermined value or bound.

Following Ignizio's terminology (Ignizio, 1976), we seek the lexicographic minimum of the achievement vector:

$$\min\left\{\sum_{k=1}^{k}(P_k + N_k), P_{k+1} + N_{k+2}, \quad \sum_{k=1}^{K}\sum_{j=1}^{M}\left(w_{kj}n_{kj} + w_{kj}P_{kj}\right)\right\},$$

such that

$$RF_k(c_{1i},x_j) + n_{kj}\text{-}P_{kj} = 0, \ j = 1,...,M,$$
$$k = 1,...,K, \ k,j = \text{a pair of indices}, \tag{10a}$$
$$RB_k(c_{1i}) + N_k - P_k = 0, \ k = 1,...,K, \tag{10b}$$
$$RIB(c_{1i}) + N_{k+1} - P_{k+1} = \varepsilon, \tag{10c}$$
$$RIB(c_{1i}) + N_{k+2} - P_{k+2} = -\varepsilon, \tag{10d}$$

where

n_{kj}, P_{kj} = deviational variables representing negative
and positive deviations from goal level k,j,
N_α, P_α = negative and positive deviational variables,
$\alpha = 1,2,...,K+2$,
ε = values of the aspiration level associated with
the objective (goal), $RIB(c_{1i})$,
w_{kj} = weights of under- and over-achievement
of the kjth goal in the third priority,
$k = 1,2,...,K; j = 1,...,M.$

In step 5 the weights w_{gj} are increased for all values of j, j = 1,2,..,M. However, the methodology is equally applicable to increasing the value of only one w_{gj} for $J \in \{1,2,..,M\}$. Also, in choosing the goal or aspiration level for $RIB(c_{1i})$, if the ε were set too low, a suboptimal and even dominated solution might be computed. On the other hand, too high an ε might degrade the accuracy of the solution. The value of ε should be such that it reflects an upper bound for the integral boundary condition. Experience with the goal programming method on other differential systems indicates that reasonably accurate solution is normally obtained whenever $0.005C < \varepsilon < 0.015C$.

In the goal programming formulation, different weighting factors w_{kj} are assigned to the under- and over-achievement of the goals. The essence in minimizing the absolute value deviations of the weighted integrals of the residuals with preference is to confine the r.m.s.$_k$, for all $_k$, to lie within a certain tolerance limit with a minimum number of trial functions. For example, it is unacceptable to have the r.m.s. for equation A, say, very small; whereas the r.m.s. for a coupled equation B within the same differential system exceeds the prespecified tolerance limit. This can be remedied by increasing the number of trial functions in the assumed mode. An alternative is to achieve a smaller r.m.s.$_B$ at the expense of r.m.s.$_A$ by having a stronger preference (larger weighting factor) in the minimum of |RF$_B$| than the minimum of |RF$_A$|. It goes without saying that the r.m.s.$_A$ must also lie within the given tolerance limit. From our goal programming experience in solving differential systems, weighting factors of values 3 or 4 are the most effective.

2.2 Optimal control problems

Consider the simple scalar version of the optimal control problem (the results can easily be generalized into the vector case), where the state and control variables x(t) and u(t) are sought such that the performance function

$$J = \int_{t_o}^{t_f} \varnothing(x(t),u(t),t)dt$$

is minimized subject to the differential system equation

$$\frac{dx}{dt} = f(x(t),u(t),t)$$

$$x(t_0) = x_0$$

The above nonlinear goal programming problem is solved by means of the modified pattern search algorithm described in Ignizio (1976). The procedure for solving the nonlinear boundary value problem, including the determination of the number of trial functions used, is as follows. We define

$$\text{r.m.s.}_k = \text{root mean square residual of differential}$$
$$\text{equation } F_k, \, k = 1,...,K,$$
$$W = \text{upper bound for } w_{kj}, \, k = 1,...,K, j = 1,...,M,$$

and the steps of the goal programming model for system (1-3) are:

Step 1. Choose W and an appropriate aspiration level ε for $RIB(C_{1i})$, equation (10c). Assume an educated guess on the number of trial functions I_l, $l = 1,2,...,L$ in $\hat{u}(t)$, equation (4).

Step 2. Set $w_{kj} = 1, k = 1,2,...,K; j = 1,2,...,M$. Go to Step 3.

Step 3. Solve for c_{li} by means of the modified pattern search algorithm, equation (10). Go to Step 4.

Step 4. Evaluate the root mean square residuals r.m.s.$_k$ for F_k, $k = 1,2,...,K$. If r.m.s.$_k$ is less than or equal to a prespecified tolerance factor for all k, then stop and the solution is given by $\hat{u}(t)$ with c_{li} evaluated in Step 3. Otherwise, go to Step 5.

Step 5. If r.m.s.$_g$ > prespecified tolerance factor for F_g for at least one g, g $\in \{1,2,...,K\}$ and w_{gj} < W for all j, $j = 1,2,...,M$; then set $w_{gj} = w_{gj} + 1, j = 1,2...,M$. Go to Step 3. Otherwise, go to Step 6.

Step 6. Set $I_1 = I_1 + 1$ (or any appropriate increment). $l = 1,2,...,L$. Go to Step 2.

and the control and state variables constraints,

$$g_k(x(t),u(t),t) \geq 0, \qquad k = 1,2,...,l$$

$$= 0, \qquad k = l+1,...,K.$$

Since the state and control variables constraints normally represent physical limits or restrictions, they are thus rigid or hard constraints and must be closely satisfied. They are assigned absolute priority. Also, the assumed modes for the state and control variables might not satisfy the differential system equation at each $t, t \in [t_0, t_f]$. Our intention is to have the assumed modes satisfy the differential system equation (interpreted as "soft" constraints) as closely as possible without violating the physical limits of the variables, namely the control and state variables (hard) constraints. Finally, the minimization of the performance function occupies the lowest priority. The weighted integrals of the residuals are next defined. (Since they resemble those defined in the previous section and are therefore omitted.) The sum of the absolute value deviations of the weighted integrals is then minimized over the collocation points, with the weighted integrals of the residuals for the state and control variables constraints, for the differential system equation and the performance function ranked at priorities 1, 2 and 3 respectively.

Again following Ignizio's terminology, we seek the lexicographic minimum of the achievement vector,

$$
\min \left\{ \sum_{j=1}^{M} \left[\sum_{k=1}^{l} n_{kj} + \sum_{k=l+1}^{K} \left(n_{kj} + P_{kj} \right) \right], \right.
$$

$$
\left. \sum_{j=1}^{M} \left(P_{KM+j} + n_{KM+M+j} \right), n_{KM+2M+1} + P_{KM+2M-1} \right\}
$$

such that

$$R_{gk}(a_i, b_i, t_j) + n_{kj} - P_{kj} = 0,$$

$$\text{for all } k, k = 1, 2, \ldots, K$$

$$\text{and for all } j, j = 1, 2, \ldots, M$$

$$R_f(a_i, b_i, t_j) + n_{KM+j} - P_{KM+j} = \varepsilon,$$

$$\text{for all } j, j = 1, 2, \ldots, M$$

$$R_f(a_i, b_i, t_j) + n_{KM+M+j} - P_{KM+M+j} = -\varepsilon,$$

$$\text{for all } j, j = 1, 2, \ldots, M$$

$$R_\varnothing(a_i, b_i, t_j) + n_{KM+2M+1} - P_{KM+2M+1} = 0,$$

$$\text{for all } j, j = 1, 2, \ldots, M$$

where

n_α, P_α deviational variables representing negative and positive deviations from goal level α, respectively (α denotes the subscripts ranging from 1,2,..., kj,..., KM,...,$KM+j$,...,$KM+M$,...,$KM+M+j$,..., $KM+2M+1$),

ε value of the aspiration level associated with objectives (goals) $R_f(a_j, b_j, t)$ at collocation points t_j, $j = 1, 2, \ldots, M$.

The steps for solving the optimal control problem are similar to those of solving the nonlinear boundary value problem. For simplicity we omit the details and refer the reader to the paper by Ng (1987).

3 EXAMPLES

3.1 We consider the goal programming solution of a viscoelastic boundary layer equation, which arises in the description of a second order fluid near a two-dimensional stagnation point (Ng, 1981),

$$f''' + ff'' + 1 - f'^2 + k\left(ff'''' - 2f'f''' + f''^2\right) = 0,$$

$$f(0) = f'(0) = 0. \qquad \lim_{x \to \infty} f'(x) = 1.$$

Here f is a dimensionless stream function, k is a non-negative elastic parameter and x is a similarity variable. On physical grounds we also assume that the limit of the fourth derivative of f as x tends to ∞ exists and is finite. Thus the evaluation of the above system at x = 0 also gives

$$f'''(0) = -\left(1 + kf''^2(0)\right).$$

Throughout this example we will assume k to be equal to 0.2, a case corresponding to the study of a "weakly" viscoelastic fluid. Approximate solutions to the aforementioned system have been obtained by the Karman-Pohlhausen method, the perturbation method and the orthogonal collocation method(see references in Ng, 1981). Standard finite difference techniques such as the Runge-Kutta and the predictor-corrector methods have been shown to be highly unstable when applied to the above system (Ng, 1981). We choose the trial function to be the set of Laguerre functions which constitutes a complete orthonormal system in $(0,\infty)$,

$$\upsilon(x) = -1 + x + \exp(-x) + x^2 \exp\left(-\frac{x}{2}\right) \sum_{i=1}^{N} c_i L_i(x),$$

where

$$L_i(x) = (1/i!) \exp(x)(d/dx)^i \exp(-x)x^i$$

in the Laguerre polynomial of degree i (Abramowitz and Stegun, 1966). We let the tolerance factor for the achievement function to be 2×10^{-4}, the number of terms in the trial function to be 5, the weighting factors $w_{kj}=1$, for all k and j, and the initial guess for c_i be $c_1=c_3=c_5=1.0$, $c_2=c_4=-1.0$. The collocation points are taken over 16 uniformly distributed points from 0 to 1.5. The results are compared with those obtained by means of the orthogonal collocation method, in terms of the wall shear stress f''(0). The results are summarized in Table I.

Number of independent Laguerre functions	Goal Programming solution	Orthogonal collocation solution
	f''(0)	f''(0)
5	1.58788	1.56640
12		1.58678
16		1.58800
24		1.58719

TABLE I - Wall shear stress comparison

It is interesting to find out that with the trial function containing only 5 independent functions, the goal programming technique provides a solution (with r.m.s. residual equal to 9.646×10^{-2}) that is comparable in accuracy to the orthogonal collocation method employing a trial function involving 12 or 16 independent functions. The difficulty that arises in the use of orthogonal collocation method to solve boundary value problems over a semi-infinite domain is that the collocation points (roots of orthonormal polynomials) are spread so far out in the space domain that the approximate solution relies primarily on information outside the diffusion layer where only minimal variations in the dependent variables take place. On the other hand, the fact that the goal programming solution with only a low order trial function compares favourably to the results of the orthogonal collocation method hardly comes as a surprise. The success is attributed to the flexibility of the collocation points for the goal programming approach, namely, there is no restriction on where to place the collocation points, as well as the number of points used in our method. For example, the collocation points can be equally spaced, or concentrated in areas where the solution increases rapidly. This striking characteristic of our method accounts for the superb accuracy achieved on the wall shear stress f''(0), as compared to the orthogonal collocation method.

3.2   Consider the flow around a flat disk which rotates about an axis perpendicular to its plane with a uniform angular velocity, in a fluid otherwise at rest. The layer near the disk is carried by it through friction and is thrown outwards owing to the action of centrifugal forces. This is compensated by particles which flow in an axial direction towards the disk to be in turn carried and ejected centrifugally. The governing system of equations is given by (Schlichting, 1968)

$$2F + H' = 0,$$
$$F^2 + F'H - G^2 - F'' = 0,$$
$$2FG + HG' - G'' = 0,$$
$$P' + HH' - H'' = 0,$$

with boundary conditions

$$F(0) = H(0) = P(0) = 0, \ G(0) = 1,$$

$$\lim_{x\to\infty} F(x) = \lim_{x\to\infty} G(x) = 0,$$

where F, G, H, P are the dimensionless dependent (similarity) variables related to the radial velocity, tangential velocity, axial velocity of the flow and the pressure, respectively; x is the similarity variable.

We choose a set of linearly independent functions w_{11}, w_{12},..., w_{21}, w_{22},.. which fulfill all the boundary conditions of the problem. The approximate solutions u_1, u_2, u_3 of F, G, H are given by:

$$u_1 = c_{11} w_{11} + c_{12} w_{12},$$

$$u_2 = w_{21} + c_{22} w_{22} + c_{23} w_{23},$$

$$u_3 = c_{11} w_{31} + c_{12} w_{32},$$

where

$$w_{11} = 0.5xe^{-x}, \ w_{12} = 2xe^{-2x}$$

$$w_{21} = e^{-3x}, \ w_{22} = e^{-x} - e^{-3x}, \ w_{23} = e^{-2x} - e^{-3x},$$

$$w_{31} = 1 + (1 + x)e^{-x},$$

$$w_{32} = -1 + (1 + 2x)e^{-2x}.$$

Applying the goal programming formulation with 14 suitably chosen collocation points, $x_k = 0 + 0.2k$, $k = 1,2,..,10$; $x_{11} = 2.5$, $x_{12} = 3$, $x_{13} = 3.5$, $x_{14} = 4$ and initial guess $c_{11} = c_{12} = c_{22} = c_{23} = 0.1$, the r.m.s. residual taken over 21 uniformly distributed points from 0 to 4 is found to be 2.38×10^{-2}. The results are compared with those obtained by the extermal point collocation method (see Ng, 1991; Fletcher, 1984; and Schlichting, 1968) in Table II.

$$F_k(x,u,u^{(1)},...,u^{(i)},...,u^{(n)}) = 0, \quad x \in [x_0, x_f], \quad k = 1,2,...,K,$$

(1)

$$B_k(x,u,u^{(1)},...,u^{(m)}) = 0, \quad x = x_0 \text{ and } x = x_f, \text{ respectively,} \quad (2)$$

$$IB(x,u,u^{(1)},...,u^{(p)}) = \int_{x_o}^{x_j} \varnothing(x,u,...,u^{(p)})dx - C = 0. \quad (3)$$

where

$u = (u_1, u_2,...,u_L)^t$, t = transpose, $u^{(t)} = d^t u/dx^t$,

F_k = nonlinear differential operator,

B_k = boundary operator,

IB = integral boundary condition and C = constant.

An assumed mode is taken in the form

$$\hat{u} = \left(\sum_{i=1}^{l_1} c_{1i}v_{1i}, \sum_{i=1}^{l_2} c_{2i}v_{2i},..., \sum_{i=1}^{l_L} c_{lL}v_{lL} \right), \quad (4)$$

where c_{li}, $1 = 1,2,..,L$; $i = 1,2,..,I_L$, are unknown parameters and V_{li} are monomials or orthorgonal polynomials. The trial functions (4) are substituted into the differential system (1) and boundary conditions (2), (3) to form the residuals:

$$RF_k(c_{li},x) = F_k(x,\hat{U},\hat{U}^{(1)},...,\hat{U}^{(n)}), \tag{5}$$

$$RB_k(c_{li}) = B_k(x,\hat{U},\hat{U}^{(1)},...,\hat{U}^{(m)}), \tag{6}$$

$$RIB(c_{li}) = IB(x,\hat{U},...,\hat{U}^{(p)}) = \int_{x_o}^{x_f} \emptyset(x,\hat{U},...,\hat{U}^{(p)})dx - C. \tag{7}$$

If the trial functions were the exact solution, the residuals RF_k, RB_k, $k = 1,2,...,K$ and RIB would be zero. In the approximation (4), the coefficients c_{li}, are chosen so that $\hat{U}$ satisfies the associated boundary conditions (2), (3) and $\hat{U}$ satisfies the differential system (1) in the weighted residual sense.

	F(x)			G(x)		
x	Schlichting	Goal[a] programming	Extremal point collocation[b]	Schlichting	Goal programming	Extremal point collocation
0.2	0.084	0.0845	0.0828	0.878	0.8764	0.8753
0.4	0.136	0.1366	0.1354	0.762	0.7591	0.7574
0.6	0.166	0.1662	0.1660	0.656	0.6510	0.6483
0.8	0.179	0.1802	0.1811	0.561	0.5537	0.5503
1.0	0.180	0.1833	0.1851	0.468	0.4677	0.4640
1.2	0.173	0.1792	0.1817	0.404	0.3928	0.3891
1.4	0.162	0.1704	0.1735	0.341	0.3284	0.3248
1.6	0.148	0.1590	0.1623	0.288	0.2735	0.2702
1.8	0.133	0.1460	0.1494	0.242	0.2271	0.2241
2.0	0.118	0.1325	0.1359	0.203	0.1880	0.1854
3.0	0.058	0.0726	0.0749	0.083	0.0715	0.0703
4.0	0.026	0.0355	0.0367	0.035	0.0266	0.0262

[a] 24.6 s CPU time on a Honeywell Multics System.
[b] 28.2 s CPU time on a Honeywell Multics System.

TABLE II - The functions for the radial and tangential velocity distribution in the neighbourhood of a disk rotating in a fluid at rest.

3.3   This example concerns the numerical solution to free convection problems due to heat sources. Examples of the buoyancy-induced flows abound in nature, such as dustdevils, firewhirls, tornadoes and the swirling columns above volcanic eruptions, see for example Ng (1978), Fendall and Coats (1967), Carrier et al (1971) to mention only a few. Buoyancy-induced flows also occur in our enclosures and devices, for example building insulation often consists of air gaps in multilayered walls. The heat transfer then involves natural convection in enclosures filled with either an ordinary fluid or a fluid-saturated porous material (Gebhart, 1988). In this example, we consider a simple model describing the free convection due to a line source of heat. The governing boundary layer equations can be expressed in the following form (Ng, 1991),

$$f''' + 3ff'' - (f')^2 + \theta = 0,$$

$$f\theta' + f'\theta + \theta/3\sigma = 0,$$

$$\int_{-\infty}^{\infty} f'\theta d\xi = 125,$$

subject to

$$f(0) = f'(0) = \theta'(0) = 0,$$

$$f'(\pm\infty) = \theta(\pm\infty) = 0,$$

where σ is the Prandtl number, f is a dimensionless dependent (similarity) variable related to the stream function and ξ is a dimensionless dependent (similarity) variable related to the above-ambient temperature.

The inherent characteristics of the above system enable closed-form solutions for Prandtl numbers $\sigma = 5/9$ and 2 to be found (Yih, 1969). However, numerical methods have to be utilized in order to yield solutions to other Prandtl numbers. A close examination on the above system reveals that f is odd and θ is even in ξ. By virtue of the boundary conditions, let v_{1i}, v_{2i} in the trial function (eq.(4)) be given by:

$$v_{11} = \tanh(\xi), \; v_{12} = \tanh(\xi/2), \; v_{13} = \tanh(2\xi),$$

$$v_{14} = \tanh(\xi/3), \; v_{15} = \tanh(3\xi),...,$$

$$v_{21} = \{\text{sech}(\xi)\}^{3\sigma}, \; v_{22} = \{\text{sech}(\xi/2)\}^{3\sigma},$$

$$v_{23} = \{\text{sech}(2\xi)\}^{3\sigma},$$

$$v_{24} = \{\text{sech}(\xi/3)\}^{3\sigma}, \; v_{25} = \{\text{sech}(3\xi)\}^{3\sigma},...$$

Since the trial functions satisfy the boundary conditions at 0 and ∞ exactly, the first priority in the goal programming formulation for the system becomes the satisfaction of the integral (soft) constraint. The second priority minimizes the absolute value deviations of the residuals Rf and Rθ at collocation points evenly distributed between $\xi = 0$ and $\xi = 1$ at intervals of 0.05 apart. For Prandtl number σ = 5/9 and σ = 2, the approximate solutions (with the number of trial functions $I_1 = 5$, $I_2 = 3$) are reasonably accurate with r.m.s. residuals less than 0.04 (taken over 100 uniformly distributed points from 0 to 1).

Our concern is to compute the goal programming solution for σ = 0.733, corresponding to the case for air under normal conditions. Choose ε to be 0.01, W equal to 4 and the tolerance factor for the r.m.s. residual to be 1.5×10^{-1}. Following the steps outlined in the previous section, we initially set $I_1 = 5$ and $I_2 = 3$. The r.m.s. residuals obtained for f and θ are between 0.2 and 0.25 for w_{kj} ranging from 1 to 4 inclusively. If we set $I_1 = 7$ and $I_2 = 5$, the r.m.s. residuals for f and θ with $w_{kj} = 1$ (for all k and j) are 0.19513 and 0.06692.

$$\hat{x}(t) = 1 + a_1 t(1-t) + a_2 t^2(1-t)^2 + \ldots$$

$$= \sum_{i=0}^{N} a_i t^i (1-t)^i, \text{ with } a_o = 1$$

$$\hat{u}(t) = \sum_{i=0}^{N} b_i t^i$$

with the boundary conditions satisfied. It is found that with only 4 terms for the state and control variables, evenly distributed collocation points from 0 to 1.0 and initial guess given by $a_1 = -a_2 = a_3 = -1$, $b_0 = b_1 = 1$, $b_2 = -b_3 = -2$, the r.m.s. residual is 3.7×10^{-2} (taken over 100 evenly distributed points from 0 to 1.0). The optimum value of the performance function is 1.6968 (a 2.42% deviation from the solution obtained by the sequential gradient restoration algorithm).

4   OBSERVATIONS AND CONCLUSION

In this paper, we have only compared the goal programming solution to the collocation and/or orthogonal collocation methods. The reason is that all other MWR techniques requires the evaluation of integrals of the trial function; whereas, in the collocation method it is only necessary to evaluate the residual at the collocation points. The problem is thus reduced to the solution of a set of nonlinear algebraic equations. Nonetheless, the ambiguity in the choice of collocation points has severely dampened the usefulness of the collocation method.

The orthogonal collocation method provides a viable alternative by evaluating the weighted integrals of residuals at the roots of the orthogonal polynomials. It has been used extensively and successfully to solve problems over a finite domain. However, application of the orthogonal collocation method to problems where the independent variable extends over a semi-infinite domain has been limited (Ng, 1991; Caban and Chapman, 1981).

The flexibility of the goal programming approach, which allows one to choose as many collocation points as one desires but without increasing the dimension of the nonlinear system of equations to be solved, is undoubtedly an advantage over the collocation methods. This is evidenced by the results on the worked examples in the previous section. In addition, the collocation points can be chosen to be concentrated in areas where the solution increases rapidly, as in the study of boundary layer equations.

Next we increase the weighting factors in increments of one until $w_{ij} \leq 4$, where the r.m.s. residuals satisfy the prespecified tolerance factor of 1.5×10^{-1}. The effect of different weights on the r.m.s. residuals of f and θ are summarized in Table III (taken over 100 uniformly distributed points from 0 to 1).

			r.m.s. Residuals[a]	
	Values of weighting factors		f	θ
1.	$w_{kj}=1, k=1,2$	$j=1,2,..,M$	0.19513	0.06692
2.	$w_{1j}=2, w_{2j}=1,$	$j=1,2,..,M$	0.19011	0.06883
3.	$w_{1j}=3, w_{2j}=1,$	$j=1,2,..,M$	0.19088	0.06908
4.	$w_{1j}=4, w_{2j}=1,$	$j=1,2,..,M$	0.13228	0.06737

[a]The CPU times for the solutions range between 100 and 120 s on the Honeywell Multics System.

TABLE III - Effect of weighting factors on the r.m.s. residuals

3.4   The last example deals with a simple scalar optimal control problem (Ng, 1987),

$$\min J = \int_0^1 \left(x^2 + u^2 \right) dt$$

subject to

$$\dot{x} = x^2 - u \quad x(0) = x(1) = 1 \quad \dot{x} \equiv \frac{d}{dt} x(t)$$

$$x - 0.9 \geq 0.$$

The solution to this problem has already been obtained by Miele et al. (1974) using the sequential gradient-restoration algorithm with optimum $J = 1.6567$. It should be noted that the above system can be easily simplified by substituting u(t) into the performance function J. However, in this example, we shall treat the system as if no simplification is possible so that the pre-emptive goal programming technique can be evaluated.

Last but not least, the goal programming formulation is independent of the form of the field equations or system. A general purpose software has been developed, in which problem changes affect only short subprograms and input data. All the examples were run with the same computer program, thus changes from one problem to another is minimal.

On the other hand, the goal programming approach is not without its limitations. First, the speed of convergence for the technique hinges on how well one can provide the initial approximations. A more important concern is that the goal programming approach, like all other methods of weighted residuals, is troubled by the choice of trial functions for the solution. The accuracy of the solution for the field equations is dependent on one's ability in providing the appropriate assumed modes. At times, the construction of the assumed modes can benefit from physical intuition as well as linearization techniques. Recent advances in finite element analysis, such as the h, p and spectral type elements, can also greatly enhance our ability to define the appropriate trial functions. Finally, it is the author's belief that hybrid goal programming-finite element, hybrid goal programming-perturbation approaches offer viable and/or promising alternatives in the construction of the trial functions. These research proposals are currently under investigation. The findings will be reported in subsequent papers.

On the application front, technically all problems that are solvable by the method of weighted residuals can be tackled by the goal programming technique. This opens the door for the goal programming approach to a myriad of potential areas of application. Partial differential equations will undoubtedly be testbeds for our goal programming model. The feasibility of utilizing our method to solve improperly posed or inverse problems, bifurcation problems can exploit the potential of goal programming to the limit. The endeavour to solve optimal control and its related problems are equally challenging, for example model reduction, estimation, identification and differential games problems are but a few of these applications. The integrodifferential equations that arise in the studies of elasticity and electromagnetics provide further ground towards interesting applications for the goal programming technique.

In summary, the goal programming model described herein is by no means a panacea to the numerical solution of nonlinear field equations. In fact, it is only in the infancy stage. For example, the comparison of the goal programming solution to other method of weighted residuals solutions on some benchmark test problems, such as the solution to the 2-dimensional wall-driven enclosure flow etc., has yet to be established. In my opinion, the breakthrough of the model relies on devising more systematic and accurate methodology in the construction of the trial functions. Until then, the goal programming approach will suffer the same kind of criticism that has plagued the method of weighted residuals for the past decades.

REFERENCES

1. Abramowitz M. and I.A. Stegan, Handbook of Mathematical Functions. National Bureau of Standards, 1964.

2. Caban R. and T.W. Chapman, Solution of boundary layer transport problems by orthogonal collocation. Chem. Engng. Science, vol. 36, 849-861, 1981.

3. Carrier G.F., A.L. Hammond and O.D. George, A model of mature hurricane. J. Fluid Mechanics, vol. 47, 145-170, 1971.

4. Faghri M. and E.M. Sparrow, Forced convection in a horizontal pipe subject to nonlinear external natural convection and to external radiation. Int. J. Heat Mass Transfer, vol. 23, 1980.

5. Fendell F. and D. Coats, Natural convection flow above a point heat source in a rotating environment. Proc. 1967 Heat Transfer and Fluid Mechanics Institute, Vol. 17, 341-360, 1967.

6. Finlayson B.A., The Method of Weighted Residuals and Variational Principles, Academic Press, N.Y., 1972.

7. Fletcher C.A.J., Computational Galerkin Methods, Springer-Verlag, N.Y., 1984.

8. Gebhart B. et al., Bouyancy Induced Flows and Transport, Hemisphere, N.Y., 1988.

9. Ignizio J.P., Goal Programming and Extensions, Lexington Books, Lexington, MA, 1976.

10. Kendall J.W., Hard and soft constraints in linear programming, OMEGA, vol.3, 709-715, 1975.

11. Minkowycz W.I., E.M. Sparrow, G.E. Schneider and R.H. Pletcher, Handbook of Numerical Heat Transfer, Wiley-Interscience, N.Y., 1988.

12. Morton B.R., Geophysical Vortices, in Progress in Aeronautical Sciences, Vol. 7, (Kuchemann ed.), 165-208, Pergamon Press, Oxford, 1966.

13. Ng K.Y.K., On the natural convection of a point heat source in a rotating environment, ZAMM, vol. 58, 391-396, 1978.

14. Ng K.Y.K., Solution of Navier-Stokes equations by goal programming, J. Computational Physics, vol. 39, 103-111, 1981.

15. Ng K.Y.K., Goal programming, method of weighted residuals and optimal control problems, IEEE Trans. on Systems, Man and Cybernetics, vol. 17, 102-106, 1987.

16. Ng K.Y.K., Goal Programming - from Management Science to Physical Science, The 13th Int. Symposium on Mathematical Programming, Tokyo, Japan, 1988.

17. Ng K.Y.K., Goal programming solutions for heat and mass transfer problems - a feasibility study, Computers and Chemical Engineering, vol. 15, 539-547, 1991.

18.   Salomatov V.V. and Puzyrez, Heat transfer in laminar flow of liquids in radiation channels, <u>Heat Transfer-Soviet Res.</u>, vol. 6, 128, 1974.
19   Schlichting H., Boundary Layer Theory, McGraw-Hill, N.Y. 1968.
20.   Yeh L.T. and B.T.F. Chung, Thermally developing laminar flow in a duct with external radiation and convection, <u>AIAA J.</u>, vol. 22, 727-729, 1984.
21.   Yih C.S., <u>Fluid Mechanics</u>, McGraw-Hill, N.Y., 1969.

Incorporating the Decision-Maker's Preferences in the Goal Programming Model with Fuzzy Goal Values : A new Formulation[1]

Jean-Marc Martel and Belaïd Aouni

Faculté des sciences de l'administration, Université Laval, Sainte-Foy (Québec). G1K 7P4. Canada.

Abstract. The goal programming model is probably the most known in mathematical programming with multiple objectives. Available in various versions, this model has been applied in much varied fields. It has also been the target of many criticisms among which are those related to the difficulty of determining precisely the goals as well as those concerning the decision-maker's near absence in this modelling process. In the actual paper, we focus mainly on the explicit integration of the decision-maker in this model, especially when he may not express his goals in a precise way. To this end, we are building up with the help of the decision-maker his functions of satisfaction related to the type of deviation in accordance with each of his imprecise goals.

Keywords. Fuzzy Goal Programming, Goal Programming with intrevals, satisfaction function, Decision maker preferences.

---

[1]This research was partially supported by Le Fonds FCAR and by the National Sciences and Engireering Research Council (NSERC).

1 Introduction

The paradigm of multicriterion decision aid processes remains in the fact that the decision-makers try to incorporate in their decisions many factors of quite diversified nature but not to optimize one single objective; for example, profit maximizing or cost minimizing. Practically, this is expressed by searching the most satisfying compromise among several objectives which are often conflicting (Romero, 1991). In fact, the **GP** model is based on a satisfaction philosophy and may be seen as an expression technique of the human being, known as intelligent, often marred of ambiguity; this differs considerably from the optimization philosophy generally adopted in mathematical programming (Zeleny, 1981; Min and Storbeck, 1991).

In other terms, the **GP** model enables the decision-maker to take simultaneously many goals into account in a problem where he has to choose the most satisfying action among a set of acceptable actions. According to Hwang et al. (1980), **GP** comes in the category of the multiple objective programming models with an a priori articulation of the decision-maker's preferences. However, Hannan (1985) rather places it between an a priori articulation and no articulation of the decision-maker's preferences since his task is only to determine the goals value.

During the last two decades, the **GP** model has been frequently used as an aid to the decision with multiple objectives and this in much varied application fields. On one hand, its popularity is certainly due to the fact that it is simple and easy to understand and, on the other hand, that it is easy to apply since it is an extension of the linear mathematical programming for which very fast performing solution algorithms are available. Charnes and Cooper (1977) do not hesitate to qualify this model as very powerful and easy to use.

In the great majority, the variants of the **GP** model deal with the determined goals for each objective as deterministic values; it is to say that we assume that the decision-maker is in a position to determine precisely his aspiration levels. However, Zeleny (1981) raises the difficulty of determining precisely the goal value associated with each objective. Often, in practice, the decision-maker has only partial information on the parameters of such a problematic situation. This lack of information may be expressed, in part, by imprecision or uncertainty as to these goal values.

In literature, we find formulation attempts of the **GP** model in a Stochastic environment (**SGP**) (Contini, 1966), in a Fuzzy environment (**FGP**), namely the works of Narasimhan (1980) and Hannan (1981-a;1981-

b). These two authors are inspired by a membership function notion introduced by Zimmermann (1978). More recently, Inuiguchi and Kume (1991) formulated the **GP** model in a context where goals and technologic parameters are expressed with intervals.

However, these formulations do not take into account, in an explicit way, the preference structure of the decision-maker. So, in this paper, we present first a brief survey of some works related to the **FGP** and the **GP** model with interval (**GPI**). Then, we will reformulate the **GP** model with imprecise goals (fuzzy or interval) incorporating explicitly the decision-maker's preference structure.

2 The GP model in an imprecise environment

The GP model has been developed in a perspective to respond to the decision-maker's desire to satisfy many goals at the same time while exploiting the optimization potential in mathematical programming. This model was originally developed by Charnes and Cooper (1961) under the following form:

Minimize
$$z= \sum_{i=1}^{p} (W_i^+ \delta_i^+ + W_i^- \delta_i^-)$$

Subject to :

$$\sum_{j=1}^{n} a_{ij} x_j - \delta_i^+ + \delta_i^- = g_i \quad \text{for } i=1,2,\dots,p.$$

$$Cx \leq c \quad \text{(system constraints)}$$

$$x_j, \delta_i^+, \delta_i^- \geq 0 \quad \text{for } i=1,2,\dots,p \text{ and } j=1,2,\dots,n.$$

Where :

g_i: the goal associated to the objective;

$x = (x_1, x_2, \dots, x_n)$: a n-dimensional vector of decision variables;

a_{ij} : technological parameters related to the system constraints;

c : the resources available;

C : the coefficients related to the system constraints;

W_i^+ : the importance coefficient associated with the positive deviations;

W_i^- : the importance coefficient associated with the negative deviations.

Then, many varieties of this model were developed to solve particular problems in various application fields. This model knows many reformulations and this in function of an application context, such as environments of imprecise and stochastic nature, it is to say that the goals g_i are not known with certainty. Despite this fact, this model is always the object of many criticisms (Hannan, 1985). Our interest in this paper is limited to the formulations in the context whereas goals are expressed in an imprecise way, i.e. in a fuzzy way or with intervals.

3 The Fuzzy Goal Programming model (FGP)

Narasimhan (1980) and Hannan (1981-a; 1981-b) were the first to give a FGP formulation. Their formulation is inspired by the concept of membership functions introduced by Zadeh (1965) to model imprecision related to the coefficients of the economic function and to the technologic parameters within constraints. The membership functions are defined on the interval [0,1]. So, the membership function for the the i th goal has a value of 1 when the i th is attained precisely; otherwise the membership function assumes a value between 0 and 1. Rao and al. (1988) have also used these types of functions to reformulate the GP model with fuzzy aspiration levels.

We present here Hannan's formulation (1981-a) since it is more simple and efficient than that of Narasimhan (1980).

Maximize $z = \lambda$

Subject to :

$$(\sum_j a_{ij} x_j / \Delta_i) - \delta_i^+ + \delta_i^- = g_i / \Delta_i \quad \text{for } i=1,2, \dots , p;$$

$$\lambda + \delta_i^- + \delta_i^+ \leq 1 \qquad \text{for } i=1,2, \dots , p;$$

$$Cx \leq c \qquad \text{system constraints}$$

$$\lambda, \delta_i^+, \delta_i^-, x_j \geq 0 \qquad \text{for } j=1,2, \dots , n \text{ and } i=1,2, \dots , p.$$

where :

Δ_i : the constant of deviation in relation to the aspiration levels g_i. The values of Δ_i are subjectively chosen by the decision-maker.

Hannan's formulation is more efficient than Narasimhan's since it needs the introduction of only $2p$ additional constraints, while Narasimhan's requires the solution of 2^p sub-problems and each sub-problem introduces $3p$ additional constraints to the system constraints.

This reduction in the number of constraints and number of sub-problems will have a positive impact on the time of resolution.

Ignizio (1982) questions Narasimhan and Hannan's developments concerning the analytical form of the membership functions. Hence, he stresses the fact that the authors have only considered the case where the decision-maker has membership functions of particular forms. In fact, these membership functions are often established without the decision-maker and are, by their nature, very different from the preference functions of this latter.

Ignizio (1982) also raises Narasimhan and Hannan's contribution in the matter of imprecision modelling in the process of a decision aid with multiple objectives. He judges that Zimmermann's formulation is more efficient since it requires a very reduced number of additional constraints. Hannan (1982) partially responded to these criticisms by bringing out the difference between the **FGP** formulation of Narasimhan (1980) and that of Zimmermann (1978). According to Hannan, there is no imprecision modelled in this last formulation; this formulation is equivalent to a "max.min." program and does not really introduce imprecision. Besides, the developed mathematical model does not take into account the imprecise inspiration levels of the decision-maker. But, this is against the philosophy of any decision aid where one tries as often as possible to involve the decision-maker in different stages of the decisional process.

However, as mentioned earlier, according to us, the membership functions introduced by Narasimhan and Hannan, do not express the decision-maker's preference structure. In fact, their effort is far from responding adequately to the decision-maker's expectations which are to give him the place due to him during the stage of the problem formulation by expressing his degrees of satisfaction faced with observed deviations in relation to imprecise goals.

4 The GP model with intervals (GPI)

Charnes and Cooper (1977) were the first to develop a **GP** formulation where goals are expressed with intervals. Deviations in relation to intervals are penalized with the aid of linear penalty functions having different slopes. Precisely, these functions change their slope from an interval to another. Therefore, these are linear penalty functions defined on several intervals.

The concept of penalty functions in the GP model was explored more in the works of Kvanli (1980), Romero (1984) and Can and Houck (1984) who propose new formulations with imprecise goals. These formulations do not deal in an equal way with the positive and negative deviations compared to the **GP** standard formulation. This may be considered as an interesting contribution since a positive or a negative interval in relation to any goal does not necessarily have the same effect on the decision-maker. It is interesting to be able to make the difference at the level of the decision-maker's preferences when a determined goal on a given objective is exceeded or not reached.

In their paper, Inuiguchi and Kume (1991) present a **GP** formulation whereas technological parameters and goals are determined by intervals. These latters reflect in part imprecision associated to these parameters. In other words, the decision-maker does not have enough information to determine precisely the values associated to the problem parameters. The goals g_i are then expressed with the aid of an interval having an inferior limit g_i^L and a superior limit g_i^U. This formulation is made up of four mathematical programs. As an illustration, we present a program called, in their paper, "NES-UPP":

Minimize $\quad Z = \lambda \sum_{i=1}^{p} W_i \mu_i + (1 - \lambda) \mu^U$

Subject to :

$$\sum_{j=1}^{n} a_{ij}^U x_j + \delta_i^{U-} - \delta_i^{U+} = g_i^U ;$$

$$\sum_{j=1}^{n} a_{ij}^L x_j + \delta_i^{L-} - \delta_i^{L+} = g_i^L ;$$

$Cx \leq c; \quad$ (system constraints);

$\delta_i^{L-} + \delta_i^{L+} \leq \mu_i ;$

$\delta_i^{U-} + \delta_i^{U+} \leq \mu_i ;$

$\mu_i \leq \mu^U ;$

Despite the fact that the decision-maker judges that his goals are imprecise and that he is not in a position to define them, this formulation proposed by Inuiguchi and Kume (1991) favors central values of the intervals. Therefore, it is as if the goals associated to various objectives were deterministic and equal to the central value of each interval (Figure 1 (a)). Since the goal value is fuzzy and expressed with the aid of an interval, the decision-maker should be indifferent between the solutions contained between both limits defining the interval (Figure 1 (b)). Moreover, this formulation does not add anything to the level of the decision-maker involvement in the modelling and solution process. In other words, the decision-maker's preference structure has not been introduced yet in an explicit way in this formulation of the GPI.

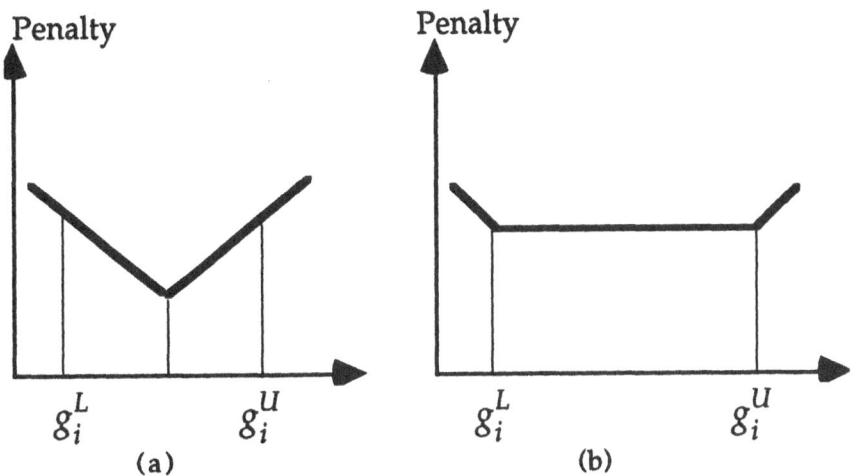

Fig. 1. Penalty functions

In a previous paper (Martel and Aouni, 1990), we proposed a reformulation of the GP model in which the decision-maker's preferences are explicitly introduced. This reformulation reads as followed:

Maximize $Z = \sum_{i=1}^{p} [W_i^+ F_i^+(\delta_i^+) + W_i^- F_i^-(\delta_i^-)]$

Subject to :

$$\sum_{j=1}^{n} a_{ij} x_j - \delta_i^+ + \delta_i^- = g_i; \quad \text{for } i=1,2, \dots ,p.$$

$Cx \leq c;$ system constraints

$\delta_i^+, \delta_i^-, x_j \geq 0.$ for $i=1,2, \dots , p$ and $j=1,2, \dots , n$

Where :

$F_i(\delta_i)$: is the Decision-Maker satisfaction function according to the deviation δ_i. This function may be different from one objective to another and for whether it is positive or negative deviation (see Figure 2 (a) and (b)).

In this formulation, the goals g_i are any point within the interval, and indifference thresholds of satisfaction functions enable the characterization of imprecision expressed by the interval. These functions may be different from one objective to another and depending on whether it is a positive or a negative deviation. Within the interval, the satisfaction function of the decision-maker is at its maximum level which is 1. Outside this interval, these functions are monotonic decreasing and they may take various forms.

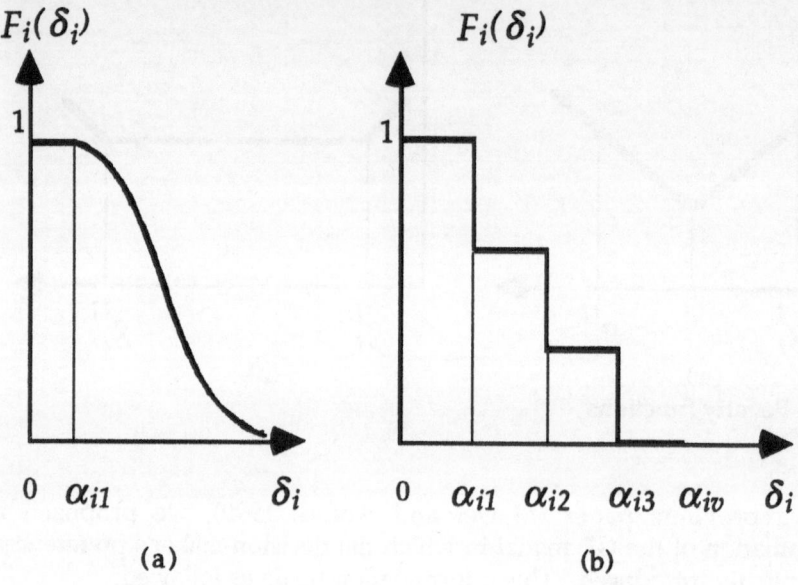

Fig. 2. The satisfaction functions.

With these functions, the decision-maker may keep an area of indifference α_{i1} in which the determined goal deviation is not penalized, it is to say that its degree of satisfaction for an action showing such deviation takes its maximum value of 1. Moreover, each action with a deviation bigger than α_{iv} (α_{iv} is a veto threshold) would be rejected by the decision-maker.

Building up these satisfaction functions is the decision-maker's right, which is very different from the **FGP** formulation where we have membership functions (in the sense of fuzzy sets) or Inuiguchi and Kume's (1991) where the mathematical model favors the central value of the interval. Besides, they take the same preference structure for all decision-makers. But, for example, in situations where the decision variable is discrete the step functions can be more appropriate (Figure 2 (b)).

The relative importance of objectives is represented in large part by the weighting vector W_i and partially by thresholds in the satisfaction functions $F_i(\delta_i)$ which contributes to overcome the difficulties raised by Hannan (1981-b) in the resolution stage of the FGP model in the presence of objective priorities.

Finally, we consider that the GP model reformulation proposed in Martel and Aouni (1990) is more general than the FGP and GPI formulations concerning the imprecise goal processing and it enables the decision-maker to introduce explicitly his preferences by the satisfaction functions. In Appendix A, we have some satisfaction functions equivalent to the following situations : 1) FGP with triangular membership function, 2) the formulation of UPP in GPI and 3) the formulation of LOW in GPI.

A sensitivity analysis may enable him to review and modify his satisfaction functions. The proposed model could be qualified as a decision aid tool since it enables the decision-maker to get more involved in his decision process.

5 Conclusion

Penalty functions and membership functions introduced in various GP model formulations do not really allow the decision-maker to express his satisfaction degrees in relation to the different expectation levels of his objectives. Therefore, these formulations are far from filling the gap in order to put at the decision-maker disposal a tool processing with imprecision related to goals g_i while taking his preference structure into account. We cannot qualify these models as tools of multicriterion decision aid at least as Roy conceives it (1987).

The integration of the decision-maker's preference structure in the GP with the aid of the satisfaction function gives the decision-maker the place due to him in various modelling stages of his problem. Hence, during the problem formulation stage, the decision-maker is required to express his satisfaction degrees in function of observed deviations in relation to the goals which are often of imprecise nature. With the aid of a sensitivity analysis, the decision-maker could verify the robustness of the results or choices proposed by the mathematical model developed, on one hand, and revising his satisfaction functions in order to make sure these latter reflect, on the other hand, his preference structure in the best way.

6 References

Can, E.K. and M.H. Houck, "Real-time Reservoir Operations by Goal Programming", *Journal of Water Ressources Planning Management*, Vol. 110, 1984 (297-309).

Charnes, A. and W.W. Cooper, " Management Models and Industrial Applications of Linear Programming", *Wiley, New-York,* 1961.

Charnes, A. and W.W. Cooper, " Goal Programming and Multiple Objectives Optimisations", *European Journal of Operational Research*, No. 1, 1977 (39-54).

Contini, B., "A stochastic Approach to Goal Programming", *Operation Research*, Vol. 16, No. 3, 1968 (576-586).

Hannan, E. L.,"On Fuzzy Goal Programming", *Decision Sciences*, Vol. 12, 1981-a (522-531).

Hannan, E. L., "Some Further Comments on Fuzzy Priorities", *Decision Sciences*, Vol. 12, 1981-b (539-541).

Hannan, E. L.,"Constrasting Fuzzy Goal Programming and Fuzzy Multicriteria programming", *Decision Sciences*, Vol. 13, 1983 (331-336).

Hannan, E. L., "An Assessment of Some Criticisms of Goal Programming", *Computer and Operation Research*, Vol. 12, No. 6, 1985 (525-541).

Hwang, C.L., S.R. Paidy, K. Yonn and A.S.M. Maud, "Mathematical Programming with Multiple Objectives : A Tutorial", *Computer and Operation Research,,* Vol. 7, 1980 (5-31).

Inuiguci, M. and Kume, Y., "Goal Programming Problems with Interval and Coefficients and Target Intervals", *European Journal of Operational Research*, No. 52, 1991 (345-360)

Ignizio, P. J., "Notes and Communications of the (Re)Discovery of Fuzzy Goal Programming", *Decision Sciences*, Vol. 13, 1982 (331-336).

Kvanli, A. H., "Financial Planning Using Goal Porgramming", *Omega,* Vol. 8, 1980 (207-218).

Martel, J-M. and B. Aouni, "Incorporating the Decision-Maker's Preferences in the Goal Programming Model", *Journal of Operational Research Society*, Vol. 41, No. 12, 1990 (1121-1132).

Min, H. and J. Storbeck, "On the Origin and Persistence of Misconceptions in Goal Programming", *Journal of Operational Research Society,,* Vol. 42, 1991 (301-312).

Narasimhan, R., "Goal Programming in a Fuzzy Environment", *Decision Sciences*, Vol. 11, 1980 (325-336).

Narasimhan, R., "On Fuzzy Goal Programming : Some Comments", *Decision Sciences*, Vol. 12, 1981 (532-538).

Rao, J. R., R. N. Tiwari and B. K. Mohanty, "A Preference Structure on Aspiration Levels in a Goal Programming Problem : A Fuzzy Approach", *Fuzzy Sets and Systems*, Vol. 25, 1988 (175-182).

Romero, C., "Handbook of Critical Issues in Goal Programming", *Pergamon Press*, 1991.

Romero, C., "A Note : Effects of Five-Side Penalty Functions in Goal Programming", *Omega*, Vol. 12, 1984.

Roy, B., "Des critères multiples en recherche opérationnelle: Pourquoi?" *Cahier de LAMSADE* No. 80, septembre 1987.

Zadeh, L. A., "Fuzzy Sets", *Information and Control*, Vol. 8, 1965 (338-353).

Zeleny, M., "Multiple Criteria Decision Making", *Mc Graw-Hill*, *New-York*, 1982.

Zimmermann, H.J., "Fuzzy Programming and Linear Programming with Several Objevtive Functions", *Fuzzy Sets and Systems*, Vol. 1, 1979 (45-55).

Appendix A

The satisfaction functions $F_i(\delta_i)$ equivalent to :

1. FGP with triangular membership function

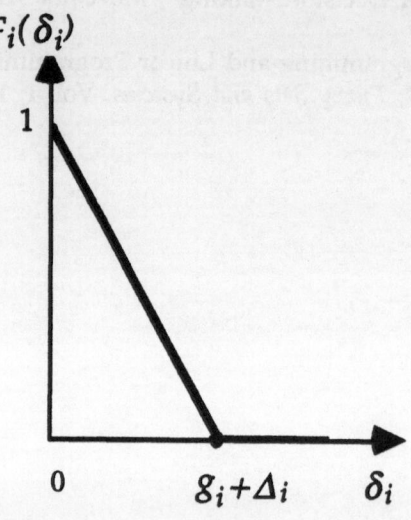

2 The formulation of UPP in GPI

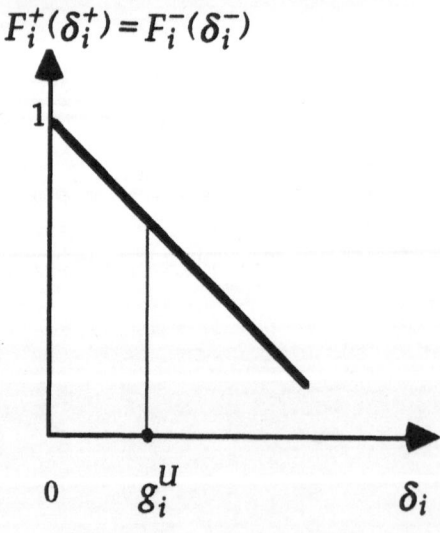

3 The formulation of LOW in GPI

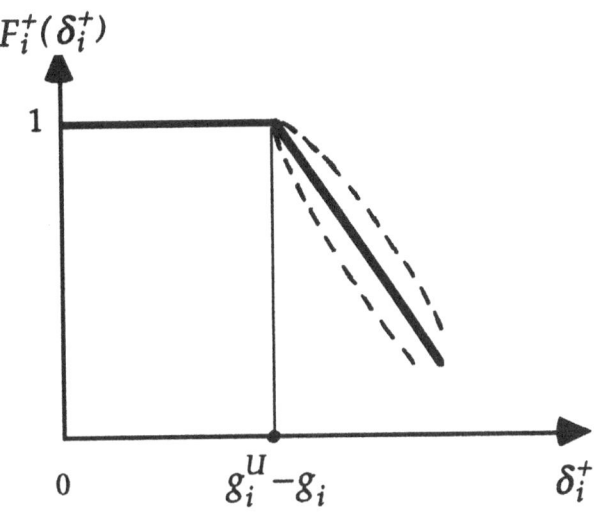

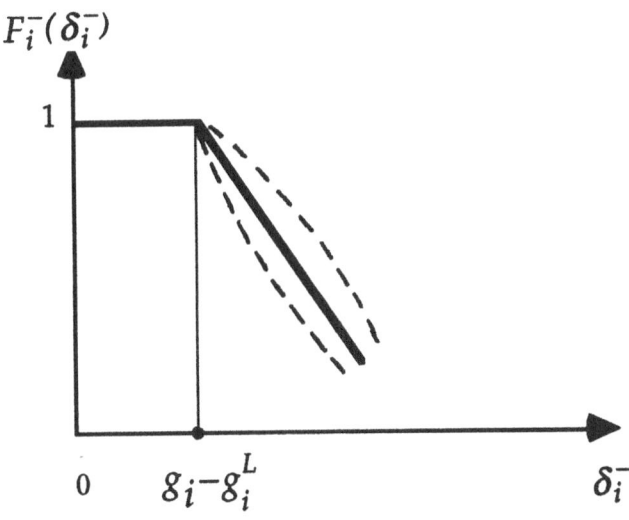

A FORMULATION OF A FUZZY LINEAR GOAL PROGRAMMING PROBLEM WITH FUZZY CONSTRAINTS AND FUZZY TARGET VALUES

ARENAS PARRA, M. MAR
RODRIGUEZ URIA, M. VICTORIA

Dpto. Matemáticas. Universidad de Oviedo. Asturias. Spain.

Abstract. In conventional Mathematical Goal Programming problems, the coefficients of objetive functions, constraints and target values are expressed as crisp numbers. In real world problems, however, it is no frequent for coefficients and target values to be known precisely. In such case, the parameters of the model should be represented by fuzzy sets for reflecting imprecision.

Our object is the formulation of a fuzzy linear goal programming problem to obtain a reasonable solution under consideration of the ambiguity in the achievement level desired for each attribute and to solve that fuzzy linear goal programming problem. The problem is transformed into a parametric one, and the solution is given in fuzzy terms; a numerical example is provided.

KEYWORDS: Goal Programming, Fuzzy Sets, Fuzzy numbers

1 Introduction

Fuzzy set theory, which was first proposed by Zadeh, has been understood to be useful for dealing with vagueness or ambiguity. Since Bellman and Zadeh [1970] proposed the concept of "decision making in a fuzzy environment", fuzzy set theory gives the conceptual and theoretical framework for handling complexity, imprecision and vagueness and has become a helpful tool in dealing with ambiguity of human judgment quantitatively. This approach has been combined with mathematical programming by Tanaka-Okuda-Assai [1974] and Negoita-Sularia [1976]. The fuzzy programming approach to multiobjetive linear programming problems was irst introduced by Zimmerman [1978] and further developed by Hannan, Yager and other authors. In many situations the decision maker has imprecise information about the real problem being studied, then he/she usually expresses his/her preference in terms of ambiguity, no randomness but fuzzyness; therefore, modelling decision-making problems in terms of fuzzy sets may be possible and a lot of fuzzy linear programming approaches have been developed.

Goal programming (G.P.) is perhaps the oldest approach within the field of multicriterion decision making (M.C.D.M.) because many decissors might follow a satisfaction criterion rather than the criterion of maximizing an objective function; and the satisfaction criterion leads to the concept of goal. When attributes and/or goals are in an imprecise environment and they cannot be stated with precision, we work in fuzzy goal programming.

Since Narasinhan [1980] a number of fuzzy linear goals programming models have been developed.

Our main concern in this paper is to formulate a fuzzy linear goal programming problem (F.L.G.P.) where the aspiration level of the decision maker for each goal —target values— is represented by fuzzy numbers and also the constraints of the model define the feasible set as a fuzzy set, described by its membership function. The problem is transformed into a parametric one, and the solution is given in fuzzy terms.

2 Basic Concepts

2.1 α-level set (α-cut)

The α-level set (α-cut) of a fuzzy set $\bar{A}$ is a crisp subset of $\Re$ and is denoted by:

$$A_\alpha = \{x \in \Re, \ \mu_{\bar{A}}(x) \geq \alpha\}, \qquad \alpha \in [0, 1] \tag{1}$$

There is a characterization of any fuzzy set $\bar{A}$ through this α-cuts using the next theorem:

2.2 Decomposition theorem

Any fuzzy set $\bar{A}$ may be associated to a totally ordered family of α-cuts:

$$\{A_\alpha, \quad \alpha \in [0, 1]\} \tag{2}$$

and decomposed in the following form:

$$\forall x \in \Re, \quad \mu_{\bar{A}}(x) = sup\{\alpha \ \mathcal{X}_{A_\alpha}(x) : \alpha \in [0, 1]\} = sup\{\alpha \in [0, 1] \ / \ x \in A_\alpha\} \tag{3}$$

where $\mathcal{X}_{A_\alpha}(x) = \begin{cases} 1 & if \ x \in A_\alpha \\ 0 & else \end{cases}$ and $\alpha \ \mathcal{X}_{A_\alpha}(x)$ means the product.

2.3 Fuzzy numbers

A *fuzzy number* is a convex fuzzy subset of the real line $\Re$ with a normalized membership function. Thus, a fuzzy subset $\tilde{a}$ of $\Re$ with the membership function:

$$\mu_{\tilde{a}} : \Re \longrightarrow [0, 1] \qquad (4)$$

is a fuzzy number, if for any $\alpha \in [0, 1]$ the α-level set

$$\tilde{a}_{\alpha} = \{u \in \Re : \mu_{\tilde{a}}(u) \geq \alpha\} \qquad (5)$$

is a convex subset of $\Re$ and there is $\lambda \in \Re$ such that

$$\mu_{\tilde{a}}(\lambda) = 1$$

This is equivalent to the condition that the membership function $\mu_{\tilde{a}}$ is non-decreasing on the interval $]-\infty, \lambda]$ and nonincreasing on the interval $[\lambda, +\infty[$ for certain $\lambda \in \Re$, where λ is the "*main value*".

To distinguish a fuzzy number from a crisp (nonfuzzy) one, the former will always be denoted by a tilde $\sim$.

We restrict ourselves to a class of fuzzy numbers, called triangular fuzzy numbers which are characterized by an ordered triad $\tilde{a} = (\lambda, \sigma_1, \sigma_2)$ and call the value λ the main value, and the value σ_1 (σ_2) the left (right) spread of the triangular number fuzzy, $\tilde{a}$. In a particular case σ_1 can be the same that σ_2, then we have *symetric triangular fuzzy numbers*.

2.4 Fuzzy parameter

A *fuzzy parameter* $\tilde{A} = \left(\tilde{A}_1, \tilde{A}_2, \cdots, \tilde{A}_n\right)$ is defined by fuzzy sets, $\tilde{A}_i$, and can be represented through:

$$\mu_{\tilde{A}}(\mathbf{x}) = \min_i \left[\mu_{\tilde{A}_i}(x_i)\right] \qquad (6)$$

If $\tilde{A}_i$, $i = 1, \ldots, n$ are symetric triangular fuzzy numbers, $\tilde{A}_i = (\lambda_i, \sigma_i)$, where

$$\mu_{\tilde{A}_i}(x_i) = \begin{cases} 1 - \dfrac{|\lambda_i - x_i|}{\sigma_i} & \lambda_i - \sigma_i \leq x_i \leq \lambda_i + \sigma_i \\ 0 & \text{otherwise} \end{cases} \qquad (7)$$

the fuzzy parameter $\tilde{A}$ can be denoted in vectorial form $\tilde{A} = \{\lambda, \sigma\}$ where

$$\lambda = (\lambda_1, \lambda_2, \ldots, \lambda_n)^t$$
$$\sigma = (\sigma_1, \sigma_2, \ldots, \sigma_n)^t$$

and means *approximately* λ, which is described by center "λ" and width "σ".

2.5   Fuzzy function: The extension principle

The extension principle, introduced by Zadeh, is one of the most important and powerful tools in fuzzy sets theory. Owing to this principle, models, functions and algorithms involving nonfuzzy variables can be extended to the case of fuzzy variables.

The principle may be so stated: given $\bar{A}_i$ a fuzzy subset of X_i, $i = 1, \ldots, n$, and a (nonfuzzy) function f

$$
\begin{aligned}
f : X_1 \times X_2 \times \cdots \times X_n &\longrightarrow Y \\
(x_1, x_2, \ldots, x_n) &\longrightarrow y = f(x_1, x_2, \ldots, x_n)
\end{aligned}
\tag{8}
$$

the fuzzy image $\bar{A}$ of $\bar{A}_1, \bar{A}_2, \ldots, \bar{A}_n$ through f, has the following membership function:

$$
\mu_{\bar{A}}(y) = \max_{\substack{(x_1, x_2, \ldots, x_n) \in \bar{A}_1 \times \bar{A}_2 \times \ \times \bar{A}_n \\ y = f(x_1, x_2, \ldots, x_n)}} \mu_{\bar{A}_1 \times \cdots \times \bar{A}_n}(x_1, \ldots, x_n)
\tag{9}
$$

for each $y \in Y$, where the Cartesian product $\bar{A}_1 \times \bar{A}_2 \times \cdots \times \bar{A}_n$ is defined as

$$
\mu_{\bar{A}_1 \times \cdots \times \bar{A}_n}(x_1, \ldots, x_n) = \min_i \left(\mu_{\bar{A}_i}(x_i) \right)
\tag{10}
$$

2.6   Almost positive

The fuzzy function $\bar{G}$ is said to be **almost positive** and denoted by $\bar{G} \gtrsim 0$ and is defined by

$$
\bar{G} \gtrsim 0 \iff \mu_{\bar{G}}(0) \leq 1 - h \quad \text{and} \quad \min_x \{\mu_{\bar{G}}(x) = 1\} > 0
\tag{11}
$$

where h stands for the degree of $\bar{G} \gtrsim 0$ and the larger h is, the stronger the meaning of *"almost positive"* is.

3   A Formulation of Fuzzy Linear Goal Programming

Multiple criteria decision making (M.C.D.M.) attempts to give the decision maker a mathematical support to decision making processes that involve several conflicting objectives subjected to a set of constraints. In real world problems, the concept of goal may be more useful than the concept of optimization of an objective function because, as we said in the introduction, many corporations follow a satisfaction criterion, which can be described by an acceptable level of achievement for any attributes considered, that is by setting goals.

In classical Goal Programming (G.P.), the optimal realization of multiple goals is sought within the constraints imposed by the decision environment. Goals are stated precisely and algebraic equations are formulated to correspond to the stated goals. If the decision environment is fuzzy then the goals cannot be stated precisely.

In this paper we intend to propose how fuzzy or imprecise goals of the decision maker and also uncertainty on the restrictions of the problem, may be incorporated into standard goal programming formulation by considering the coefficients and target values as fuzzy numbers.

The general formulation of a linear goal programming problem usually is:

Find the optimal decision $X^* = (x_0^*, x_1^*, \ldots, x_n^*)^t$, such that:

Goals $\qquad G_r \equiv c_r^1 x_1 + c_r^2 x_2 + \ldots + c_r^n x_n \geq g_r \qquad r = 1, 2, \ldots, m$

Constraints $\qquad AX \leq b, \qquad X \geq 0$

We assume now that goals and constraints are given by fuzzy sets $\tilde{G}_r$ and $\tilde{A}_s$ which are characterized by the membership function $\mu_{\tilde{G}_r}(x)$ and $\mu_{\tilde{A}_s}(x)$, let a fuzzy decision $\tilde{D}$ be defined as the fuzzy set resulting from the intersection of $\tilde{G}_r$ and $\tilde{A}_s$, that is,

$$\mu_{\tilde{D}}(x) = \min_{r,s} \left\{ \mu_{\tilde{G}_r}(x), \mu_{\tilde{A}_s}(x) \right\} \qquad (12)$$

where $\bigwedge$ denotes minimum.

Fuzzy mathematical programming problems are to obtain an optimal decision X^* such as:

$$\max_x \mu_{\tilde{D}}(x) = \mu_{\tilde{D}}(\mathbf{x}^*) \qquad (13)$$

Notice that in fuzzy linear goal programming problems, the conventional distinction between goals $\tilde{G}_r$ and constraints $\tilde{A}_s$ no longer applies; both enter in the expression for a fuzzy decision $\tilde{D}$ precisely in the same manner. One formulation for fuzzy linear goal programming problem with fuzzy constraints and fuzzy target values can be written as:

Find the "better" decision $X^* = (x_0^*, x_1^*, \ldots, x_n^*)^t$ that verifies:

Fuzzy Goals

$$\tilde{G}_r \equiv \tilde{g}_r x_0 + \tilde{c}_r^1 x_1 + \tilde{c}_r^2 x_2 + \cdots + \tilde{c}_r^n x_n \gtrsim 0, \quad r = 1, 2, \ldots, m$$

Fuzzy Constrains

$$\tilde{A}_s \equiv (Ax)_s \lesssim b_s, \qquad X \geq 0$$

where $\tilde{g}_r = (\lambda_r, \sigma_r)$ is a symetric triangular fuzzy numbers and means approximately λ_r representing the fuzzy aspiration level –target value– associated with the r-objetive that transforms it into ı fuzzy goal, and $\tilde{c}_r^i = (\lambda_r^i, 0)$ represents the crisp coefficients; the fuzzy inequality $\tilde{G}_r \gtrsim 0$ means "almost positive" (definition 2.6) and where A and b are a convenient matrix and vector, $X = (x_1, x_2, \ldots, x_n)^t$, $x_0 = 1$ and $\lesssim$ is a fuzzy inequality represented by membership functions[1]:

$$\mu_{\tilde{A}_s}(X) = \begin{cases} 1 & \text{if} \quad (AX)_s < b_s \\ 1 - \dfrac{(AX)_s - b_s}{p_s} & \text{if} \quad b_s \leq (AX)_s \leq b_s + p_s \\ 0 & \text{if} \quad (AX)_s > b_s + p_s \end{cases} \tag{14}$$

The fuzzy inequality $\tilde{G}_r \gtrsim 0$ may be written in the vector form

$$\tilde{G}_r = \tilde{C}_r X \gtrsim 0 \tag{15}$$

where

$$\tilde{C}_r = (\tilde{g}_r, \tilde{c}_r^1, \tilde{c}_r^2, \ldots, \tilde{c}_r^n) = \{\lambda^r = (\lambda_r, \lambda_r^1, \lambda_r^2, \ldots, \lambda_r^n), \sigma^r = (\sigma_r, 0, 0, \ldots, 0)\}$$

is a fuzzy parameter and

$$X = (x_0, x_1, \ldots, x_n)^t, \quad X \geq 0$$

Proposition 3.1

The fuzzy linear function $\tilde{G}_r$, that describe the fuzzy goals, is defined through the following membership function[2]:

$$\mu_{\tilde{G}_r}(y) = \begin{cases} 1 - \dfrac{|y - (\lambda_r + \sum_{i=1}^{n} \lambda_r^i x_i)|}{\sigma_r} & \text{if} \quad X > 0 \\ 1 & \text{if} \quad X = 0, y = 0 \\ 0 & \text{if} \quad X = 0, y \neq 0 \end{cases} \tag{16}$$

[1] Verdegay, J.L.(1982): "Fuzzy mathematical programming", in Gupta, M.M. and E. Sanchez (eds.) *Approximate Reasoning in Decision Analysis*. North-Holland, Amsterdam, 231-236.

[2] Arenas, M.M.; Rodríguez, M.V.: Fuzzy Linear Goal Programming with Fuzzy Numbers. DECISION MAKING:TOWARDS THE 21st CENTURY, Madrid 1993.

The decision maker usually establish preemptive priorities over the goals, $\tilde{G}_r$, then they are ranked into the priority levels; let $I = \{1, \ldots, k, \ldots, K\}$ be the set of index of priorities:

$$\mathbf{P}_1 \equiv \left\{ \tilde{G}_1, \ldots, \tilde{G}_{l_1} \right\}$$

$$\mathbf{P}_2 \equiv \left\{ \tilde{G}_{l_1+1}, \ldots, \tilde{G}_{l_2} \right\}$$

$$\vdots$$

$$\mathbf{P}_K \equiv \left\{ \tilde{G}_{l_{k-1}+1}, \ldots, \tilde{G}_m \right\}$$

We call

$$\mathcal{A} = \{ X \in \Re^n \ / \ AX \overset{\leq}{\approx} b, \ X \geq 0 \} \tag{17}$$

the fuzzy constraint set of the problem and we manipulate it through its $\alpha - cuts$:

$$\mathcal{A}_\alpha = \left\{ X \in \Re^n \ / \ \min_i \mu_{\tilde{A}_i}(X) \geq \alpha, \ X \geq 0 \right\} \tag{18}$$

to obtain the solution.

From Decomposition Theorem, as we know, the fuzzy set, $\mathcal{A}$, can be uniquely represented by all such $\alpha - cuts$, then $X \in \mathcal{A}$ is equivalent to $X \in \mathcal{A}_\alpha$ for some $\alpha \in [0, 1]$

With the lexicographic goal programming approach, the achievement function is made up of an ordered vector whose dimension coincides with the k number of priority levels established in the model. Each component in this vector represents the deviation variables that must be maximized in order to make sure that the goals ranked in this priority come closest to the established achievement levels.

From the above considerations we can set the following results:

Proposition 3.2

The better decision $X^* = (x_0^*, x_1^*, \ldots, x_n^*)^t$ that verifies:

$$\tilde{G}_r \equiv \tilde{g}_r x_0 + \tilde{c}_r^1 x_1 + \tilde{c}_r^2 x_2 + \cdots + \tilde{c}_r^n x_n \overset{>}{\approx} 0$$

$$\tilde{A}_s \equiv (Ax)_s \overset{\leq}{\approx} b_s$$

$$X \geq 0$$

can be obtained by solving the following parameteric Fuzzy Linear Goal Programming Problem:

Find X that:

$$\text{Lexicographically maximize } \mathbf{h} = \{(h_1), (h_2), \ldots, (h_K)\}$$

subject to

$$\mathbf{P_1} \begin{cases} \mu_{\tilde{G}_1}(0) \leq 1 - h_1 \\ \vdots \\ \mu_{\tilde{G}_{l_1}}(0) \leq 1 - h_1 \end{cases}$$

$$\mathbf{P_2} \begin{cases} \mu_{\tilde{G}_{l_1+1}}(0) \leq 1 - h_2 \\ \vdots \\ \mu_{\tilde{G}_{l_2}}(0) \leq 1 - h_2 \end{cases}$$

$$\vdots$$

$$\mathbf{P_K} \begin{cases} \mu_{\tilde{G}_{l_{K-1}+1}}(0) \leq 1 - h_K \\ \vdots \\ \mu_{\tilde{G}_m}(0) \leq 1 - h_K \end{cases}$$

$$X \in \mathcal{A}_\alpha$$

$$\alpha \in [0, 1]$$

$$0 \leq h_k \leq 1 \qquad k = 1, 2, \ldots, K$$

The problem now is reduced to obtain the largest h_k that is compatible with the expression of each priority $\mathbf{P_k}$ in fuzzy terms.

Proposition 3.3.- Sequential Algorithm.

A methodology for solving the above problem can be the following sequential algorithm for linear goal programming:

STEP 1.- Let $k = 1$.

STEP 2.- Determine the value of h_1 such that solve the single-objetive no linear programming model associated to the first priority, $\mathbf{P_1}$:

$$\left.\begin{array}{l} \max \; h_1 \\[2mm] \text{such that} \\[2mm] \lambda_i - \sigma_i \, h_1 + \displaystyle\sum_{j=1}^{n} \lambda_i^j x_j \geq 0 \qquad i \in \mathbf{P_1} \\[4mm] X \in \mathcal{A}_\alpha \\[2mm] 0 \leq h_1 \leq 1; \qquad \alpha \in [0,1] \end{array}\right\} \qquad (19)$$

Let the optimal solution to this problem be given as h_1^* .

STEP 3.- If $k < K$ then $k = k + 1$

STEP 4.- Solve the problem asociated with priority level k, $\mathbf{P_k}$:

$$\left.\begin{array}{l} \max \; h_k \\[2mm] \text{such that} \\[2mm] \lambda_i - \sigma_i \, h_1 + \sum_{j=1}^{n} \lambda_i^j x_j \geq 0 \qquad i \in \mathbf{P_1} \\[2mm] \lambda_i - \sigma_i \, h_2 + \sum_{j=1}^{n} \lambda_i^j x_j \geq 0 \qquad i \in \mathbf{P_2} \\[2mm] \vdots \\[2mm] \lambda_i - \sigma_i \, h_k + \sum_{j=1}^{n} \lambda_i^j x_j \geq 0 \qquad i \in \mathbf{P_k} \\[2mm] h_s = h_s^* \qquad s = 1, 2, \ldots, k-1 \\[2mm] X \in \mathcal{A}_\alpha \\[2mm] 0 \leq h_k \leq 1; \qquad \alpha \in [0,1] \end{array}\right\} \qquad (20)$$

STEP 5.- Go back to step 3 and continue to the last priority, K.

STEP 6.- The fuzzy solution $x^*(\alpha)$, associated with the last single-objetive model solved, is the solution vector for the original fuzzy linear goal programming model.

In order to illustrate the previous formulation, let us consider a numerical example:

4 Example

A company manufactures four products x_1, x_2, x_3, x_4. The share of each one in the output is shown in **table 1**:

Table 1. Monetary unit is 100.000 pesetas.

PRODUCT	PRICE/UNIT	COST
x_1	0.40	0.12
x_2	0.045	0.014
x_3	0.06	0.018
x_4	0.06	0.01

In the manufacturing of each x_i two professionals P_1 and P_2 and two machines, M_1 and M_2 participate; their contribution to each product is given in **table 2**:

Table 2. Time in minutes.

PRODUCT	P_1	P_2	M_1	M_2
x_1	60	–	15	30
x_2	20	–	40	10
x_3	–	20	–	10
x_4	10	–	–	10

Let us suppose that constraints on working time per week, for professionals and machines are given by the membership function below:

$$\mu_{P_1}(x) = \begin{cases} 1 & \text{if} \quad P_1(x) < 1620 \\ 1 - \dfrac{60x_1 + 20x_2 + 10x_4 - 1620}{1080} & \text{if} \quad 1620 \le P_1(x) \le 2700 \\ 0 & \text{if} \quad P_1(x) > 2700 \end{cases}$$

$$\mu_{P_2}(x) = \begin{cases} 1 & \text{if} \quad P_2(x) < 540 \\ 1 - \dfrac{20x_3 - 540}{360} & \text{if} \quad 540 \le P_2(x) \le 900 \\ 0 & \text{if} \quad P_2(x) > 900 \end{cases}$$

$$\mu_{M_1}(x) = \begin{cases} 1 & \text{if} \quad M_1(x) < 540 \\ 1 - \dfrac{15x_1 + 40x_2 - 540}{360} & \text{if} \quad 540 \leq M_1(x) \leq 900 \\ 0 & \text{if} \quad M_1(x) > 900 \end{cases}$$

$$\mu_{M_2}(x) = \begin{cases} 1 & \text{if} \quad M_2(x) < 1260 \\ 1 - \dfrac{30x_1 + 10x_2 + 10x_3 + 10x_4 - 1260}{900} & \text{if } 1260 \leq M_2(x) \leq 2160 \\ 0 & \text{if} \quad M_2(x) > 2160 \end{cases}$$

The company manager wants the turnover to be around 12.5 million pesetas and the expenses between 35% and 40% of the turnover; it is also convenient to restructure production so that X_4 will increase its share in the the global turnover.

The decision maker, as we see, expresses his aspiration level for each goal not very precisely; then we represent the target values by fuzzy numbers, and the problem formulation is:

Find a $x^t = (x_1, x_2, x_3, x_4)$, such that

$$G_1 \equiv 0.4x_1 + 0.045x_2 + 0.06x_3 + 0.06x_4 \gtrsim (12.5, 6)$$
$$G_2 \equiv 0.06x_4 \gtrsim (2.8, 0.3)$$
$$G_3 \equiv 0.12x_1 + 0.018x_2 + 0.03x_3 + 0.03x_4 \lesssim (4.69, 0.3)$$

$$\mu_{P_1} \geq \alpha$$
$$\mu_{P_2} \geq \alpha$$
$$\mu_{M_1} \geq \alpha$$
$$\mu_{M_2} \geq \alpha$$
$$\alpha \in [0, 1]$$
$$xi \geq 0, \ i = 1, 2, 3, 4$$

that is:

Find a $x^t = (x_1, x_2, x_3, x_4)$, such that

$$-\tilde{12.5} + 0.4x_1 + 0.045x_2 + 0.06x_3 + 0.06x_4 \gtrsim 0$$
$$-\tilde{2.8} + 0.06x_4 \gtrsim 0$$
$$4.\tilde{69} - 0.12x_1 - 0.014x_2 - 0.018x_3 - 0.01x_4 \gtrsim 0$$

$$1 - \frac{60x_1 + 20x_2 + 10x_4 - 1620}{1080} \geq \alpha$$

$$1 - \frac{20x_3 - 540}{360} \geq \alpha$$

$$1 - \frac{15x_1 + 40x_2 - 540}{360} \geq \alpha$$

$$1 - \frac{30x_1 + 10x_2 + 10x_3 + 10x_4 - 1260}{900} \geq \alpha$$

$$\alpha \in [0, 1]$$

$$xi \geq 0, \ i = 1, 2, 3, 4$$

From the definition of "almost positive" and using the membership function proposed by us, we have the following formulation of the problem:

Find a $\mathbf{x}^t = (x_1, x_2, x_3, x_4)$, such that

Lexmax h

subject to:

$-12.5 - 6h + 0.4x_1 + 0.045x_2 + 0.06x_3 + 0.06x_4 \geq 0$

$-2.8 - 0.3h + 0.06x_4 \geq 0$

$-4.69 + 0.3h + 0.12x_1 + 0.014x_2 + 0.018x_3 + 0.01x_4 \leq 0$

$60x_1 + 20x_2 + 10x_4 \leq 1620 + 1080(1 - \alpha)$

$20x_3 \leq 540 + 360(1 - \alpha)$

$15x_1 + 40x_2 \leq 540 + 360(1 - \alpha)$

$30x_1 + 10x_2 + 10x_3 + 10x_4 \leq 1260 + 900(1 - \alpha)$

$0 \leq h \leq 1$

$x = (x_1, x_2, x_3, x_4) \geq 0,$

$\alpha \in [0, 1]$

The problem has a parameter, α. We can apply the above sequential algorithm and then we obtain the following results given in **table 3**, that is provided for the decision maker:

Table 3.

α	h	G_1	G_2	G_3	P_1	P_2	M_1	M_2
0	.9266	18.06	7.56	4.41	2700	300	360	2130
0.2	.8133	17.38	5.82	4.438	2470	520	375	1980
0.4	.6866	16.62	3.84	4.474	2260	660	405	1780
0.6	.4666	15.32	2.94	4.204	2050	660	390	1600
0.8	.1633	13.48	3.06	3.636	1830	540	330	1440
1	.0000	11.7888	2.8	3.16333	1620	433.3333	288.3333	1260

If now the decision maker ranks the goals in terms of importance:

$$\mathbf{P}_1 \equiv \left\{ \tilde{G}_2, \tilde{G}_3 \right\}$$

$$\mathbf{P}_2 \equiv \left\{ \tilde{G}_1 \right\}$$

the problem will be stated as:

Find a $\mathbf{x}^t = (x_1, x_2, x_3, x_4)$, such that

$$\mathbf{P}_1 \begin{cases} -\tilde{2}.8 + 0.06x_4 \gtrsim 0 \\ 4.\tilde{6}9 - 0.12x_1 - 0.014x_2 - 0.018x_3 - 0.01x_4 \gtrsim 0 \end{cases}$$

$$\mathbf{P}_2 \left\{ -\tilde{1}2.5 + 0.4x_1 + 0.045x_2 + 0.06x_3 + 0.06x_4 \gtrsim 0 \right.$$

$$1 - \frac{60x_1 + 20x_2 + 10x_4 - 1620}{1080} \geq \alpha$$

$$1 - \frac{20x_3 - 540}{360} \geq \alpha$$

$$1 - \frac{15x_1 + 40x_2 - 540}{360} \geq \alpha$$

$$1 - \frac{30x_1 + 10x_2 + 10x_3 + 10x_4 - 1260}{900} \geq \alpha$$

$$\alpha \in [0, 1]$$

$$xi \geq 0, \; i = 1, 2, 3, 4$$

From the definition of "almost positive" and using the membership function proposed by us, we have the following formulation of the problem:

Find a $x^t = (x_1, x_2, x_3, x_4)$, such that

Lexmax $\quad h = (h_1, h_2)$

subject to:

$$P_1 \begin{cases} -2.8 - 0.3h_1 + 0.06x_4 \geq 0 \\ -4.69 + 0.3h_1 + 0.12x_1 + 0.014x_2 + 0.018x_3 + 0.01x_4 \leq 0 \end{cases}$$

$$P_2 \{-12.5 - 6h_2 + 0.4x_1 + 0.045x_2 + 0.06x_3 + 0.06x_4 \geq 0$$

$$60x_1 + 20x_2 + 10x_4 \leq 1620 + 1080(1 - \alpha)$$
$$20x_3 \leq 540 + 360(1 - \alpha)$$
$$15x_1 + 40x_2 \leq 540 + 360(1 - \alpha)$$
$$30x_1 + 10x_2 + 10x_3 + 10x_4 \leq 1260 + 900(1 - \alpha)$$

$$0 \leq h_1 \leq 1$$
$$0 \leq h_2 \leq 1$$

$$x = (x_1, x_2, x_3, x_4) \geq 0$$

$$\alpha \in [0, 1]$$

We can apply the above sequential algorit'um and then we obtain the following results given in **table 4**:

Table 4.

α	h_1	h_2	G_1	G_2	G_3	P_1	P_2	M_1	M_2
0	1	0.92	18.02	7.92	4.35	2700	300	345	2160
0.2	1	0.7833	17.2	5.82	4.4384	2470	460	375	1950
0.4	1	0.66	16.46	4.2	4.378	2260	620	390	1790
0.6	1	0.4533	15.22	3.18	4.142	2030	680	375	1620
0.8	1	0.1266	13.26	3.42	3.522	1830	480	315	1440
1	0.8	0	11.6422	3.04	3.0873	1620	393.33	278.33	1260

We offer the Decision Center the solutions of the parametric goal programming problems with one priority level (Table 3) and with two priority levels (Table 4) in order to permit it to adopt the "best" solution.

5 Concluding Remarks

In this paper we have tried to formulate a linear Goal Programming problem whose elements are subjectively perceived an 1 represented into the fuzzy set frame. This formulation is performed by the fuzzy function and the α-cuts of the fuzzy -soft- constraints.

We can observe from our formulation that we have dealt with soft constraints and flexible target values of the goals modelled by symetric triangular fuzzy numbers. We have transformed the problem into a parametric one and we have proposed a sequential algorithm to solve it. One type of fuzzy numbers was discussed here but other types of fuzzy numbers can be treated with a similar concept as in this paper.

Now we are working in the design of an algorithm for fuzzy linear goal programming problems with fuzzy cost. We are also exploring the set of G.P.-efficients solutions.

6 References

Bellman, R. E.; Zadeh, L.A. (1970): "Decision-making in a fuzzy environment", *Management Science 17*, B141-B164.

Carlsson, C.; Korhonen, P. (1986): "A parametric approach to fuzzy linear programming", *Fuzzy sets and Systems 20*, 17-30.

Chanas, S. (1989): "Fuzzy programming in multiobjective linear programming - a parametric approach", *Fuzzy sets and Systems 29*, 303-313.

Delgado, M.; Verdegay, J.L.; Vila, M.A. (1989): "A general models for fuzzy linear programming", *Fuzzy sets and Systems 29*, 21-29.

Dubois,D.; Prade,H. (1980): "Systems of linear fuzzy constraints", *Fuzzy Sets and Systems 3*, 37-48.

Fedrizzi, M. (1987): "Introduction to Fuz:,y Sets and Possibility Theory", *Optimization Models using fuzzy Sets and Possibility Theory*, 13-26.

Hannan, E. L. (1980): "Nondominance in goal programming". *INFOR, Canadian Journal of O.R. and Inf. Processing, 18*, 300-309.

Hannan, E. L. (1981): "Linear programming with multiple fuzzy goals". *Fuzzy Sets and Systems, 6*.

Ignizio, J.P. (1976): *Goal Programming and Extensions.* Lexington Books. Lexington.

Lai, Y.; Hwang, C. (1992): *Fuzzy Mathematical Programming. Methods and Aplications*, Springer-Verlag, Berlin Heidelberg.

Narasimhan (1980): "Goal Programming in a fuzzy environment". *Decision Sciences, 99.* 325-336.

Negoitia, C.V.; Sularia, M. (1976): "On fuzzy mathematical programming tolerances in planning". *Econom. Comput. Econom. Cybernet. Stud. Res. 1,* 3-15.

Negoitia, C.V.; Ralescu, D.A. (1977): "On fuzzy optimization". *Kybernetes, 6,* 193-196.

Rodríguez , M. V.; Arenas, M. M. (1992): *Programación Multiobjetivo: Goal Programming*, Servicio de publicaciones del Dpto. Matemáticas, Universidad de Oviedo.

Tanaka, H.; Okuda, T.; Asai, K. (1974): "On fuzzy-mathematical programming", *Journal of Cibernetics 3,* 37-46.

Tanaka, H.; Asai, K., (1984): "Fuzzy linear Programming Problems with fuzzy numbers", *Fuzzy sets and Systems 13,* 1-10.

Verdegay, J. L. (1982): "Fuzzy mathematical programming", *Approximate Reasoning in Decision Analysis*, eds. M. M. Gupta and E. Sanchez, North-Holland, Amsterdam, 231-236.

Verdegay, J. L. (1984): "A dual approach to solve the fuzzy linear programming problems", *Fuzzy sets and Systems 14,* 131-141.

Zadeh, L. A. (1965): "Fuzzy sets". *Information and Control, 8,* 338-353.

Zimmermann, H. J., (1976) : "Description and optimization of fuzzy systems", *International Journal General Systems 2,* 209-215.

A Two Staged Goal Programming Model for Portfolio Selection

M.Tamiz [1], R.Hasham, D.F Jones, B. Hesni, E.K Fargher,
University of Portsmouth, UK.

Abstract

The basic philosophy underlying investor portfolio stems from economic utility theory. Many mathematical models have been applied to portfolio selection, however a major drawback of these methods is that a vast majority of input data is needed which requires a large amount of computation.

The aim of this paper is to investigate the multi-objective approach of Goal Programming(GP) and its application to portfolio evaluation and selection. A two stage GP model is proposed. The first stage predicts the sensitivity of the shares to specific indicators. The second stage of the model selects a portfolio based on the decision maker's priorities and goals together with the information produced by the first stage.

Keywords: Goal Programming, Portfolio Analysis,

1 Overview of Modern Portfolio Theory

This paper illustrates the selection of a portfolio of shares in the Companies included in the FTSE 100 share index using Goal Programming. Generally speaking, a portfolio is a combination of risky and risk free assets held together. However, in the context of this paper, a portfolio refers only to a combination of shares in the FTSE 100 Companies. Expressing it in mathematical terms, the portfolio can be represented by:

$$x = (x_1, x_2, ..., x_{100})^T$$

, $x_i \geq 0$, $i = 1, \ldots 100$ where $\sum x_i$ = Total amount of investment, x_i = the amount invested in share i. This follows the Modern Portfolio Theory (MPT) which dates from the 1952 paper by Harry Markowitz on portfolio selection which is subjecting a sum of money to risk in order to obtain a higher future

---

[1] Address for correspondence : Dr M.Tamiz, School of Mathematical Studies, University of Portsmouth, Mercantile House, Hampshire Terrace, Portsmouth, UK. E-mail(Int) TAMIZM@cv.port.ac.uk

reward. We also note that an efficient capital market is an arena in which many particpants with similar investment objectives and access to the same information actively compete. These investors have strikingly similar objectives in that each prefers a high rate of return to a low one, certainty to uncertainty, low risk to high risk etc.

Modern Portfolio Theory has shown that the investment selection process requires far more than just assembling a portfolio of what is believed to be the "best" available shares on the market. MTP has shown that investors undertaking either the construction or analysis of a portfolio must address the relationships between the individual securities which comprise the aggregate portfolio in order to strike the desired balance between risk and return. Furthermore, Markowitz suggests that it is not sufficient to just pick the "best securities", but that the goal of a modern portfolio should be to maximise the expected utility.

Utility can be viewed as the degree of satisfaction percieved by the consumer. Thus in the context of an investment, each investor's preferred portfolio depends on his/her preference for increase in returns relative to his/her distaste for a corresponding increase in risk. It can therefore be suggested that the goal of all rational investment decisions is to maximise investor "satisfaction" or "utility".

Clearly any technique which is capable of measuring utility and determining precisely how much utility someone would obtain from a given amount of consumption would be very desirable. Unfortunately, however, such a technique has not yet been developed and therefore it is impossible to directly gauge marginal utility - the additional unit of satisfaction that the consumer achieves for each additional unit of consumption. It may be argued that this is not a problem for although utility cannot be measured on an absolute scale, it is possible to eveluate it on a relative scale. For example, these relative judgements can be put in the form of indifference curves, first introduced in the 1880's by Francis Y. Edgeworth. However, although the theory seems sound, there are certainly many questions on its application in practice.

The Markowitz's model, suggests that a portfolio's risk and return depends on three variables:

1. The expected return for each share,

2. The variance of the expected return for each share, and

3. The covariance of return of shares within the portfolio.

This in turn means that these variables might be brought into an efficient portfolio by means of an optimisation method (using quadratic programming techniques) to produce an efficient portfolio in terms of:

1. The shares to be held

2. The proportion of available funds to be allocated to each share.

A simplification to portfolio theory was proposed by Sharpe [9] in 1963. He postulated that the returns of securities are interrelated only to some index (i.e common market response), thereby eliminating the need to compute all the covariances. The model expresses the return of any security as a linear function of the market as a whole. The expected return of the market can be approximated by using the return on a suitable stock market index. This gives an equation of the form for each security i

$$E(R_i) = a_i + b_i E(R_m)$$

where $E(R_i)$ is the expected return on security i
a_i and b_i are constants specific to the security.
$E(R_m)$ is the expected return on the market as a whole.

However in practice the return R_i would not necessarily be equal to to its expected value, $E(R_i)$. Thus using past data on returns leads to the equation of the form:

$$R_i = a_i + b_i R_m + e_i$$

where R_i and R_m are the actual returns for security i and the market as a whole. e_i is the residual term for the difference between the actual and expected result. The expected value of e_i is assumed to be zero. However as stated by Rutterford [8] such a model as yet has no theoretical foundation.

2 A Critical Overview of Portfolio Models

The Markowitz model [6] as referred to in section 1 is ideal in theoretical content, however its pratical application presents a number of problems. Firstly, a large amount of data is required which leads to a large amount of computation. The second problem is that the results are given in the form of alternative optimal portfolios from which the fund manager has to choose. Finally the size of the optimal portfolio leads to the concept of undiversifiable market risk, since the more securities are held in the portfolio, the closer the fund manager gets in creating a portfolio which exhibits risk charateristics similar to that of the stock market, which cannot be diversified.

The above shortcomings led Sharpe [9] to develop a revised model in 1963. However Sharpe's model [9] makes some strong assumptions. For example the model assumes that the only common factor affecting all securities is the return on the market and ignores factors such as industry or economic influences. Although the linear approximation brings a powerful analytical model closer to practical implementation, there is still the process of having to derive and evaluate numerous portfolios until a suitable one is found.

This paper proposes GP as a more versatile and faster tool for selecting a portfolio based on the decision maker's goals and priorities.

3 Goal Programming

Goal Programming is a branch of multi-objective decision making. It is based around the Simonan [10] concept of satisfaction of a number of goals rather than the optimisation of a single objective function as is the case in linear programming. Upon the identifaction of a set of objectives which are relevent to the problem, the decision maker is asked to specify a set of targets or goals which satisfy the objectives. Unwanted deviations from the goals are then penalised to give a satifisficing solution. The nature of this penalisation depends on the type of goal programming used. The unwanted deviations can be given weights according to their relative importance to the decision maker, and their summation minimised. This is known as weighted goal programming(WGP) and is algebraically expressed as:

$$MIN \ z = \sum_{i=1}^{k} (u_i n_i + v_i p_i)$$

Subject to,

$$f_i(x) + n_i - p_i = b_i \qquad i = 1 \ldots k$$

$$x \in C_s$$

Where n_i, p_i represent the negative and positive deviations from the target value b_i for the $i'th$ objective. u_i, v_i represent the non-negative weights attached to these deviations, a zero weight being used for any deviation that the decision maker does not wish to minimise, such as positive deviation from a profit goal which represents the level of surplus profit. C_s is an optional set of 'hard' constraints which must be satisfied at the final solution point.

In the other main type of goal programming, objectives are separated into a number of priority levels, where the satisfaction of goals placed in a higher priority level is regarded as infinitely more important than the satisfaction of goals at a lower priority level. The ordered vector of priority levels is then minimised in a lexicographic manner. This is known as lexicographic goal programming. A standard lexicographic goal programme is algebraically expressed as:

$$Lex \ Min \ \mathbf{a} = [(g_1(\mathbf{n}, \mathbf{p})), g_2(\mathbf{n}, \mathbf{p}), \ldots, g_l(\mathbf{n}, \mathbf{p})]$$

Subject to,

$$f_i(x) + n_i - p_i = b_i \qquad i = 1 \ldots k$$

Where $\mathbf{a}$ represents the achievement function, $g_i(\mathbf{n}, \mathbf{p})$ represents the deviations to be minimised in the i'th priority level. It has the standard form:

$$g_i(\mathbf{n}, \mathbf{p}) = \sum_{j=1}^{k} (u_j n_j + v_j p_j)$$

Where u_j, v_j are the inter priority weights attached to the negative and positive deviations of the j'th objective in the i'th priority level. These weights are set to zero if the objective is not considered in that priority level.

A lexicographic goal programming can be solved as a series of linear programmes with variable fixing at the end of each priority level to protect the optimal value in the following priority level minimisations [2].

The reader is referred to recent textbooks by Romero [7] and Ignizio [2] for a comprehensive review of goal programming theory and modelling practice.

4 Applications of Goal Programming to Portfolio Selection

There have been several applications of GP to Portfolio Selection. The study by Lee and Chesser [4] presents a lexicographic GP model representing the investor's priorities. A similar model was discussed by Levary and Avery [5] which compared the use of linear programming to goal programming for the selection of optimal portfolios. Hallerbach and Spronk [1] in 1985 proposed an Interactive Multiple Goal Programme to construct a portfolio which best meets a given decision makers preferences.

4.1 Lee's Portfolio Model

The models descibed above are hypothetical and were not tested with real life data. The only model to do so is that of Lee [3] in 1972.

Lee's model is formulated as a Lexicographic Goal Programme with three objectives.

1. Maximisation of expected return E_p

2. Minimisation of portfolio risk B_p

3. Maximisation of dividend yield D_p

The following 'hard' constraints insured the integrity of the solution:

1. no more than 5% of the fund may be invested in a single security.

2. no more than 25% of the fund is to be invested in any one sector.

Two possible models were formulated, the income model and the growth model.

Achievement function for growth portfolio:

$$Min \ a \ = [(n_1), (p_2, n_3)]$$

Achievement function for income portfolio:

$$Min \ \mathbf{a} \ = \ [(n_3), (p_2, n_1)]$$

GOALS:

$$\sum_{i=1}^{53} E_i x_i + n_1 - p_1 = 1.391$$

$$\sum_{i=1}^{53} B_i x_i + n_2 - p_2 = 0.403$$

$$\sum_{i=1}^{53} D_i x_i + n_3 - p_3 = 0.048$$

Some hard constraints:

$$\sum_{i=1}^{53} x_i = 1$$

$$x_i \leq 0.05 \quad i = 1, \ldots, 53$$

$$\sum_{i=1}^{4} x_i \leq 0.25 \ Automation$$

$$\sum_{i=5}^{12} x_i \leq 0.25 \ Cosmetics$$

$$\sum_{i=47}^{53} x_i \leq 0.25 \ Utilities$$

Using the model Lee constructed a portfolio of shares in 1968 and a year later compared the actual performance with the expected performance.

In reality both Lee's models degenerate into Linear Programmes. This is because the optimal solution of priority level one has to be obtained and held whilst achieving the solution for the other priority level.

Therefore the growth model selects the 20 stocks with the highest E_p and the income model picks the 20 stocks with the highest D_p values.

Since further improvement to the model requires trade off to take place between growth, risk and dividend, a WGP approach to the portfolio problem is proposed.

5 WGP Version of Lee's Model

The data was used from Lee's model by using the percentage normalisation method as suggested by Romero [7]. The following model was obtained:

$$Min\ z = W_1 n_1 + W_2 p_2 + W_3 n_3$$

subject to

$$\sum_{i=1}^{53} \frac{E_i x_i}{\frac{1.391}{100}} + n_1 - p_1 = 100$$

$$\sum_{i=1}^{53} \frac{B_i x_i}{\frac{0\ 403}{100}} + n_2 - p_2 = 100$$

$$\sum_{i=1}^{53} \frac{D_i x_i}{\frac{0\ 048}{100}} + n_3 - p_3 = 100$$

$$x \in C_s$$

C_s: other constraints from Lee's model.

RESULTS OBTAINED

These results were obtained using heuristic methods to determine suitable weighting policies. A selection of these are given in the table below:

Policy	Return	Risk	Dividend
(1,0,0)	0. 68	1. 36	0. 009
(0,1,0)	0. 18	0. 71	0. 029
(0,0,1)	0. 11	0. 89	0. 033
(1,1,1)	0. 22	0. 72	0. 029
(1,1,0)	0. 22	0. 72	0. 028
(1,0,1)	0. 55	1. 07	0. 024
(0,1,1)	0. 17	0. 71	0. 030
(3,1,2)	0. 38	0. 80	0. 027
(8,1,3)	0. 56	0. 92	0. 022

The first and third results correspond to Lee's model with the growth and income policies optimised respectively. Thus proving the hypothesis that Lee's model degenerated into an LP. The actual optimal portfolio is a function of the preferences of the fund analyst.

In our opinion it is necessary to assess the risk by considering a number of economic and financial factors which are likely to affect the share price instead of only concentrating at a single factor. The globalisation of financial markets and the ease with which capital can be moved from one market to another provide further justification for this approach. Furthermore, since the majority of the companies, particurlarly those included in the FTSE 100 Index are multinational companies, it is inappropriate to ignore the factors which affect the major foreign markets and concentrate only on the UK economic indicators. For example, there is little dispute that, amongst other things, a rise in the US interest rate not only affects the share prices of the companies listed on the US stock markets but also has an indirect affect on the share prices of the British listed companies. The reason for this is the fact that almost inevitably a rise in interest rates by the US has to be reflected in the UK interest rate by an appropriate rise to avoid an outflow of capital by investors in the UK markets. Further, a rise in the UK interest rate will mean both higher interest payments by companies with debt capital and higher expected return from equities by investors which in return will mean lower share prices.

Clearly the process of both identifying and determining the relative importance of all the factors which affect the UK share prices either directly or indirectly is quite complicated. Moreover, not all these factors are quantitative factors and capable of being measured in a subjective manner. Despite these facts, however, it is widely believed that there are only a few quantifyable factors which significantly affect the share prices.

Once these factors are identified, a sensitivity analysis needs to be carried out to determine the extent to which a particular share price is affected by changes in these factors. The classical approach for predicting the sensitivity of the shares to such economic factors would be to use conventional regression analysis. However Ignizio [2] states that under certain circumstances a goal programming based regression approach can be used as an alternative to find the predictive function.

A goal programming model is proposed for this type of analysis as it is computationally faster than conventional regression and does not give overdue importance to outliers and dubious data points. A two stage model is presented which chooses an optimum portfolio based on one or more economic scenarios. A full description of the models are outlined in the following sections.

6 A Sensitivity Based GP: Two-Staged Approach

6.1 First Stage: Sensitivity Analysis of the Shares

In the first stage a GP model is formulated which predicts the sensitivity of the shares to specific economic indicators. The data used is taken from the FTSE 100 index. The share price for all of the shares in the index were collected for each quarter from 1988-1992 (giving 19 sets of share prices). Only the data for 97 shares were available due to changes in the FTSE 100 index constituents. The data was cleaned by adjusting the share prices for any rights issues, script issues and dividends. The twelve economic factors used in the model are:

1. UK interest rate

2. US interest rate

3. German interest rate

4. US inflation rate

5. German inflation rate

6. Dow Jones index

7. Nikkei Average

8. Hang Sang index

9. Oil price

10. Gold price

11. House price

12. Sterling index

6.2 Mathematical Representation of Sensitivity Analysis

Sensitivity for each individual share for the 19 time periods is given by the solution to the following GP (Model A):

$$Min \ z = \sum_{i=1}^{19}(n_i + p_i)$$

subject to

$$\sum_{j=1}^{12} C_{ij}y_j + n_i - p_i = P_i \quad i = 1, \ldots, 19$$

$$y_j \ free$$

Similarly sensitivity for FTSE index is given by (Model B):

$$Min \ z = \sum_{i=1}^{19}(n_i + p_i)$$

subject to

$$\sum_{j=1}^{12} C_{ij}y_j + n_i - p_i = F_i \quad i = 1,\ldots,19$$

$$y_j \ free$$

where
C_{ij}=change in factor j for period i
P_i=Share price movement in period i.
F_i=movement of FTSE in period i.
y_j=predicted sensitivity of share i or FTSE to factor j

Model A is solved for each of the 97 shares and model B is solved for FTSE index, using a goal programming package GPSYS [11]. Solving this series of models gives the sensitivity of each share and the FTSE index to the economic factors. Model A is used to predict the sensitivity of a given share to the economic factors. Thus the model must be processed 97 times. Each set of outcomes (y_j, $j = 1,\ldots 12$) is stored in a matrix (MATRIX.DAT in fig 1) which when complet forms the sensitivity matrix S_{ij} used in the second stage model. Model B is processed once to obtain the sensitivity of the FTSE against the economic factors. The outcome (y_j, $j = 1,\ldots 12$) (FTSE.DAT in fig 1) represented by vector F_j used in the second stage model.

6.3   Second Stage: Selection of Optimal Portfolio

Using the results obtained from the the above sensitivity analysis a subsequent GP model is formulated to select an optimal portfolio from amongst the given set of shares.

Mathematical Representation of the Model

$$Min \ z = W_1 \sum_{j=1}^{12}(nfact_j + pfact_j) + W_2 \sum_{k=1}^{N}(nscen_k) + W_3 \sum_{i=1}^{97}(ppen_i) \quad (1)$$

subject to

$$x_i + npen_i - ppen_i = 0.05 \quad i = 1,\ldots,97 \quad (2)$$

$$\sum_{i=1}^{97} \frac{S_{ij}x_i L_{FTSE}}{F_J L_i} + nfact_j - pfact_j = 1 \quad j = 1,\ldots,12 \tag{3}$$

$$\sum_{j=1}^{12} M_{jk} \sum_{i=1}^{97} \frac{100 \times S_{ij}x_i}{L_i} + nscen_k - pscen_k = T_k \quad k = 1,\ldots,N \tag{4}$$

$$\sum_{i=1}^{97} x_i = 1$$

$$sector \ \ constraints \le 0.25 \tag{5}$$

$$0 \le x_i \le 0.09 \quad i = 1,\ldots,97$$

where N is the number of scenarios and S_{ij} is the sensitivity obtained for share i against factor j. F_j is the sensitivity of the FTSE to factor j, M_{jk}, represent the changes in the factors for scenario k and T_k represents the percentage profit desired for scenario k. There is a strict upper bound of 9% of the portfolio that can be invested in any individual share. Also the set of objectives in (2) states that ideally this figure should not exceed 5% of the total funds available for investment. The extent to which shares take values under 5% rather than values in the range 5% to 9% is given by the weight placed on the minimisation of the positive deviations from these objectives in the achievement function, W_3. A further set of constraints (5) concerning unsystematic risk are that at most 25% of the portfolio can be invested in any one sector. These are taken directly from the model outlined by Lee[3] in section 4.1.

The objective set (3) deals with the control of the systematic risk associated with the portfolio. Each objective in this set represents the specific risk of the portfolio to one of the economic factors included in the sensitivity analysis above. The individual sensitivity of a share to the factor divided by the FTSE sensitivity to that factor gives a measure of the relative sensitivity of the share to the factor compared with the market. Due to the different magnitude of the share prices involved this coefficient is divided by a scaling factor of the ratio of the last price of the share against the last value of the FTSE. The sum of the percentage invested in each share multiplied by these coefficients then gives the sensitivity of the portfolio to the factor relative to that of the market. Thus, by setting the goal of these objectives to a value of one, the modeller is expressing the desire to follow the market as regards these factors. If the goal for a factor was set to zero this would indicate a wish to de-sensitise the portfolio towards that factor.

The second stage of the analysis of the model is based on the concept of a number of scenarios. A scenario is defined as a set of values assigned to the economic factors included in the sensitivity analysis. For each scenario the sensitivity for each share and the amount invested in the share is multiplied by the factor changes for the scenario and summed to give the total profit for the portfolio under that scenario. The target value expresses the modeller's required profit level for that particular scenario. This is represented by the objective set (4).

A summary of the overall model is illustrated in Figure 1.

7 Results

As an experiment the model was solved four times, each time using a single scenario based on the actual values of the twelve factors referred to in section 6.1 at the end of each quarter in 1993. As shown in the following table, in each sum, the optimal portfolio selected by model outperformed the FTSE 100 index quite significantly.

	Nov 92	Jan 93	Mar 93	June 93	Sep 93
FTSE 100 Index	2697	2950	2886	3100	3237
Cum increase(FTSE 100)		9.4%	7.0%	14.9%	20.0%
Cum increase(Portfolio)		21.5%	46.0%	45.3%	30.8%
Outperformed by		129.1%	556.7%	203.3%	53.6%

8 Conclusions and Further Research

From the previous section it can be seen that the model selects satisfactory portfolios based on actual changes in the twelve risk factor which took place in 1993. It may be suggested that such extraordinary results were achieved because the values of the twelve economic factors were known to us with certainty and that had it not been for this fact, the results would have been less impressive. Such criticism may only be justified if the reader were to believe that despite access to a vast amount of financial and non financial data, investment analysts and fund managers are still unable to make any reasonable forecast of the short term changes in the twelve economic factors included in the model. Besides, as it is explained in section 5, a number of possible scenarios can be incorporated in the model.

However what is clear from this research is that the first stage of the sensitivity analysis is accurate, and furthermore, providing the decision maker's predictions regarding the risk factors used in this model are reasonably reliable, the model is capable of selecting an optimum portfolio which satisfies the desired goals.

Clearly further research is needed to determine the most suitable method of predicting the appropriate scenarios. We believe the use of neural networks for this purpose is one option which is worth considering.

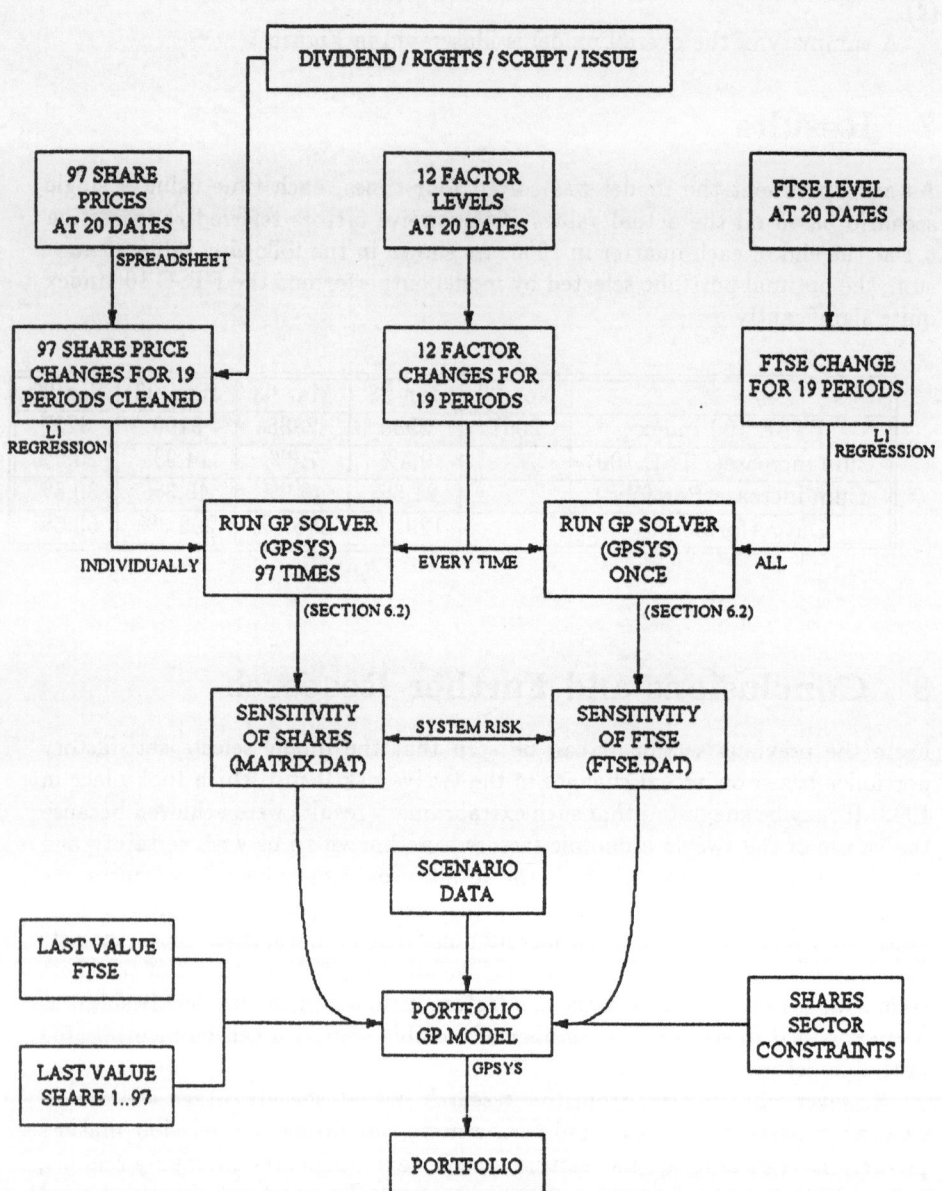

Figure 1: Flow Chart of the Portfolio Selection Process using GPSYS.

A further idea would be to reduce the number of risk factors which should help ease the task of scenario building. This can be achieved by vigorous testing of the correlation between the factors. For this purpose multi-regression analysis tools may be used to identify the dominant factors.

References

[1] HALLERBACH, W. and SPRONK, J. (1986) An Interactive Multi-factor Portfolio Model.

[2] IGNIZIO, J.P. and CAVALIER, T.M. (1994) *Linear Programming*, Prentice Hall.

[3] LEE, S.M. (1972) *Goal Programming for Decision Analysis*. Auerback, Philadelphia.

[4] LEE, S.M. and CHESSER, D.L. (1980) Goal Programming for Portfolio Selection, *Journal of Portfolio Management*, 23-25.

[5] LEVARY, R.R. and AVERY, M.L. (1984) On the Practical Application of Weighting Equities in a Portfolio via Goal Programming. *Opsearch*, **21**, 246-261.

[6] MARKOWITZ, H.M. (1959) *Portfolio Selection, Efficient Diversification of Investments*, Wiley, New York.

[7] ROMERO, C.(1991) *Handbook of Critical Issues in Goal Programming*, Pergamon Press.

[8] RUTTERFORD, J. (1983) *Introduction to Stock Exchange Investment*, The Macmillan Press Ltd.

[9] SHARPE, W.F.(1963) A Simplified Model for Portfolio Analysis. *Management Science*,**9**,277-293.

[10] SIMON, H.A.(1955) *Models of Man*, J.Wiley & Sons, New York.

[11] TAMIZ, M., JONES, D.F.(1994) GPSYS : Preliminary User Guide, University of Portsmouth, UK.

Application of MOP and GP to Wildlife Management (Deer). A Case in the Mediterranean Ecosystem

Julio Berbel and Ricardo Zamora

ETS Ingenieros Agrónomos y Montes, Universidad de Córdoba, Apdo. 3048,
14080 Córdoba, Spain.

Abstract. This paper presents an application of MOPGP to wildlife management. The case study is the development of a game management operational plan for deer in Southern Spain. As there are two species, ecologic and economic conflicts are in conflict. The solution procedures start with a Multiobjective analysis of the efficient set, and is complemented by using Lexicographical Goal Programming to obtain a "satisfactory" operational plan. This research proves that management models can be improved by using multicriteria approaches, especially when a satisfactory combination of methods is employed.

Keywords. Wildlife management, Multicriteria Decision Making Techniques, Semi-arid ecosystems.

1. Introduction

Multicriteria Decision Making (MCDM), and especially mathematic programming approaches, Multiobjective Program (MOP), Goal Programming (GP) and Interactive MCDM (IMCDM), are presently reaching maturity in the theoretical development. But as Ignizio (1994) affirms, two important problems faced by MCDM paradigm are:
- lack of successful applications
- internal fight between two approaches (GP vs. MOP).

Following this line of reasoning, we will try to illustrate a new field of application for MCDM and an example of complementary use of MOP and GP. The research will focus on wildlife management by studying a real-case based on the development of an operational management plan facing conflicting ecologic and economic objectives.

Wildlife management problems are subsets of natural resource management problems. Generally, these type of problems involve a search for ecological, economic and social objectives, usually in conflict when real systems are managed. Most natural resource applications are in the field of land-use planning. Natural resource management problems have been tackled by some authors during the last two decades with the use of Multiple Criteria Decision Making (MCDM) techniques. A complete review can be found in Romero & Rehman (1987).

The main definitions of MCDM such as objectives, criteria, goals, etc. can be found elsewhere; we recommend the book of Romero & Rehman (1989) as a complete review of MCDM techniques, approaches and examples of application to agricultural problems.

2. Agroforestry and wildlife management

Within the field of natural resources we wish to focus on wildlife management as an activity closely related to forestry management. We can quote some examples of MCDM in forestry systems such as Zadnik M. & Stirn, L. (1990) which is an adaptive dynamic model for optimal forest management. Romero (1989) uses compromise programming to obtain an efficient set of operational plans. Wildlife management cannot be treated independently from other criteria in forest management such as timber yield sustainability, timber production maximization, outdoor recreation and other economic and social factors.

It is frequent to find wildlife as a single criterion among others of an economic, social or ecological nature; then, the satisfaction of a certain level of the population is usually the main goal of the model. There are not many examples in the literature dealing with wildlife populations as the main object of the model. Operational management plans have not been tackled within the MCDM paradigm. The most important criteria in wildlife management are: a) preservation of endangered species; b) optimization of game uses of certain species; c) maximizing bio-diversity.

This paper deals mainly with the second aspect but taking into account ecological and endangered species protection. The model focuses on wildlife management itself but it does not incorporate other forest management objectives, which can be done at a different planning stage because of the different time-span of wildlife systems (10 to 20 years) and forest plans (40 to 80 years).

We have not found in the literature many examples of game management. Davis (1967) expounds a dynamic linear program for deer population management in New Zealand, in a LP context; an example of MCDM model is Jordi & Peddie (1988), who manage a natural park in South Africa where twenty different species are simultaneously considered for commercial purposes, and different preservation goals are sought. This model has a static nature and is related mainly to land use plans.

Bare & Mendoza (1988) proposed that the objective of the model should be the optimization of certain wildlife populations that are considered as an index for ecological wildlife quality and simultaneously optimizing timber yield as the other objective.

It is difficult to find a good index for wildlife quality, and it depends on the ecosystem characteristics. In Mediterranean ecosystems, the number of rabbits has been proposed by Zamora (1994) as a good index for the protected species system capacity, because in countries such as Spain, a typically Mediterranean one, the number of rabbits is the main source of food for the most important endangered species (*Lynx pardina, Aquila spp.*, etc.)

On the other hand, a more complex task is to define "biodiversity"; this can be treated with the help of definitions related to the entropy of the system, number of species, etc. There are not many applications in that field, and it is probably one of the more interesting research problems in wildlife management.   As we define above, our model is focused on medium term planning of game resources by means of a set of criteria representing the game economic value and simultaneously maintaining an acceptable population level for two different species of deer.

3. The case study

The most important output of Mediterranean forest areas is the protection of the environment. The scant importance of marketable forest products is a consequence of the low productivity of lands under Mediterranean conditions. At the same time, the use for game marketing as an outdoor recreation has become very important by exceeding the economic value of the forest products themselves.

The case study is a typical piece of land in a Mediterranean forest in Southern Spain. The land has 3,600 ha with the following vegetation cover: pinewoods, groups of *Quercus spp.*, *maqquia* and range lands under *Quercus* cover. Consequently it has clear game potential. On this land grows two main species for hunting purposes: roe deer (*Capreolus capreolus*), and red deer (*Cervus elaphus*).

The management of game commercial farms is based upon operational plans defining, for each species, the number of captures classified by sex and age for every year in the planning period. An operational plan should take care of economic considerations and simultaneously preserve the population's sustainability and demographic equilibrium.

Regarding the planning horizon, it should be of between 10 and 20 years according to Berbel (1992) for deer management in Spain. This time the horizon is explained by the population dynamics which needs longer periods for stabilization. On the other hand, a too lengthy planning period does not permit practical considerations, consequently we have selected a planning period of 10 years. A complete description of the case study can be found in Zamora (1994).

Nutritional requirements are satisfied by the vegetation cover each year by defining nutritional constraints considering nitrogen (proteins) and energy demand/supply. There is also a set of equations defining the dynamic relations between different ages and sex for the species. The complete model is composed of the objectives section and the constraint and dynamic relationship section.

A considerable effort was devoted to finding good models of the supply of nutrients by the natural cover and the demand of nutrients and energy by the population. A previous paper by Berbel & Zamora (1992) deals with the demographic evolution of red deer populations in Southern Spain Mediterranean ecosystems (Sierra Morena area) and the main parameters of these species were taken from the works of López-Giménez (1972) and Zamora-Lozano et al (1976).

Zamora-Lozano et al (1979) published a simple linear program of red deer management under Southern Spain conditions.

4. The model

The model introduced in the last section has complex dynamics. There are variables for both deer species (red and roe); these population variables were divided into age and sex each year during the period of analysis. As in any other system model, a crucial decision to be made is the selection of the relevant criteria to be optimized. In our case we see quite clearly the need for an economic performance criterion and an ecological value criterion.

The capitalized value of all yearly income of individuals hunted is the economic attribute to be maximized. Market value for each species and age is known from the local market, and the interest rate was 4%, which is quite usual in forest analysis and is suggested by Berbel (1992) as a rate that does not produce population distortion. Therefore, the first criteria will be to maximize Economic Income, which is:

$$\text{Maximize } Z_1 = \sum_{t=1,10} r_t V_t \tag{1}$$

where V_t is the income for year t (t=1,10) both for roe deer and red deer. The coefficient r_t will take the values:

$$r_t = 1 / (1+i)^t$$

$$V_t = \sum_{i=2,5} (V_i^{cc} HM_{ti}^{cc} + V_i^{ce} HM_{ti}^{ce}) \tag{2}$$

The V_t will be computed for years 1 to 10. Model allows only adults to be hunted (both sexes older than 2 years). Only hunted males have economic value. Variables in the equation (2) mean:

V_i^{cc} = Value of roe deer (*Capreolus capreolus*) of age "i"
V_i^{ce} = Value of red deer (*Cervus elaphus*) of age "i"
HM_{ti}^{cc} = Hunted males of roe deer (age "i", year "t")
HM_{ti}^{ce} = Hunted males of red deer (age "i", year "t")

With this objective the selection of the appropriate rate of interest for the present value formula is an important decision. We based our decision in the work of Berbel (1992), whose main finding is shown in figure 1. In this figure, we can see the effect on the population levels of different interest rates for a model with dynamic relationship similar to these.

We selected an intermediate rate as 4% based on figure 1 to avoid extreme behavior. The dynamic relationships and nutritional constraints used to generate these plans are the same as those set up in this work.

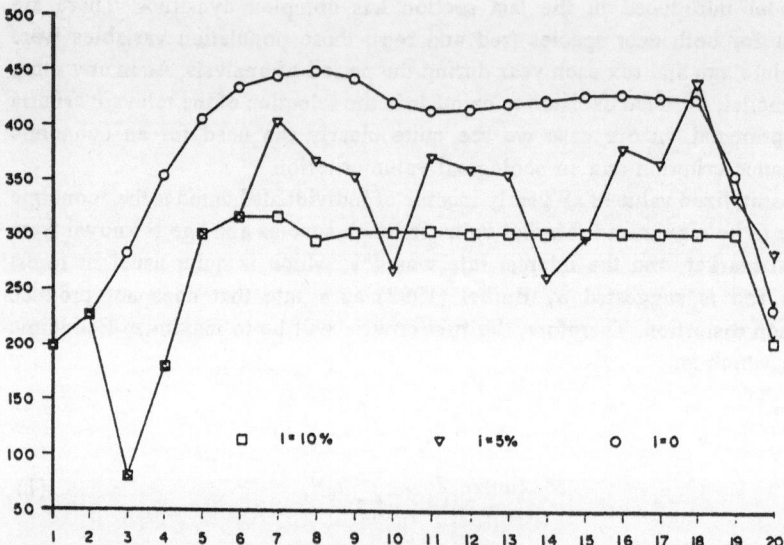

Figure 1. Evolution of deer population for different interest rates.

Roe deer in Southern Spain is an endangered species, especially when in competition with red deer. Therefore the population will be the ecological attribute to be maximized. The ecology of roe deer implies that it needs almost ten times more area for survival than the red deer although the value of hunted pieces is about the same. Obviously when only economic performance is considered, the model will eliminate roe deer from the land.

Therefore, the second objective will be to maximize the number of roe deer at the end of the planning period. This criterion will be done as:

$$\text{Maximize} Z_2 = \sum_{i=2,5} LM_{10,i}^{\infty} \qquad (3)$$

where $LM_{10,i}^{\infty}$ is the amount at the end of the planning period (year t=10) of male roe deer adults.

Sex ratio constraints are enforced during the planning period for each year including the last one. Therefore, the maximization of males implies the maximization of the balanced population. The schema of the program is as follows.

Efficient (Z_1, Z_2)

$$HM_{t,i} + LM_{t,i} = K1 - LM_{t-1,i-1} \quad ;i=2,5 \tag{4}$$
$$HF_{t,i} + LF_{t,i} = K2 - LF_{t-1,i-1} \quad ;i=2,5 \tag{5}$$
$$LM_{t,1} = K3 \; LM_{t-1,0} \tag{6}$$
$$LF_{t,1} = K4 \; LF_{t-1,0} \tag{7}$$
$$LM_{t,0} = K5 \; LF_{t-1,*} \tag{8}$$
$$LF_{t,0} = K6 \; LF_{t-1,*} \tag{9}$$

HM stands for hunted male during the year, LM live male at the end of the year, HF is hunted female and LM live female: the subscript stands for year "t", age "i" (each of the population classes). The initial population, year $t=1$ starts from the known values for $t=0$.   Subscript "*" stands for ages $i=2$ to 5 years. Parameters K1 to K6 defining population dynamics are taken from Zamora (1994) and they are different for roe deer and red deer. The system is fully defined with equations (4) to (9). We have included some constraints for the sex ratio as suggested by Davis (1967).

The system is constrained by the nutritional requirements by age, sex and species that have to be smaller or equal to supply (computed from land cover ecosystems). We have included nutritional constraints for both energy and nitrogen requirements. There are additional constraints to set the minimum level at the end of the planning period for both species defined by adult males, being more than five for roe deer and 80 for red deer. The model was solved by an IBM-PC with HIPERLINDO[c].

$$\sum_{i=1,5} NR_1(HM_i + LM_i) + \sum_{i=1,5} NR_2(HF + LF) \leq ND_i) \tag{10}$$
$$(year \; t=1,10)$$

$$0{,}5 \; LM \leq LF \leq 2 \; LM \tag{11}$$
$$LM^{ce} \geq 80 \tag{12}$$
$$LM^{cc} \geq 5 \tag{13}$$

Model was firstly designed to be solved by Multiobjective Programming. The use of MOP allows the analyst to understand the conflicts and tradeoffs between the different objectives in conflict, which is the first step in analyzing a system. The solution of this problem gives us a small efficient set defined only by two corner points.

Table 1. Efficient set for MOP problem

Point	Income (Z_1)[1]	Population (Z_2)
A	110,7	5
B	103,0	61

(1) Million Sp. Ptas.

The tradeoff analysis shows that we can substantially increase the number of roe deers giving up only a little income. On a purely theoretical basis there is no way to incite us to adopt "B" as the best solution because, by definition, the efficient set is formed by all non-dominated solutions, therefore both solutions are efficient and non-dominated.

Nevertheless, we decided to analyze further this system, based on our previous experience of game management via spreadsheet analysis. Thus, the second step in the analysis of this problem was to redefine it as a Lexicographical Goal Program to be described in the next section.

5. Goal Program for Wildlife Management

The first phase of the system analysis showed that ecological and economic objectives have only a small conflict, this outcome is explained when we consider the following system characteristics: a) the red deer population is fully exploiting the resources; therefore, avoiding hunting some roe deer does not have a significant impact on economic performance, b) as the value of hunting red is equal to hunting roe deer, the substitution of the endangered species for the commercial one has a scant influence on income.

We decided to use a lexicographical goal programming as follows:

$$\text{Lex } \{(n1,n2) ; (\sum_{t=1,10} n_t+p_t)\} \tag{14}$$

subject to

$$Z_1+n_1-p_1 = 100 \tag{15}$$
$$Z_2+n_2-p_2 = 60 \tag{16}$$
$$LM^t_{ce2}+LM^t_{ce3}+LM^t_{ce4}+LM^t_{ce5}+n_t-p_t = 400; \ t=1,10 \tag{17}$$
constraints (4) to (13)

Equation (14) is the objective function, (15) sets the economic goal (in million Sp.Ptas.), (16) is the number of roe deer adults living at the end of year 10.

The second level of goals (eq.17) attempted to obtain a scheme in which every year the number of adult males of red deer is as close as possible to 400. The target "400" is the average obtained in the MOP program explained in the previous section.

A solution to the first problem achieved the fulfilment of the first set of goals (i.e. $n_1=0$; $n_2=0$). Therefore, there are alternatives and we went on to minimize the second level of goals that gave us a unique solution. The analysis of the solution is interesting (table 2 and figure 2). We can see how this solution has achieved a steady level of population, satisfying the first level of goals.

Table 2. Population parameters for different solutions.

Year	Sol. A		G.P. Solution		Sol. B	
	Red Deer	Roe Deer	Red Deer	Roe Deer	Red Deer	Roe Deer
1	424	20	400	20	425	20
2	426	15	400	26	426	26
3	465	11	400	29	465	32
4	470	6	400	34	460	39
5	400	9	400	41	381	45
6	437	16	400	49	406	52
7	450	23	400	57	411	59
8	443	30	400	66	405	68
9	324	26	400	75	298	76
10	199	19	212	84	190	86

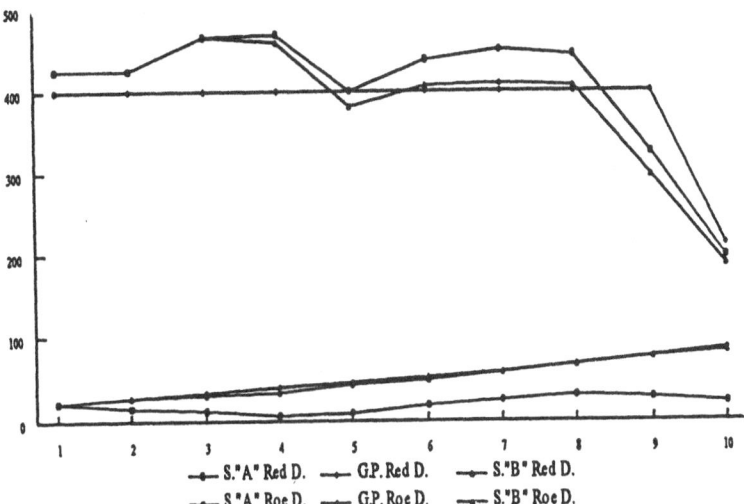

Figure 2. Population of deers (Cervus elaphus) for different criteria.

6. Concluding remarks

The aim of this paper was to enlarge the field of applications of MOP and GP paradigm. We describe a case study of wildlife management in Southern Spain where commercial and endangered species compete for the land. The model includes an analysis of conflict between economic and ecological objectives and suggests criteria to measure both dimensions of the managed system.

The main finding of the model is the successful application to this relatively new field, and the complementarity between GP and MOP. Also, we have defined several different criteria related to wildlife management (bio-diversity, endangered species preservation, or economic performance) some of them with applications scarcely found in the literature.

We suggest that the field of natural resources management in general, and wildlife management in particular, should be treated with the help of MCDM paradigm. It is important to remark that MOP and GP should not be treated as conflicting approaches but treated as a complementary model for solving the real world decision problems.

References

Bare, B.B.; Mendoza, G. (1988) Multiple Objective Forest Land Management Planning: An Illustration. European Journal of Operational Research 34: 44-55.

Berbel, J. (1992) Gestión de explotaciones cinegéticas (ciervos) en Sierra Morena: influencia de la tasa de interés. Investigación Agraria. Economía, 7 (1) 1-23.

Berbel, J. & Zamora, R. (1992) Modelo de evolución de poblaciones de ciervo en Sierra Morena. Evolución de una finca sin caza de hembras ni restricciones alimenticias. Montes, 27: 39-44.

Davis (1967) Dynamic Programming for Deer Management Planning. Journal of Wildlife Management, 21: 667-679.

Ignizio, M (1994) "Who killed GP" MOPGP94. Portsmouth. UK (unpublished)

Jordi, K. & Peddie, D. (1988) A Wildlife Management Problem: a Case Study in Multiple-Objective Linear Programming. Journal of the Operational Research Society, 39: 1011-1020.

López Giménez, R. (1972) Aportaciones al estudio de las explotaciones de caza mayor en Sierra Morena occidental. Archivos de Zootecnia 21, 82:167-91.

Romero, (1989) "Modelos de planificación forestal: una aproximación desde el análisis multicriterio" Revista de estudios Agro-Sociales. num. 147 (en-mar.1989).

Romero, C. & Rehman, T. (1987) Natural Resources Management and the Use of Multiple Criteria Decision-Making Techniques: a Review. Eur. R. Agric. Econ. 14, 61-89.

Romero, C. & Rehman, T. (1989) Multiple Criteria Analysis for Agricultural Decisions, Elsevier, Amsterdam.

Zadnik M. & Stirn, L. (1990). Adaptive dynamic model for optimal forest management. Forest Ecology and Management, 31, 167-188.

Zamora, R. (1994) Modelos decisionales multicriterio en planificación forestal. Una aplicación a la ordenación integral de montes. Tesis Doctoral. Universidad de Córdoba.

Zamora-Lozano, M., Barasona Mata, J. y Rodríguez Berrocal, J. (1976). Contribución al estudio del potencial productivo y cinegético de áreas marginales de la provincia de Córdoba. Bases técnicas para un estudio económico. Boletín de la Estación Central de Ecología 5: 31-43.

Zamora-Lozano, M., Medina Carnicer, R. y Barasona Mata (1979) Establecimiento de un Modelo de Programación Lineal para ordenar cualitativamente la población de Ciervos. Archivos de Zootecnia 28, 161-66

Flight Trajectory Optimization by Goal Programming with Fuzzy Objectives

Shinji Suzuki

Department of Aeronautics and Astronautics, the University of Tokyo, 3-7-1, Bunkyo-ku, Hongo, Tokyo 113,Japan

Abstract. A sequential goal programming (GP) approach is considered for nonlinear optimal control problems. This paper puts emphasis on flight trajectory problems which have no feasible solutions satisfying all constraints. The prioritization of multiple goals and fuzzy objectives are utilized to deal with unsatisfied constraints and multiple performance requirements. In order to apply a conventional simplex algorithm for a nonlinear problem, a linearized problem with respect to a set of discrete control variables is solved sequentially. A rocket ascending problem is shown as numerical examples of a well-defined problem. As a practical ill-defined problem, a takeoff trajectory of a jet airplane through a severe wind called a microburst is investigated.

Keywords. Optimal Control, Goal Programming, Fuzzy Set

1 Introduction

Optimal control is one of the most active research areas of applied mathematics and is utilized in a variety of fields. A flight trajectory problem is a typical application in aerospace engineering. This problem is formulated to determine the admissible control variables which maximize or minimize an objective function while satisfying boundary conditions at the initial and at the terminal time and path constraints on the state variables and control variables.[1] Although numerous solutions approaches (for example, calculus of variations, maximum principle, and dynamic programming) have been developed, recent investigations have revealed the advantages of using a mathematical programming approach. Mathematical programming algorithms handle inequality constrains in an efficient manner, and do not have as excessive computer-storage requirements as dynamic programming.[2] Whereas theoretical studies consider mathematically well-defined problems, real-world problems may have no admissible control variables satisfying all constraints. In these ill-defined cases, some constraints must be relaxed to accommodate other objectives using various decision making methods.[3] This paper will explore the solution method for ill-defined flight trajectory problems by using a goal programming formulation with fuzzy objective functions.

In order to formulate an optimal control problem as an optimization problem, the state variables and control variables are discretized. The methods of discretization are mainly divided into two: 1) the state equations are approximated by difference equations which are regarded as constraints in nonlinear problems,[2,4] and 2) the control variables are approximated by using a set of time functions with discrete parameters and the state equations are integrated numerically.[5] This paper uses the latter approach. The time scale is divided into a set of elements and a control variable in each time element is approximated as a linear or constant function.

As a mathematical programming algorithm, a Goal Programming (GP) formulation is employed. GP was developed to apply linear programming to linear multiple criteria optimization problems.[6~9] By introducing target values (goal values) of achievement for each objective or constraint, GP can find solutions even if a problem is unfeasible. This characteristic of GP is suitable for the present study since the initial presumption of admissible control variables is very difficult, and since GP can offer solutions for ill-defined problems. Linear GP problems can be solved efficiently by a conventional simplex algorithm; for this reason, a successive linearization approach is used for nonlinear GP problems.[10]

Ill-defined problems require multi-criteria decision making method to manage a design trade-off among the unsatisfied constraints and performance requirements. This study utilizes both the prioritization of target goals in a lexicographic GP formulation and a fuzzy set theory. Fuzzy set theories was applied to consider the decision maker's vague ideas or the fuzzy characteristics of problems.[11,12] This study applies the Zimmermann's formulation[12] which uses membership functions with respect to multiple objective functions to present competing goals in the same priority class in GP.

As numerical examples, an ascent trajectory of a rocket is demonstrated as a well-defined problem, and a flight trajectory of a jet transport plane is investigated as an ill-defined problem. The aircraft has no admissible flight paths satisfying terminal conditions due to a severe wind called a microburst.

2 Mathematical Formulation

2.1 Optimal Control Problem

An optimal control problem is defined to search the control variables which minimize or maximize a performance index under several kinds of constraints. This problem is formulated in the following manner.

A system is subjected to the state equations

$$\frac{dx(t)}{dt} = \dot{x}(t) = f[x(t), u(t), t] \tag{1}$$

where $x(t)$ and $u(t)$ are the vector of states and the vector of controls, respectively. At the initial time, t_I, and the final or terminal time, t_F, some states may be specified as

$$\psi_I[x(t_I)] = o, \quad \psi_F[x(t_F)] = o \tag{2}$$

Equations (2) are called boundary conditions.

The upper limits or the lower limits of the state or control variables may be determined during, $t_I \leq t \leq t_F$. This type of constraint is called a path constraint and defined as

$$L_{PC}(t) \leq S[x(t), u(t)] \leq U_{PC}(t) \tag{3}$$

The control variables $u(t)$ is sought to minimize or maximize a given performance index

$$J = \phi[x(t), u(t), t] \tag{4}$$

2.2 Nonlinear Programming Formulation

In order to transform an optimal control problem to a parameter optimization problem, the control variables are approximated by defining discrete parameters. When a set of time functions are introduced, the control variables are

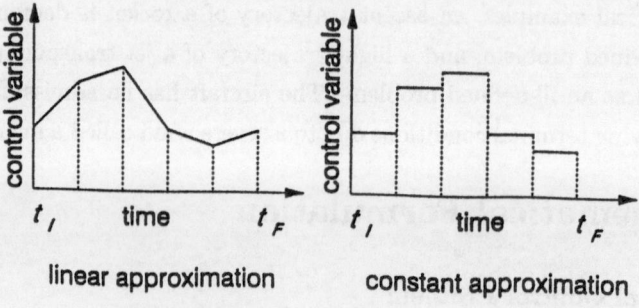

Fig. 1: Approximation of control variable.

represented as

$$u(t) \approx n_1(t)p_1 + n_2(t)p_2 + ... + n_N(t)p_N = N(t)p \tag{5}$$

While any kind of approximation can be utilized, this paper employs a piecewise linear or constant approximation. As shown in Fig. 1, the time scale is divided into elements, and a control variable in each time element is approximated as a linear or constant function.

When all the initial boundary conditions are specified and the terminal time, t_F, is determined, the state variables, $x(t)$, are calculated by integrating the state equations (Eqs. (1)) combined with Eqs. (5). As a numerical integration method, the modified Euler method[3] was applied in this study. Therefore, the terminal boundary condition, the path constraints, and the performance index are written as the functions of the discrete parameters, p. When some of the initial states are unknown, and when the terminal time, t_F, is not given, these unknown variables are incorporated into the parameter vector, p.

Consequently, a trajectory optimization problem can be formulated in the following nonlinear programming problem:

$$\begin{aligned} \text{minimize}: \quad & \phi(p) \\ \text{subject to}: \quad & \psi_I(p) = o \\ & \psi_F(p) = o \\ & L_i \leq S_i(p) \leq U_i \\ & i = 1, 2, ..., N_t \end{aligned} \tag{6}$$

where the path constraints are imposed at a set of discrete times, t_i, ($i = 1, 2, ..., N_t$).

2.3 Goal Programming Formulation

A nonlinear programming algorithm generally requires initial design variables for an iterative search process. In a flight trajectory problem, it is quite difficult to find the initial control variables satisfying both the terminal conditions and the path constraints. Furthermore, there may be no feasible solutions which satisfy all constraints in a real-world flight trajectory problem.

A Goal Programming (GP) approach is suitable for this kind of optimization problem. GP deals with both an objective function and constraints as goals to be achieved and introduces the concept of the "priority level"[7] which represents the importance of each goal. Since the "priority level" means the "preemptive" priority, goals in the high priority level must be satisfied before we consider goals in the low level. By considering constraints as goals in the higher level and by managing an objective function as a goal in the lowest level, feasible control variables are searched at the beginning, and then these variables are optimized without violating constraints. This process can accommodate a design trade-off in ill-defined problems by specifying priority levels for unsatisfied constraints and performance requirements. A penalty function approach can deal with the same situation by introducing some penalty functions to limit constraints and by adding these penalty functions to the original objective function. Note that the penalty function often leads to problems of numerical calculations, and has difficulty in determining the weights introduced in the combined objective function.

In a GP formulation, n_g goal values, g, are defined for both the constraints and objective function which are written as $f(p)$. Deviation variables d^+ and d^- are introduced to measure the overachievement and the underachievement of the target goal in the following way:

$$
\begin{aligned}
\text{subject to}: \quad & f(p) - d^+ + d^- = g \\
& d^+ \geq 0, \ d^- \geq 0 \\
& d_j^+ \cdot d_j^- = 0 \\
& j = 1, 2, ..., n_g
\end{aligned}
\tag{7}
$$

By using the deviation variables, the objective functions to be minimized are defined as

$$w_j^+ d_j^+ + w_j^- d_j^-, \quad j = 1, 2, ..., n_g \tag{8}$$

where w_j^+ and w_j^- are weights which are appropriately chosen to manipulate various types of design requirements: e.g., inequality constraints, equality constraints, and function minimization or maximization.

In many cases, goals are prioritized according to their significance. Priority levels P_l $(l = 1, 2, ..., L)$ are then introduced to make the following objective function:

$$\text{minimize} : \sum_{l=1}^{L} \{P_l[\sum_{j \in I_l}(w_j^+ d_j^+ + w_j^- d_j^-)]\} \tag{9}$$

where I_l denotes a set of goals which are categorized in the same priority level l. P_l is called the "preemptive" priority factor.[7~9] This factor satisfies the relationship of $P_i >>> P_{i+1}$, which implies that the goals at the upper priority level should be considered to be infinitely more important than goals at the lower priority level.

This kind of optimization problem is categorized as preemptive (or lexicographic) goal programming, which can be solved efficiently by a modified simplex algorithm[8] if all of the functions, f, in Eqs. (7) are linear with respect to p. In order to utilize this algorithm, a sequential linearization method is applied in the following manner: 1) the initial values of each variable, p, are assumed, 2) all of the functions in Eqs. (7) are linearized with respect to p, 3) the optimal solutions are obtained within the "move limits" of each design variable, where "move limits" are defined as the upper and lower limits of each design variable[13], and 4) the variables, p, are successively updated until all of the solutions converge. Linearization of the state equations with respect to p can be carried out analytically (see Appendix). Note that analytical deviation of the sensitivities, $\partial x/\partial p$, can eliminate numerical errors due to finite difference and reduce the computational time.

The move limits should be selected which assure the linearity assumption and converge the iteration process in a conventional sequential linear programming approach.[13] Initial values of the move limits which have large values to help reduce the number of iterations are gradually shortened when a parameter changes its sign during the iteration process.[3]

2.4 Fuzzy Programming

In the GP formulation, a design trade-off between goals in the same priority level is manipulated by scaling the weights, w_j^+ and w_j^-. However, the
decision maker may not strictly specify the weighing factors but in a fuzzy
manner. A fuzzy set theory is utilized in the GP formulation.

In a fuzzy environment, the decision maker can define the membership
function, μ_i, for each target value, g_i, $(i = 1, 2, ..., n_f)$. This function has the
value of zero for the worst possible case, the value of one for the best possible
case, and the intermediate value for those cases in between (as shown in Fig.
2). When the membership functions are mapped on a p-μ plane and when
the degree of the decision maker's satisfaction of the solution-set is defined
as a minimum value among all membership functions[12], we have

$$\lambda_D(p) = \text{Min}[\mu_1(p), \mu_2(p), ..., \mu_{n_f}(p)] \tag{10}$$

and "the maximization decision"[12] yields the following Min-Max programming problem. By introducing a independent variable, λ, which represents
the maximum value of $\lambda_D(p)$, the problem is formulated as

$$\begin{aligned}
\text{maximize}: &\quad \lambda \\
\text{subject to}: &\quad \lambda \leq \mu_i(p) \\
&\quad i = 1, 2, ..., n_f
\end{aligned} \tag{11}$$

Equations (11) are incorporated in the GP formulation to represent the fuzzy
"decisions" for goals in the same priority level.

3 Well-defined Problem: Rocket Ascent Tra-jectory

As a well-defined problem, the accent trajectory problem of a simple rocket
which has analytical solutions was studied. A rocket with mass m is guided
by controlling the direction angle, β, of a constant thrust, ma, in a vacuum
and a gravity-free sate as shown in Fig. 3.

The state equations of motion are

$$x = (x_1, x_2, x_3, x_4)^T$$

316

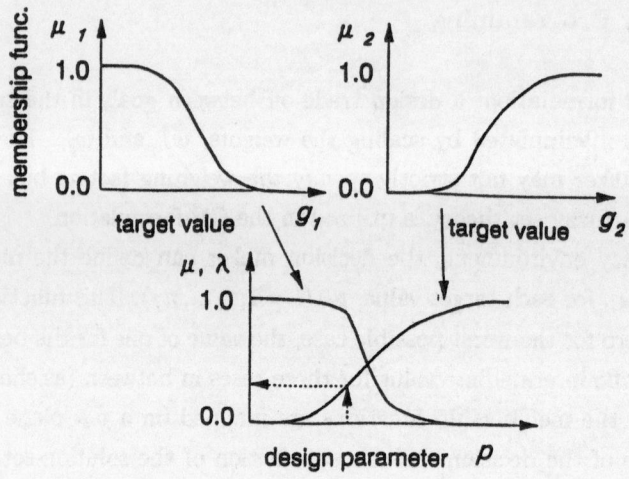

Fig. 2: Membership functions and fuzzy decision.

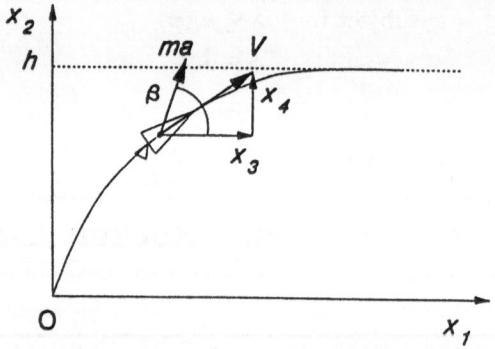

Fig. 3: Rocket ascent trajectory problem.

$$\dot{x} = \begin{bmatrix} 0 & 0 & 1 & 0 \\ 0 & 0 & 0 & 1 \\ 0 & 0 & 0 & 0 \\ 0 & 0 & 0 & 0 \end{bmatrix} x + \begin{bmatrix} 0 \\ 0 \\ a\cos\beta \\ a\sin\beta \end{bmatrix} \tag{12}$$

where x_1 and x_2 are a horizontal range and an altitude, and x_3 and x_4 are velocity components in an inertia coordinate system, respectively. The trust angle β is the control variable.

The rocket is required to attain a given altitude h, at a given time t_F, with zero vertical speed, and maximum horizontal speed. The terminal range $x_1(t_F)$ is not specified.

The boundary conditions and performance index for the problem are

$$\psi_I = x(0) = o \tag{13}$$

$$\psi_F = \begin{bmatrix} x_2(t_F) \\ x_4(t_F) \end{bmatrix} - \begin{bmatrix} h \\ 0 \end{bmatrix} = o \tag{14}$$

$$\phi = -x_3(t_F) \tag{15}$$

The time scale is divided into N elements, and the control variable $\beta(t)$ is approximated as a linear function in each time element. Therefore, the discrete control variables are represented as β_i, $(i = 1, 2, ..., (N+1))$ at nodal points of time elements. The prioritization of goals is defined in the following way: the goals associated with the side constraints of the design variables are the most important class, the goals for the boundary conditions at the terminal time are the second, and the maximization of the final horizontal speed is the third class. Note that the boundary conditions at the initial time are employed automatically in the numerical integration.

The iteration history of the flight path is illustrated in Fig. 4 where h, t_F, a, and N are designated as 100, 20, 1.123972, and 20, respectively. The initial control was a horizontal flight, i.e., the initial values of β_i were set to be 0. In an early stage, the terminal conditions are tried to be satisfied, and then the final horizontal speed is minimized.

Figures 5(a), (b) and (c) show the calculated flight path, the time histories of velocity components, and the control angle time history. In these figures, solid lines are the exact solutions, and "$\bigcirc$" and "$\triangle$" denote the calculated results.

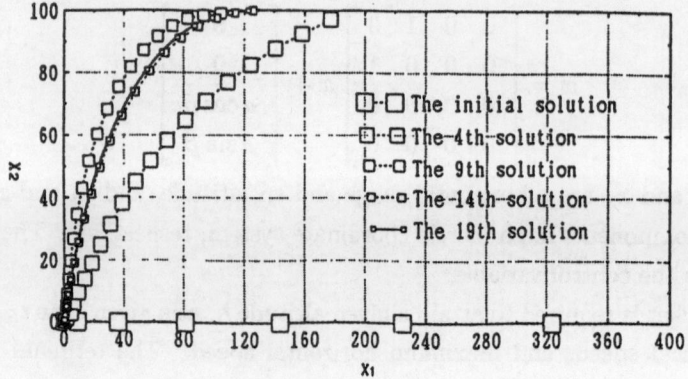

Fig. 4: Iteration history of the flight path.

The relative errors of the performance index versus the number of elements and the number of iterations are presented in Table 1. This table indicates that the presenting method is very stable because the number of iterations is almost the same independently of the number of time elements and the solution accuracy is improved as the number of time elements increases. The criterion of convergence was defined as

$$\sum_i \frac{f_i^{k+1}(p) - f_i^k(p)}{1+ \mid f_i^k(p) \mid} < 1.0E - 4, \tag{16}$$

where the upper subscript means the iteration number.

4 Aircraft Trajectory in Microburst

The problems described in the preceding section are mathematically well-defined; however, real world flight trajectory designs may be formulated as the ill-defined problems in which we can find no flight paths satisfying all boundary conditions or path constraints. Ill-defined problems require multi-criteria decision making method to accommodate objectives and unsatisfied constraints.

4.1 Formulation of Problem

The flight trajectory optimization problem of an aircraft in a microburst is

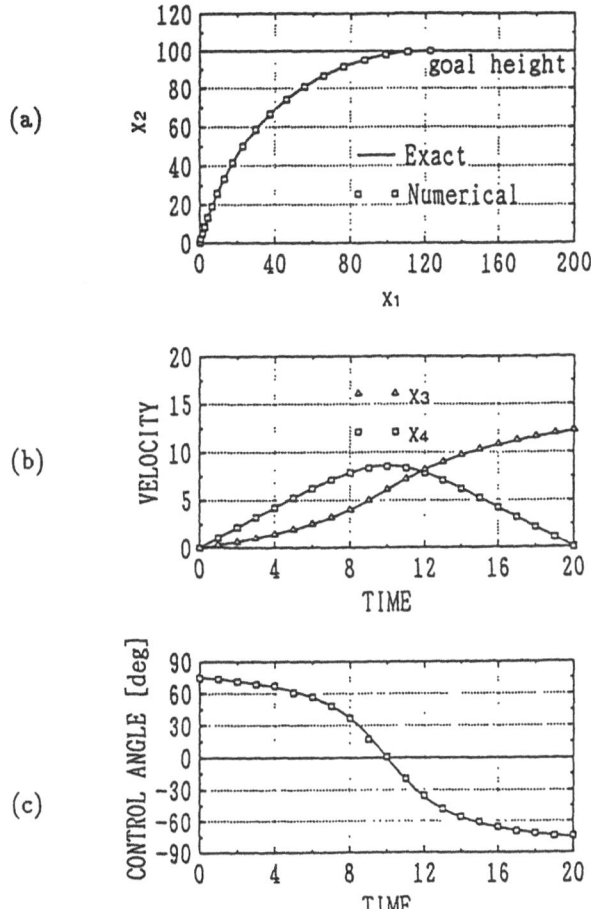

Fig. 5: Solutions of rocket problem. (a) Flight paths x_1 vs x_2. (b) Time histories of velocity components x_3 and x_4 vs t. (c) Time histories of control angle β vs t.

Table 1: Number of iterations and relative error

Number of elements	Number of iterations	Relative error
2	20	3.40E-1
4	17	1.18E-1
6	21	5.73E-2
8	17	3.34E-2
10	23	2,17E-2
16	20	8.71E-3
20	19	5.69E-3
40	28	1.41E-3

considered. Microburst is one of the most hazardous or dangerous wind for an aircraft in takeoff or landing conditions. It contains a descending air which spreads horizontally near the ground. It is reported that there were more than 30 accidents in the last 20 years in the United States. In this study, the takeoff condition of a B-727 aircraft model is investigated as shown in Fig. 6.

By considering an airplane as a point mass m, the equations of motion are written as follows:

$$\dot{V} = \frac{T}{m}\cos(\alpha+\delta) - \frac{D}{m} - g\sin\gamma - (\dot{W}_x\cos\gamma + \dot{W}_h\sin\gamma)$$
$$\dot{\gamma} = \frac{T}{mV}\sin(\alpha+\delta) + \frac{L}{mV} - \frac{g\cos\gamma}{V} + \frac{1}{V}(\dot{W}_x\sin\gamma - \dot{W}_h\cos\gamma)$$
$$\dot{x} = V\cos\gamma + W_x$$
$$\dot{h} = V\sin\gamma + W_h \tag{17}$$

where V and γ and the airspeed and the flight-path angle relative to the wind axis, x and h are the horizontal distance and the vertical height, α is the angle of attack, δ is the thrust inclination, and W_x and W_h are the horizontal and vertical wind components (Fig. 7). T, D, L are the thrust force, the lift, and the drag which are defined as

$$T = A_0 + A_1V + A_2V^2$$
$$D = \frac{1}{2}\rho V^2 SC_D(\alpha)$$

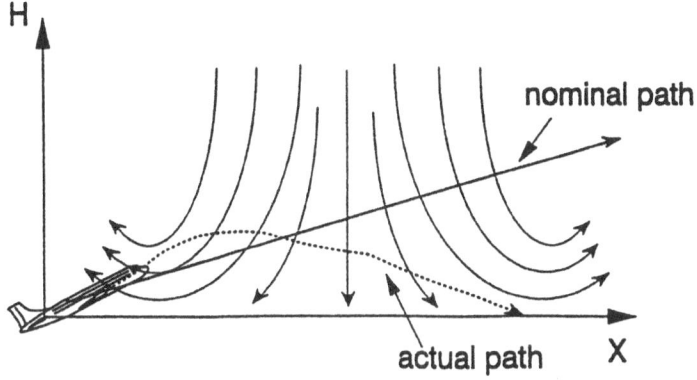

Fig. 6: Takeoff trajectory of a jet airplane in a microburst.

$$L = \frac{1}{2}\rho V^2 S C_L(\alpha)$$
$$C_D = B_0 + B_1\alpha + B_2\alpha^2$$
$$C_L = C_0 + C_1\alpha$$
$$\alpha \leq 16 \text{ degree} \tag{18}$$

where ρ and S are an air density and a wing area. There are four state variables: V, γ, x, and y, and one control variable, α. A stall condition requires the upper limit of α. A piecewise constant approximation of the control variable is employed. Numerical data of the takeoff configuration of a B-727 aircraft model are used in this study.[14,15]

Figure 8 illustrates the wind velocity profile used in this study. The maximum horizontal and down velocities are assumed to be 40 ft/s and 10 ft/s for a sever case and 20 ft/s and 5 ft/s for a moderate case.

4.2 Climb Rate Maintaining Formulation

Previous researches formulate this problem as a clime rate maintaining problem in which the maximum value of the deviation between a clime rate, $\dot{h}$, and an intended clime rate, $\dot{h}_0$, is minimized. The performance index is defined as

$$\min \max |\dot{h} - \dot{h}_0| \tag{19}$$

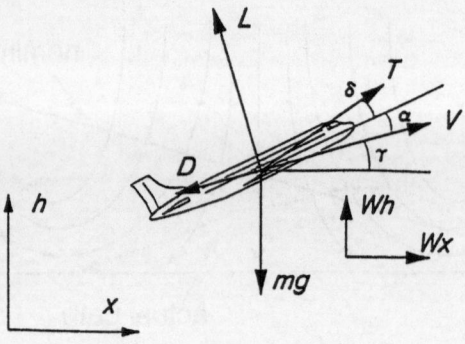

Fig. 7: Coordinate system and forces acting on an airplane.

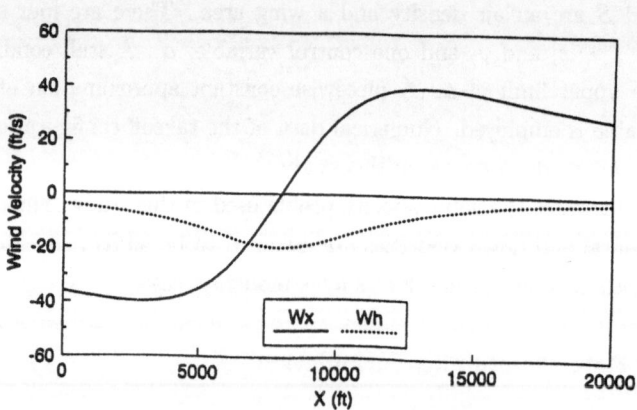

Fig. 8: Wind velocity components of a microburst.

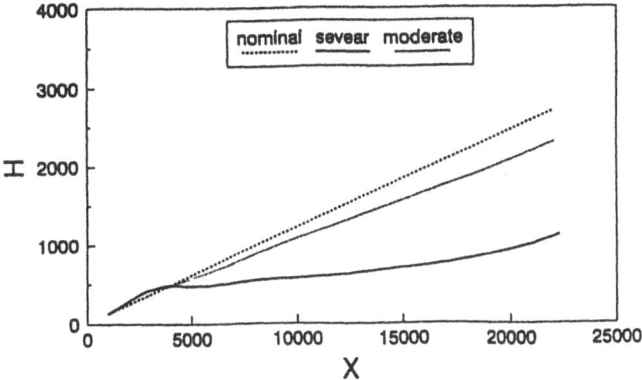

Fig. 9: Optimal trajectories of clime rate maintaining problem.

Miele et al.[14] transformed this problem into the following Bolza problem:

$$\min J = \int_0^{t_f} (\dot{h} - \dot{h}_0)^q \, dt \tag{20}$$

where q is the large positive even number.

Figure 9 shows the calculated flight paths obtained from the performance index in Eq. (20) in which the terminal time t_f is 80 s and the time axis is divided into 20 elements. This figure indicates that the clime rate maintaining strategy can produce a satisfactory performance for a moderate microburst, but an airplane lost the terminal height in a severe microburst.

4.3 GP Formulation

Since the climb maintaining strategy cannot give an adequate terminal state for a severe microburst, Zhao and Bryson[15] proposed that the maximization of the potential energy at the terminal time should be selected. This paper will apply GP formulation to this problem.

In a GP formulation, the prioritization of goals has an important role. Two cases are considered as shown in Table 2. The first three classes are the same in the two cases: the first class is related to the side constraints of control variables, the second class requires that the minimum height is greater than equal 0, and the third class demands that the terminal path angle holds the nominal value. In case 1, the fourth class requires that terminal airspeed sustains the nominal value, and the final class demands that the terminal

Table 2: Prioritization of goals in GP formulation

Class	Case 1	Case 2
1	Side constraints	←
2	Min $h \geq 0$	←
3	Terminal flight path angle $\gamma(t_f) = 0.122$ rad	←
4	Terminal speed $V(t_f) = 277$ ft/s	Terminal height $h(t_f) = 2700$ ft
5	Terminal height $h(t_f) = 2700$ ft	Terminal speed $V(t_f) = 277$ ft/s

height should reach the nominal height. In case 2, the last two classes are exchanged.

Figures 10(a), (b) demonstrate the obtained flight paths and airspeed time histories. In case 1, the upper four classes are achieved, but the terminal height as the final class does not reach the target value. In case 2, the last two classes are not satisfied: i.e., the terminal height in the class 4 tries to increase by greatly scarifying the terminal speed in the class 5. In both cases, not all terminal states achieve the target values simultaneously. Therefore, this problem is ill-defined in a mathematical sense when the terminal states are specified. Note that both flight paths lack smoothness near the ground because the clime rate performance was not considered.

4.4 GP Formulation with Fuzzy Objectives

In the final example, fuzzy "decisions" are incorporated into the GP formulation. In this study, membership functions are considered for the following three design criteria: 1) the clime-rate performance, 2) the terminal height, and 3) the terminal airspeed. Figure 11 shows the membership functions utilized in this study. While any type of membership function is available, the function used in this study is given as

$$\mu(g) = \frac{1}{2} \tanh[(g - a_0)a_1] + \frac{1}{2} \tag{21}$$

where two parameters, a_0 and a_1, determine the shape of the function. This

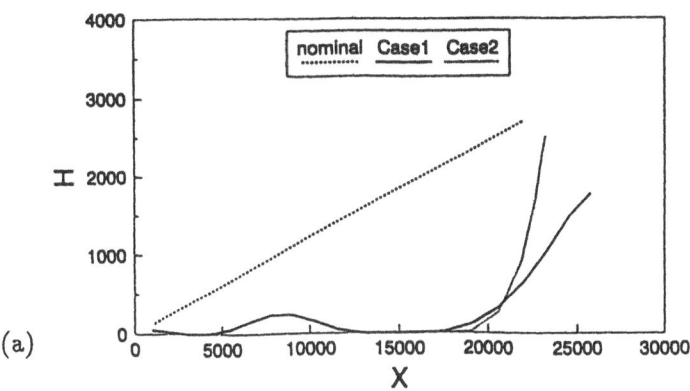

(a)

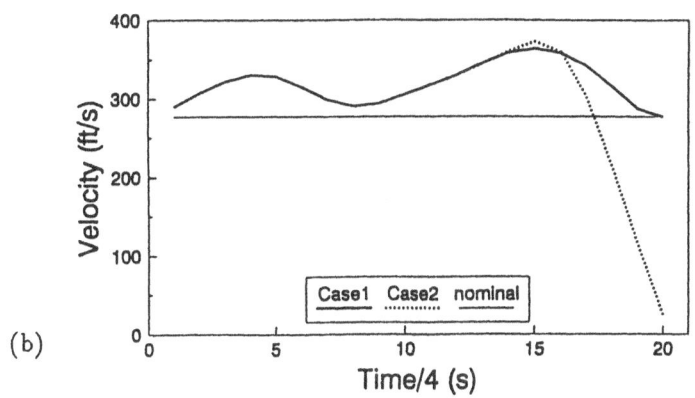

(b)

Fig. 10: Optimal trajectories of GP problem. (a)   Flight paths. (b) Flight speed histories.

326

Table 3: Prioritization of goals in GP formulation with fuzzy objectives

Class	Objective
1	Side constraints
2	Minimum height ≥ 0
3	Terminal gamma $\gamma(t_f) = 0.122$ rad
4	Fuzzy objectives
	f_1 : performance index
	f_2 : terminal height
	f_3 : terminal speed

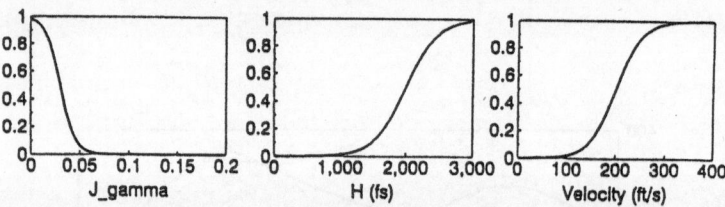

Fig. 11: Membership functions for clime rate, terminal height, and terminal speed.

function is convenient for the sequential linearization approach since it becomes a linear function as follows:

$$\mu^* \equiv \tanh^{-1}(2\mu - 1) = (g - a_0)a_1 \tag{22}$$

Table 3 indicates the prioritization in this formulation. The upper three classes are the same as the case in Table 2. The last class represents the fuzzy "decisions".

Figure 12 shows the optimal flight path and the time histories of velocity and the control variable (the angle of attack α). The obtained solutions demonstrate that the requirements for the minimum height and the terminal path angle are strictly satisfied, and the fuzzy requirements reach a compromise. The flight path is smooth and has acceptable terminal height and airspeed. This indicates that the goals in the upper classes are managed as hard constraints, and that the fuzzy objectives in the last class can represent the design trade-off of multiple criteria.

The iteration process of this problem is depicted in Fig. 13 which represents the degree of satisfaction in three fuzzy objective functions. Note that the requirement for the clime rate is the most difficult requirement, and the terminal height is the next one. These two requirements are tried to be satisfied by sacrificing the requirement for the terminal airspeed.

5 Conclusion

A sequential goal programming (GP) formulation was presented for the numerical optimization of a nonlinear optimal control problem. This paper focused on practical flight trajectory problems which have no feasible solutions satisfying all the terminal constraints due to severe disturbances. Numerical examples considered the takeoff flight trajectory of a jet airplane through a microburst. In our formulation, design goals were prioritized according their significance, and the design trade-off among vague design criteria was represented by membership functions associated with each target value. The hard constraints such as non-crash requirements were categorized into the upper classes, and the soft constraints and the performance objectives which may include fuzzy decisions were dealt with in the lower classes. The numerical examples demonstrated the practicality of the GP formulation with fuzzy objectives. Finally, although a deterministic model was used in this paper, the problem has stochastic characteristics in both disturbances and the system properties. An optimal control problem including stochastic parameters is considered worthy of further study.

Appendix: Sensitivity Matrix of State Variables

A dynamic system has the following form:

$$\frac{dx}{dt} = \dot{x} = f(x, u, t) \tag{23}$$

The control variables u are parameterized as

$$u(t) = N(t)\, p \tag{24}$$

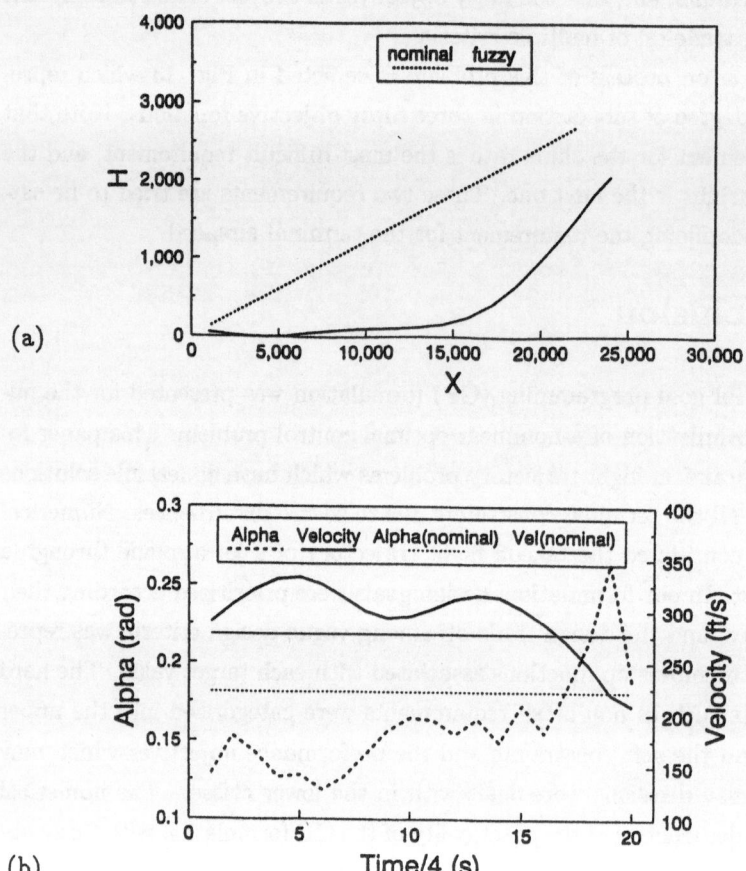

Fig. 12: Optimal trajectories of GP problem with fuzzy objectives. (a) Flight paths. (b) Flight speed and control variable histories.

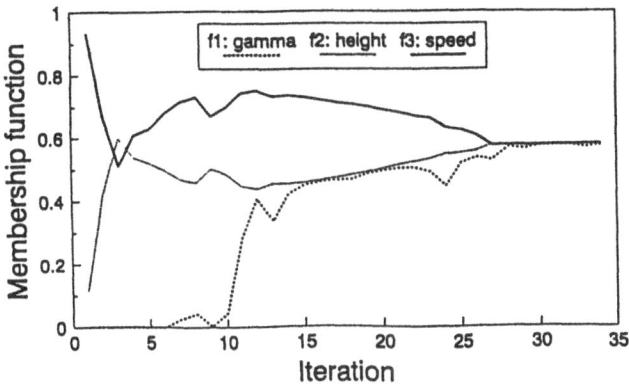

Fig. 13: Membership function values vs iteration number.

where p is the design parameter vector.

By differentiating the state equations with reference to the design parameter vector p, the differential equations associated with a matrix $x_p = [\partial x / \partial p]$ are given as

$$\frac{dx_p}{dt} = \left[\frac{\partial f}{\partial x}\right] x_p + \left[\frac{\partial f}{\partial u}\right] N \qquad (25)$$

The sensitivity matrix x_p can be obtained by integrating Eqs. (25).

References

[1] Bryson, A. E., Jr., and Ho, Y. C., *Applied Optimal Control*, Blaisdell, Waltham, MA, 1969.

[2] Tabak, D., and Kuo, B. C., *Optimal Control by Mathematical Programming*, Pentice-Hall, Englewood Cliffs, NJ, 1971.

[3] Suzuki, S., and Yoshizawa, T., "Multiobjective Trajectory Optimization by Goal Programming with Fuzzy Decisions" *Journal of Guidance, Control, and Dynamics*, Vol. 17, No. 2, 1994, pp. 297-303.

[4] Hargraves, C. R., and Paris, S. W., "Direct Trajectory Optimization Using Nonlinear Programming and Collocation," *Journal of Guidance, Control, and Dynamics*, Vol. 10, No. 4, 1987, pp. 338-342.

[5] Williamson, W. E., "Use of Polynomial Approximation to Calculate Sub-

optimal Controls," *AIAA Journal*, Vol. 9, No. 11, 1971, pp. 2271-2273.

[6] Charnes, A., and Cooper, *Management Models and Industrial Applications of Linear Programming*, Vol. 1, John Wiley & Sons, New York, 1961.

[7] Ijiri, Y., *Management Goals and Accounting for Control*, Rand McNally, Chicago, 1965.

[8] Lee, S. M., *Goal Programming for Decision Analysis*, Auerbach Publishers, Philadelphia, 1972. pp. 15-35.

[9] Ignizio, J. P., *Linear Programming in Single- & Multiple-Objective Systems*, Prentice-Hall, NJ, 1982.

[10] Suzuki, S., and Yonezawa, S., "Simultaneous Structure/Control Design Optimization of a Wing Structure with a Gust Load Alleviation System," Journal of Aircraft, Vol. 30, No. 2, 1993, pp. 268-274.

[11] Bellman, R. E., and Zadeh, L. A., "Decision Making in a Fuzzy Environment," *Management Science*, Vol. 17, 1970, pp. 141-164.

[12] Zimmermann, H. J., "Decision and Optimization of Fuzzy Systems," *International Journal of General Systems*, Vol. 2, No. 4, 1976, pp. 209-215.

[13] Vanderplaats, G. N., *Numerical Optimization Techniques for Engineering Design*, McGraw-Hill, New York, 1984. pp. 155-157.

[14] Miele, A., Wang, T., and Melvin, W. W., "Optimization and Acceleration Guidance of Flight Trajectories in a Windshear," *Journal of Guidance, Control, and Dynamics*, Vol. 10, No. 4, 1987, pp. 368-377.

[15] Zhao, Y., and Bryson Jr., A. E., "Optimal Paths Through Downbursts", *Journal of Guidance, Control, and Dynamics*, Vol. 13, No. 5, 1990, pp. 813-818.

An Application of Goal Geometric Programming to Equipment Replacement Under Fuzziness

Valentin Vitanov[1], Nikolai Mincoff[2], Tanya Vladimirova[3]

[1] Department of Mechanical and Manufacturing Engineering, University of Portsmouth, PO1 3DJ,UK

[2] Systems Optimization Laboratory,Technical University,Sofia 1756, Bulgaria

[3] Department of Electrical and Electronics Engineering,University of Surrey,Guildford,GU2 5XN,UK

Abstract. This paper presents an investigation of equipment replacement policy problem using an approach which solve geometric programming problems with fuzzy parameters and fuzzy goals. Almost all types of equipment are subject to deterioration over time through age and usage and therefore decisions regarding the need for replacement and cost reduction are required. Such an analysis is based on existing functional relationships between costs (overhaul, replacement, and operating costs) and predictor variables (replacement, overhaul, and inspection intervals). Replacement strategies are directly connected with deterioration and its permanent dynamic changes. Deterioration is continuously subjected to influence of many factors that cannot always be predicted by company engineers and experts. As a result the optimisation model is based on uncertain of data as well as target values and model parameter values are usually specified by experts, which implies for subjectivity. The approach adopted here solves a series of multiobjective linear programming problems in order to build fuzzy regression models to represent such responses. The obtained geometric programming problem with fuzzy parameters is completed with fuzzy target values of the cost objective function. Deterministic transformation of the model as well as its computational treatment are considered.

Keywords. Geometric programming, Goal Programming, Fuzzy Regression.

1 Fitting Fuzzy Regression Models

The theory of possibility first proposed by Zadeh (1978) is essentially a methodology for representing some available information (specially when it is vague or incomplete) in terms of possibility distributions $\pi_A(u)$, and for deducing from it the values of variables of interest. The quantity $\pi_A(u)$ represents the degree of possibility of the assignment $A=u$, some values of u being more possible than others, according to what is known. The *possibility measure* and its dual counterpart *necessity measure* play an important role in practical applications of the possibility theory. Following the concept of these measures here we use the well-known indices for ranking fuzzy intervals in accordance with (Dubois and Prade, 1988):

$$Pos(A \geq B) = \sup \{ \min \{ \mu_A(y), \mu_B(z) \} \, / y \geq z \};$$

$$Pos(A > B) = \sup_y \{ \inf_z \{ \mu_A(y), 1 - \mu_B(z) \} \, / y \geq z \};$$

$$Nes(A \geq B) = \inf_y \{ \sup_z \{ \max \{ 1 - \mu_A(y), \mu_B(z) \} \, / y \geq z \} \};$$

$$Nes(A > B) = \inf \{ \max \{ 1 - \mu_A(y), 1 - \mu_B(z) \} \, / y \geq z \},$$

$$Nes(A > B) = \inf \{ \max \{ 1 - \mu_A(y), 1 - \mu_B(z) \} \, / y \geq z \},$$

where $\mu_A(.)$ and $\mu_B(.)$ are membership functions of fuzzy intervals A and B, and Pos or Nes are short for possibility and necessity, respectively.

In this section, an extension for fuzzy regression analysis based on possibilistic distribution and measures is given. In contrast with existing approaches (see Sakawa and Yano, 1992) we apply different possibility and necessity indices for sample rows. This allows a generalized formalization approach to treatment of fuzzy observations and also gives an opportunity to consider inequality and set-inclusive conclusions like '*cost will increases by* **no more than about 2%**', for example. In addition we propose a new aggregation criteria for several conclusions (observations) at the same predicative basis, a situation common when dealing with a group of experts as another extension for fuzzy regression analysis. The suggested extension does not add computational difficulties to the regression model building process, however it is necessary to solve an ordinary Linear Programming (LP) problem as in the most of the former approaches.

Let us consider the problem of fuzzy regression function fitting, when cost C_i of type i is based on fuzzy observations or fuzzy expert judgements over time T.

$$\widetilde{C}_i = a_{0i} T^{\widetilde{b}_{1i}}, \quad i=1,...,n.$$

The *log*-transformed model for the cost function $\tilde{C}(T)$ can be presented as follows:

$$\tilde{y}_i = \tilde{b}_{0i} + \tilde{b}_{1i} x, \ i = 1,...,n,$$

where $\tilde{y}_i = ln(\tilde{C}_i)$, $\tilde{b}_{0i} = ln(\tilde{a}_{0i})$, $x=ln(T)$ and $\tilde{b}_{1i}$ is a fuzzy estimated power, respectively. We assume that parameters $\tilde{y}_i$ and $\tilde{b}_{ij}$ in the above functions are presented as fuzzy numbers, i.e. $\tilde{y}_i = (y_i, \gamma_i^L, \gamma_i^R)_{LR}$ and $\tilde{b}_{ij} = (b_{ij}, \beta_{ij}^L, \beta_{ij}^R)_{LR}$, respectively.

As a convenient basis to illustrate our extension of fuzzy regression analysis problem we follow the well developed modelling approach (Sakawa and Yano, 1992). Based on the indices given above and on well known transformations the following multiple objective mathematical programming problem is obtained

$$min \ J(\beta) = \sum_{k \in N_a} s_{km} w_{km} [(\beta_{0k}^L + \beta_{0k}^R) + (\beta_{1k}^L + \beta_{1k}^R) |x_k|]$$

$$max \ \alpha$$

subject to

$$Pos(\tilde{y}_{ik} = \tilde{b}_{0i} + \tilde{b}_{1i} x_{ik}) \geq \alpha, \ k = 1,..., N_{EQ}^P$$

$$Pos(\tilde{y}_{ik} \geq \tilde{b}_{0i} + \tilde{b}_{1i} x_{ik}) \geq \alpha, \ k = 1,..., N_{GE}^P$$

$$Pos(\tilde{y}_{ik} > \tilde{b}_{0i} + \tilde{b}_{1i} x_{ik}) \geq \alpha, \ k = 1,..., N_{GT}^P$$

$$Pos(\tilde{y}_{ik} \leq \tilde{b}_{0i} + \tilde{b}_{1i} x_{ik}) \geq \alpha, \ k = 1,..., N_{LE}^P$$

$$Pos(\tilde{y}_{ik} < \tilde{b}_{0i} + \tilde{b}_{1i} x_{ik}) \geq \alpha, \ k = 1,..., N_{LT}^P$$

$$Nec(\widetilde{y}_{ik} \geq \widetilde{b}_{0i} + \widetilde{b}_{1i}x_{ik}) \geq \alpha, \ k = 1,...,N_{GE}^{N}$$

$$Nec(\widetilde{y}_{ik} > \widetilde{b}_{0i} + \widetilde{b}_{1i}x_{ik}) \geq \alpha, \ k = 1,...,N_{GT}^{N}$$

$$Nec(\widetilde{y}_{ik} \supset \widetilde{b}_{0i} + \widetilde{b}_{1i}x_{ik}) \geq \alpha, \ k = 1,...,N_{SI}^{N}$$

$$Nec(\widetilde{y}_{ik} \leq \widetilde{b}_{0i} + \widetilde{b}_{1i}x_{ik}) \geq \alpha, \ k = 1,...,N_{LE}^{N}$$

$$Nec(\widetilde{y}_{ik} < \widetilde{b}_{0i} + \widetilde{b}_{1i}x_{ik}) \geq \alpha, \ k = 1,...,N_{LT}^{N}$$

$$Nec(\widetilde{y}_{ik} \subset \widetilde{b}_{0i} + \widetilde{b}_{1i}x_{ik}) \geq \alpha, \ k = 1,...,N_{SE}^{N}$$

$$\text{and} \ 1 \geq \alpha \geq 0, \ \beta_{ij}^{L}, \beta_{ij}^{R} \geq 0, \ i = 0,...,n,$$

where notations $N_{EQ}^{P}, N_{GE}^{P}, N_{GT}^{P}, N_{LE}^{P}, N_{LT}^{P}$, and $N_{GE}^{N}, N_{GT}^{N}, N_{SI}^{N}, N_{LE}^{N}, N_{LT}^{N}, N_{SE}^{N}$ in constraints denote the *Pos* and *Nec* subsamples of original fuzzy sample with N observations, such that

$$N_{EQ}^{P} \cup N_{GE}^{P} \cup N_{GT}^{P} \cup N_{LE}^{P} \cup N_{LT}^{P} \cup N_{GE}^{N} \cup N_{GT}^{N} \cup N_{SI}^{N} \cup N_{LE}^{N} \cup N_{LT}^{N} \cup N_{SE}^{N} = N$$

and s_{km} is a sign function, defined by the Pos and Nes indices (see Sakawa and Yano, 1992). The last expression supposes that there may be several fuzzy observations at the same base set and these can be treated with different indices (e.g. 'soft' and/or 'hard' expert judgements). As a first objective function we design an additive weighting utility function. The above extended treatment of the fuzzy regression model fitting process gives an additional opportunity to consider the case of several experts with different weights of experience for the fuzzy rule sample. The only restriction in this kind of fuzzy rule interpolation procedure is that consensus between experts for the structure of the modelled relationship is needed. Other approaches for instance may be considered in order to build up an utility function as well as to find compromise solutions to this multiobjective mathematical programming problem. Having obtained fuzzy estimated parameters for the set of cost functions an optimization model for matching costs/time periods for equipment replacement with fuzzy parameters can be built.

2 Multiobjective Fuzzy Geometric Programming

Based on the fuzzy extension of the *log*-linear function developed above this section will describe some basic definitions and useful transformations in the conventional and fuzzy Geometric Programming (GP). Let us consider an *exp*-transformed multiobjective GP problem as it is presented below:

$$min \ \{G_k(x,a) - H_k(x,b)\}, \ k = 1,...,q$$

subject to

$$\{G_k(x,a) \ R_k \ H_k(x,b)\}, \ k = q+1,...,p$$

$$x_i^U \geq x_i \geq x_i^L \geq 0, \ i = 1,...,m,$$

where

x is a *m*-vector of decision variables with simple bounds x_i^L, x_i^U, $i=1,...,m$; *a* and *b* are vectors of coefficients, *q* is the number of objective functions; and *p-q* is the number of constraints, R_k, $k=q+1,...,p$, represents relation operators such as $\leq$, $\geq$, $<$, $>$, $=$, $\subset$; and functions $G_k(x,a)$ and $H_k(x,b)$, $k=1,...,p$ are defined as follows:

$$G_k(x,a) = \sum_{i \in I_{G_k}} exp(\sum_{j=0}^{m} a_{ij} x_j),$$

$$H_k(x,b) = \sum_{i \in I_{H_k}} exp(\sum_{j=0}^{m} b_{ij} x_j), \ k = 1,...,p$$

where

$I_{Gk}=\{m_{Gk}, m_{Gk}+1,...,n_{Gk}\}$, $k=1,...,p$, $m_{G1}=1$, $m_{G2}=n_{G1}+1$, ..., $m_{Gp}=n_{Gp-1}+1$, $n_{Gp}=n_G$, and the same notations applies to I_{Hk}.

Here n_G and n_H are the total numbers of monomial terms in the problem, respectively.

It is assumed that parameters a_{ij} and b_{ij} in the above functions are presented as *LR* fuzzy numbers, i.e. $\tilde{a}_{ij} = (a_{ij}, \alpha_{ij}^L, \alpha_{ij}^R)_{LR}$ and $\tilde{b}_{ij} = (b_{ij}, \beta_{ij}^L, \beta_{ij}^R)_{LR}$, respectively.

2.1 Possibilistic constraints

The focal points of this approach are transition techniques of possibilistic constraints and objective functions. which are based on the above indices as well as on the monotonicity properties of $G_k(x,a)$ and $H_k(x,b)$ (following Dubois,1988) in order to propose deterministic equivalents of possibilistic constraints for different kinds of possibilistic relations R_k, $k=p+1,...,q$. This transformation allows us to use several strategies to build the δ-feasible solutions set $X(\delta)$, where δ is a common level cut for fuzzy sets.

Definition:

i)   Soft equality constraint possesses δ-weak feasibility iff:
$$WF_k(\delta) = Pos(G_k(x,A) = H_k(x,B)) \geq \delta, \, k \in I^{WF};$$

ii)   Soft inequality constraint possesses δ-very weak feasibility iff:
$$VWF_k(\delta) = Pos(G_k(x,A) \leq H_k(x,B)) \geq \delta, k \in I^{VW};$$

iii)   Soft inequality constraint possesses δ-medium weak feasibility iff:
$$MWF_k(\delta) = Pos(G_k(x,A) < H_k(x,B)) \geq \delta, k \in I^{MW};$$

iv)   Hard set-inclusive constraint possesses δ-strong feasibility iff:
$$SF_k(\delta) = Nec(G_k(x,A) \subset H_k(x,B)) \geq \delta, k \in I^{SF};$$

v)   Hard inequality constraint possesses δ-medium strong feasibility iff:
$$MSF_k(\delta) = Nec(G_k(x,A) \leq H_k(x,B)) \geq \delta, k \in I^{MS};$$

vi)   Hard inequality constraint possesses δ-very strong feasibility iff:
$$VSF_k(\delta) = Nec(G_k(x,A) < H_k(x,B)) \geq \delta, k \in I^{VS};$$

where *iff* is short for *if and only if* and $I^{WF}, I^{VW}, I^{MW}, I^{SF}, I^{MS}$, and I^{VS} denote the subsets of index set I_{pq}, such that $I^{WF} \cup I^{VW} \cup I^{MW} \cup I^{SF} \cup I^{MS} \cup I^{VS} = I_{pq}$.

In contrast to the existing approaches we consider the more realistic case, where the model builder can define different types of constraints. Such a classification of constraints in classes of feasibility is an informal problem, which can be solved, on the basis of interaction between the model builder and decision maker.

It is important to mention here, that some of constraints may belong simultaneously to different classes of feasibility, for instance, in the case of group decision making processes.

From the definition and properties of fuzzy intervals at δ-level it follows that the δ-feasible solutions set can be characterised by the following proposition.

Proposition: The set $X(\delta)$ of δ-feasible solutions to possibilistic GP problem is defined as follows:

$$X(\delta)= X^{WF}(\delta) \cap X^{VW}(\delta) \cap X^{MW}(\delta) \cap X^{SF}(\delta) \cap X^{MS}(\delta) \cap X^{VS}(\delta),$$

where $X^{WF}(\delta), X^{VW}(\delta), X^{MW}(\delta), X^{SF}(\delta), X^{MS}(\delta),$ and $X^{VS}(\delta)$ are sets, resulting from deterministic counterparts of equations in definition, as follows:

i) Set of δ-weak feasible solutions:

$$x \in X^{WF}(\delta)= \{WF_k(\delta), k \in I^{WF}\} : =$$
$$G_k\{x,[a_{ij}^R + \alpha_{ij}^R R_{a_{ij}}^{-1}(\delta)]\} \ge H_k\{x,[b_{ij}^L - \beta_{ij}^L L_{b_{ij}}^{-1}(\delta)]\}$$

$$\text{and } G_k\{x,[a_{ij}^L - \alpha_{ij}^L L_{a_{ij}}^{-1}(\delta)]\} \le H_k\{x,[b_{ij}^R + \beta_{ij}^R R_{b_{ij}}^{-1}(\delta)]\}, k \in I^{WF}.$$

ii) Set of δ-very weak feasible solutions:

$$x \in X^{VW}(\delta)= \{VW_k(\delta), k \in I^{VW}\} : =$$

$$G_k\{x,[a_{ij}^L - \alpha_{ij}^L L_{a_{ij}}^{-1}(\delta)]\} \le H_k\{x,[b_{ij}^R + \beta_{ij}^R R_{b_{ij}}^{-1}(\delta)]\}, k \in I^{VW}.$$

iii) Set of δ-medium weak feasible solutions:

$$x \in X^{MW}(\delta)= \{MW_k(\delta), k \in I^{MW}\} : =$$

$$G_k\{x,[a_{ij}^R + \alpha_{ij}^R R_{a_{ij}}^{-1}(1-\delta)]\} \le H_k\{x,[b_{ij}^R + \beta_{ij}^R R_{b_{ij}}^{-1}(\delta)]\}, k \in I^{MW}.$$

iv) Set of δ-strong feasible solutions:

$$x \in X^{SF}(\delta)= \{SF_k(\delta), k \in I^{SF}\} : =$$

$$G_k\{x,[a_{ij}^L - \alpha_{ij}^L L_{a_{ij}}^{-1}(1-\delta)]\} \ge H_k\{x,[b_{ij}^L - \beta_{ij}^L L_{b_{ij}}^{-1}(\delta)]\}$$

$$\text{and } G_k\{x,[a_{ij}^R + \alpha_{ij}^R R_{a_{ij}}^{-1}(1-\delta)]\} \le H_k\{x,[b_{ij}^R + \beta_{ij}^R R_{b_{ij}}^{-1}(\delta)]\}, k \in I^{SF}.$$

v) Set of δ-medium strong feasible solutions:

$$x \in X^{MS}(\delta)= \{MS_k(\delta), k \in I^{MS}\} : =$$

$$G_k\{x,[a_{ij}^L - \alpha_{ij}^L L_{a_{ij}}^{-1}(\delta)]\} \le H_k\{x,[b_{ij}^L - \beta_{ij}^L L_{b_{ij}}^{-1}(1-\delta)]\}, k \in I^{MS}.$$

vi) Set of δ-very strong feasible solutions:

$$x \in X^{VS}(\delta) = \{VS_k(\delta), k \in I^{VS}\} := $$

$$G_k\{x,[a_{ij}^R + \alpha_{ij}^R R_{a_{ij}}^{-1}(1-\delta)]\} \leq H_k\{x,[b_{ij}^L - \beta_{ij}^L L_{b_{ij}}^{-1}(1-\delta)]\}, k \in I^{rs}.$$

A very important from the computational point of view is that a result has been obtained: such that deterministic equivalents of possibilistic GP constraints preserve their signomial properties.

2.2 Multiobjective Geometric Programming under fuzziness

2.2.1 Fuzzy goal geometric programming

The concept of geometric programming (GP) problem with fuzzy objective functions was first considered by Verma (1990). Let the fuzzy GP model is written as follows

$$\{G_K(x,a) - H_K(x,b)\} \underset{\sim}{\leq} s_K, \quad k = 1,...q$$

subject to

$$x \in X(\delta),$$

where $X(\delta)$ presents the δ-feasible solution set as defined above in '3.1', s_K is an aspiration level for k-th objective function and notation (~) represents the fact that the model parameters are fuzzy numbers. The cost function value belongs to a fuzzy threshold defined as $S_K = (-\infty, s_K, 0, r_K)_{LR}$, where r_K is a subjectively chosen constant of admissible violation of the k-th constraint.

There exist several operators for aggregation of fuzzy objective functions. In order to choose a suitable aggregation technique decision making (DM) should take into account the specifics of the problem under consideration. Also, a sensible compromise between aggregative operator properties and computational resources for its numerical treatment is needed (Zimmermann 1991),.

If the well known 'minimum' operator is used, we yield the following crisp model

$$min \ y$$

subject to

$$G_K(x,a) \leq H_K(x,b) + r_K y - r_K \delta + s_K, \quad k = 1,...q$$

$$x \in X(\delta), \quad y > 0,$$

where the quantity δ guarantees that y is positively defined. This model is a signomial GP problem with the same degree of difficulty (see Verma, 1990). The 'product' operator has useful computational properties: it has compensatory possibilities saving at the same time the posynomial character of the original multiobjective posynomial GP problem. From application point of view, especially with regard to optimal engineering design and process planning problems, conjunction aggregation rules give an opportunity to combine closely economic and technological indices. The 'product' operator approach gives the following crisp model:

$$min \; \prod_{k=1}^{q} \frac{[G_K(x,a)-H_K(x,b)-s_K]}{r_K}$$

subject to

$$x \in X(\delta).$$

After some transformations we obtain the following program formulation:

$$min \; \prod_{k=1}^{q} y_K$$

subject to

$$G_K(x,a) \le H_K(x,b)+r_K y_K - r_K \delta_K + s_K \;, \quad k=1,...,q$$
$$x \in X(\delta), \; y_K > 0, \; k=1,...,q$$

where here the negligibly small quantities δ_K are chosen such that $s_K > \delta_K r_K$, $k=1,...,q$, and extra variables y_k, $k=1,...,q$ are constrained to satisfy $y_k > 0$ and

$$\delta_K + \frac{[G_K(x,a)-H_K(x,b)-s_K]}{r_K} \le y_K$$

The above problem is a conventional signomial GP problem with q new variables y_k and it can be solved applying some of existed numerical algorithms.

2.2.2 Objective functions with possibilistic coefficients

In this subsection we briefly discuss defuzzification approaches to GP objective functions with possibilistic coefficients presented as fuzzy intervals. Deterministic equivalents are obtained based on the monotonicity property of posynomial functions with respect to their monomial coefficients and exponents (Sotirov and Mincoff, 1994).

'Pessimistic' strategy at δ-level:

$$G_K^P(x,a(\delta)) - H_K^P(x,b(\delta)) =$$
$$\sup_a(G_K(x,A(\delta)) - \inf_b(H_K(x,B(\delta))) =$$
$$G_K(x,(a_{ij}^R + \alpha_{ij}^R R_{a_{ij}}^{-1}(\delta))) - H_K(x,(b_{ij}^L - \beta_{ij}^L L_{b_{ij}}^{-1}(\delta)))$$

where δ-level is assigned by the DM, and $A(\delta)$ and $B(\delta)$ are crisp intervals for δ-level cut.

'Optimistic' strategy at δ-level:

$$G_K^O(x,a(\delta)) - H_K^O(x,b(\delta)) =$$
$$\inf_a(G_K(x,A(\delta)) - \sup_b(H_K(x,B(\delta))) =$$
$$G_K(x,(a_{ij}^L - \alpha_{ij}^L L_{a_{ij}}^{-1}(\delta))) - H_K(x,(b_{ij}^R + \beta_{ij}^R R_{b_{ij}}^{-1}(\delta)))$$

Both above equations are signomial functions which allow to use standart GP aggregation approaches and computational techniques.

3 AN APPLICATION TO EQUIPMENT REPLACEMENT

The solution of the equipment replacement problem has been addressed by several modelling techniques in the past thrithy years. One of them, primarily applied to the solution of finite time horizon replacement problem, is dynamic programming (Bellman, 1955). An original approach exploits computational effectiveness of geometric programming (Cheng, 1992, Mincoff, 1993) to solve the infinite time horison for equipment replacement problem. In our presentation of the application of fuzzy goal geometric programming we follow the model description of the replacement problem, given in Cheng (1992), where the original objective is to minimize the total cost of replacing and operating the equipment. Comments about fuzzy extentions of the model are given, when it is necessary.

The optimal replacement policy for the deteriorate equipment components suppose a building up process of an optimization problem. The following assumption are made for the total cost of replacements and operating the equipment between the overhaul and inspection intervals (Cheng, 1992):

i)   Equipment and its components and parts do not fail but are only subject to deterioration over time due to age and wear.

ii)   Overhaul and inspection cost I is an increasing function of the inspection interval expressed as

$$I = aT^b \, ,$$

where T is the overhaul and inspection interval, a and b are fitting parameters.

iii)   Replacement cost R is a decreasing function of the inspection interval expressed as

$$R = \frac{c}{T^d} \, ,$$

where c and d are fitting parameters.

iv)   Operating and running cost M of the equipment is an increasing function of the replacement interval expressed as:

$$M = et^f \, ,$$

where t is the replacement interval, e and f are fitting parameters. In the above cost functions all fitting parameters are nonnegative defined.

The fitting process we applied for the cost elements given above is fuzzy regression analysis. Based on 'What-if" scenarios, generated from knowledge experts we collected so named 'fuzzy samples' $I(T)$, $R(T)$, and $M(t)$, with observations for I, R, as well as for M presented as fuzzy intervals. Then the LP-based procedure for building fuzzy log-linear regression models is applied (see Section 2). This way we received estimation for the fitting parameters a, b, c, d, e, and f represented as fuzzy intervals. There is only one restriction introduced by the fitting process. It is necessary that all the parameters are estimated under an equal degree α.

Since we assume that the replacement planning horizon is infinite and the replacement cycle repeats indefinitely throughout the plannig horizon, the analysis is concentrated on one typical overhaul cycle. Then the total cost $C(t,T)$ between two overhaul and inspection instants may be written as presented below (Cheng, 1992):

$$C(t,T) = aT^{b-1} + cT^{-d}t^{-1} + et^{f-1},$$

where based on Extension principle, formulated by Zadeh (1978) the total cost C is a resulted fuzzy interval.

As a first constraint we consider $t << T$, which is implicitly assumed in the above derivation that , so that the residual effect of an incomplete t at the end of T on the replacement and operating costs is minimal and could be ignored for simplicity:

$$tT^{-1} \leq E,$$

where E is a fuzzy threshold.

Also we include in the optimisation model constraints on cost components, such as

$$I(T) \leq I^U,$$

$$R(T) \leq R^U,$$

$$M(t) \leq M^U,$$

as well as sample bounds on decision variables

$$T^L \leq T \leq T^U,$$

$$t^L \leq t \leq t^U.$$

Some comments are necessary on the assigning of the bounds and ranges for the above constraints as an important step in the model building process. Analogically to the classical regression analysis if a regression-based function is optimised, then results are valid only on the decision variable range for the regression sample. This requirement defines some preliminary restrictions to our problem. For the decision variables they are given below:

$$T^L \geq max(T_I^L, T_R^L) \quad \text{and} \quad T^R \leq min(T_I^R, T_R^R)$$

$$t^L \geq t_M^L \quad \text{and} \quad t^R \leq t_M^R$$

A fuzzy extended version of Cheng's numerical example (Cheng, 1992):

Let

a = (5000, 5000, 0, 0)   b = (2, 2, 0.7, 0.7)   c = (2500, 2500, 0, 0)

d = (1, 1, 0, 0)   e = (10000, 10000, 0, 0)   f = (2, 2, 0.05, 0.05)

We have considered

Overhaul Cost as a POS α very week feasible constraint

Replacement Cost as a POS α very strong feasible constraint

Operating Cost as a NES α very strong feasible constraint

Criteria with an Pessimistic strategy at α level

The results for four different values of α are listed in Table 1.

Table 1. Simulation results for different values of α

α	0.25	0.50	0.75	0.80
I	21620	22137	22671	22780
R	11788	11725	11662	11650
M	11270	11439	11519	11535
C(t,T)	34025	35123	36221	36440
T	2.12	2.13	2.14	2.15
t	1.05	1.06	1.07	1.08

4 Concluding Remarks

In this paper we have formulated an approach for curve fitting and optimisation based on fuzzy presented sample of 'knowledge' data. An extended approach to log-linear fuzzy regression analysis recently developed is applied as a technique for knowledge interpolation and and building up models for power-type cost functions. A general case of Geometric Programming problem with fuzzy parameters as well as fuzzy goals, constructed upon fuzzy-fitted monomial terms is demonstrated. In order to illustrate a practical application an equipment replacement policy problem has been considered. Although, further detailed development would be necessary for the applied aspect, we believe that such fuzzy geometric programming would become efficient tool for analysing the real world systems in situations where the fuzziness of data and human subjective judgement influence is critical factor for the correctness of the decision making process.

References

Bellman, R. (1955) Equipment replacement policy, *SIAM Journal*, Vol.3, pp.133-136.

Cheng, T.C.E. (1992) Optimal replacement of ageing equipment using geometric programming, *International Journal of Production Research*, Vol.30, No.9, pp.2151-2158.

Dubois, D. and H.Prade (1988) Possibility Theory, Plenum Press, New York.

Mincoff, N.C. (1993) Fuzzy Geometric Programming: Model Building and Engineering Applications, Working Paper TUSOL#793, Systems Optimisation Laboratory, Technical University, Sofia, Bulgaria.

Sakawa, M. and H.Yano (1992) Fuzzy linear regression and its applications, in: *Fuzzy Regression Analysis*, Ed. by J.Kacprzyk and M.Fedrizzi, Omnitech Press, Warsaw and Physica-Verlag, Heidelberg, pp.61-80

Sotirov, G. and N.Mincoff (1994) Multiobjective Possibilistic Geometric Programming: Methods and Applications, in: *Fuzzy Optimization: Recent Advances*, Ed. by M. Delgado et al., Physica-Verlag, Heidelberg.

Verma, R.K. (1990) Fuzzy Geometric Programming with Several Objective Functions, Fuzzy Sets and Systems, Vol.35, pp.115-120.

Zadeh, L.A. (1978) Fuzzy Sets as a Basis for a Theory of Possibility, *Fuzzy Sets and Systems*, Vol.1, pp.3-28.

Zimmermann, H.-J. (1991) Fuzzy Set Theory and Its Applications, 2nd Ed., Kluwer Academic Publ., Boston.

An Exploration of Linear and Goal Programming Models in the Downstream Oil Industry

M. Tamiz[1], S. J. Mardle and D. F. Jones
University of Portsmouth, UK.

Abstract

This paper presents a comparison of Linear and Goal Programming methods applied to the downstream oil industry. A realistic, hypothetical model is developed. Results are presented and conclusions are drawn.

Key words: linear programming, goal programming, distribution

1 Introduction

Linear programming (LP) is a well recognised tool for the solution of distribution problems such as that of the downstream oil industry. In this paper we set out to explore the possibility of using goal programming(GP) techniques to model and solve this class of problems. GP is a multi-objective approach to solving real-life problems and despite being a powerful tool it is found in practice to be used comparatively little. In our opinion this is due to two main factors, (i) the general lack of knowledge required to build effective GP models, and (ii) the scarcity of commercial large-scale GP solvers.

Sear [5], describes an outline of the use of linear programming in such an industry. Using these ideas to formulate an LP and a GP, the authors have developed a realistic, but hypothetical, model to optimise a network for the distribution of fuel to customers throughout England and Wales.

The remainder of this paper is divided into six sections; section 2 gives a brief explanation of goal programming, section 3 describes the formulation of the general model, sections 4 and 5 describe the specific application of linear and goal programming to this model, section 6 gives results and section 7 draws conclusions and discusses further ideas for investigation in the models developed.

[1]Correspondence: Dr M.Tamiz, School of Mathematical Studies, University of Portsmouth, Mercantile House, Hampshire Terrace, Portsmouth, Hants. PO1 2EG.
EMail(Int): TAMIZM@cv.port.ac.uk

2   Goal Programming - A Brief Explanation

Goal programming is a powerful tool for multiple-objective decision making. The simplest approach is to give each objective a target (goal) value to be achieved. The unwanted deviations from all these target values are then minimised as a weighted euclidean sum. This is known as weighted goal programming (WGP), see [7]. The standard algebraic form for a WGP is given as:

$$Min \ z = \sum_{i=1}^{k}(w_{i_n} n_i + w_{i_p} p_i)$$

Subject to,

$$f_i(\mathbf{x}) + n_i - p_i = b_i \quad , i = 1 \ldots k$$

$$\mathbf{x} \in C_s$$

where $f_i(\mathbf{x})$ is a linear function (objective) of $\mathbf{x}$, and b_i is the target value for that objective. Variables n_i and p_i represent the negative and positive deviations from this target value, and w_{i_n} and w_{i_p} are the respective weights attached to these deviations in the achievement function z. A deviational variable which we are indifferent to is given a weight of zero. An optional set of rigid constraints, C_s, may also be included.

An amendment to this basic model is the relaxing of the goal target from a point to an interval. Any points within this interval do not incur a penalty, but deviations from either end of the interval are penalised. This technique is known as goal interval programming [1].

3   The Model

The general model of the downstream oil industry considered here is in the form of a multi-level transportation problem with constraints at the intermediate nodes.

Crude oil must first be purchased and/or collected from a *site*, from where it must be transported to a *refinery*, a task usually performed by pipeline or tanker as sites and refineries are commonly situated on the coast. Fuel types, represented primarily by petrol, diesel and fuel oils, are subsequently manufactured and transported individually to various *depots*. This is frequently done by rail, road and pipeline due to the depots' locations often being inland. It is here, at the depot, where any brand identifying chemicals are added to the products, and the finished fuel types are then distributed to the *customer zones*. A diagramatic representation of this is given below:

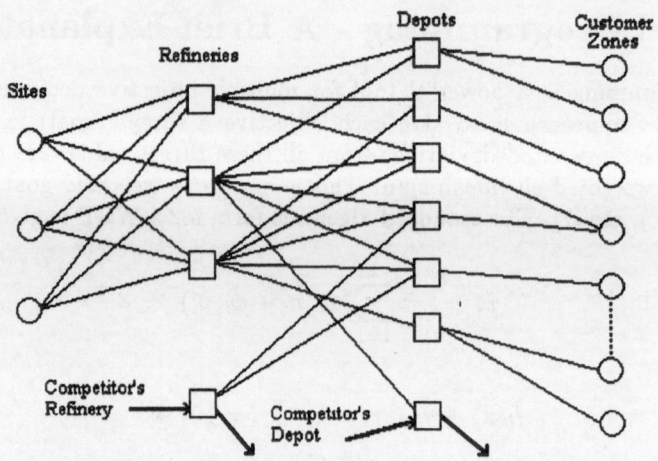

The requirements of each customer zone must be satisfied for each fuel type and they can be regarded as being the centre of a predetermined subarea of the whole distribution area. Therefore, the smaller the subarea in comparison to the complete area, the more accurate any analysis will be concerning customer zones and also the larger the model will be. It should be noted that in practice each customer will be supplied with all fuel types from the same depot.

Deals may be struck with competitors, whereby a common agreement is formed to supply a given fuel to each other's depots. The costs of these trades must be equal as this fuel is not bought and sold but simply exchanged. To do this, the relative costs for each fuel must be agreed beforehand. This is the riskiest link of the network, due to differences in company structure and oil handling. Trading must therefore be kept at a reasonable level, say around 20% of total demand.

The model used here for discussion has the following components:

- 3 sites

- 3 refineries

- 6 depots

- 20 customer zones

- 1 competitor

- 3 fuel types

All routes are made feasible except certain depot to zone distribution links, which are considered impractical from the outset.

This problem is then modelled using the MPL (mathematical programming language) system [3].

4 Linear Programming Approach

The linear programming model is concerned with just one objective, in this case to minimise total cost in the distribution network. The general form of such an LP model is:

$$Min \; z = \sum_{j=1}^{n} c_j x_j \qquad (1)$$

Subject to,

$$\sum_{j=1}^{n} a_{ij} x_j = b_i \qquad , i = 1 \ldots m \qquad (2)$$

where $x_j \geq 0$, c_j are the associated variable costs for $j = 1 \ldots n$, and a_{ij} is a set of scalar coefficients.

To maintain viability, both refineries and depots must keep their total workload between certain limits, i.e. for example between 33% and 66% in the case of refineries, and between 20% and 80% in the case of depots. This makes sure that each respective node is not overworked or underworked. These are hard constraints, since they have to be satisfied to achieve optimality.

The constraint set (2) is constructed of:

- Site purchasing constraints

- Site to refinery transportation constraints

- Refinery production constraints

- Refinery to depot transportation constraints

- Refinery to competitor's depot transportation constraints

- Competitor to depot transportation constraints

- Depot handling constraints

- Depot to zone transportation constraints

- Zone requirement constraints

The linear programming model produced is of the following size:

Constraints	Variables	Non-zeros	Density
134	231	660	2%

A more complete mathematical representation of the constraints is provided in appendix A.

5 Goal Programming Approach

The goal programming model differs to the above LP model, as it is concerned with more than just the single cost objective. Minimising total cost is important, although consideration can now be given to minimising deviations from minimum and maximum production levels at refineries and depots. Thus the model can now consider a possible overtime payment and possible undertime occurancies at the respective nodes if necessary. To do this, goal interval programming techniques [1] are introduced in order to determine the modeller's utility towards production rates. The graph below shows the case for refinery production rates, (depots are considered in the same manner);

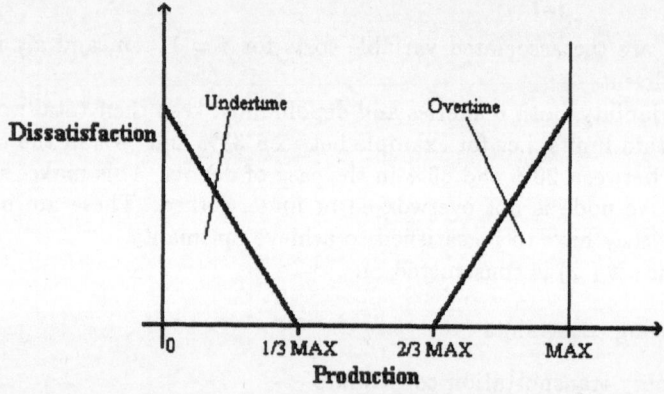

The soft constraints described above, now make the model more flexible as both over-production and under-production are allowed at a cost. Information such as this can be analysed in greater detail than that given by the simple hard constraints of LP, as can be seen in the given results (see section 6). Minimisation of competitor trading can also be included as a goal in the model.

All other constraints in the model are handled in the same way as in the linear programming case.

The goal programming model produced is of the following size:

Objectives	Variables	Non-zeros	Density
136	271	940	3%

6 Results

The linear programme was solved using the commercially available FortLP package [4], a large scale linear and integer programme solver. The goal programme was solved using GPSys, a large scale goal programme solver, currently in development by the authors at the University of Portsmouth.

Five different LP and GP models are developed, to show the effects of modelling an oil distribution network under different scenarios. These models have the characteristics shown in table 1 (concerning the supply capacities and the capacity limits at refineries and depots respectively, and the customer demand level):

Model	Refinery Supply	Refinery Limits	Depot Supply	Depot Limits	Customer Demand
A	Medium	33%-66%	High	20%-80%	Medium
B	Low	33%-66%	Medium	35%-65%	Medium
C	Medium	40%-60%	Medium	20%-80%	High
D	High	40%-60%	High	35%-65%	Low
E	Low	33%-66%	Low	20%-80%	High

Table 1: Refinery, Depot and Customer Levels

For model A, an LP and two differently weighted GP models, GP1 and GP2, are produced. GP1 weighs all the objectives equally, whereas GP2 places a greater weight on the minimisation of the competitor trading objective.

The results from the LP and the WGP models are given in table 2 and table 3.

Model	Total Cost $(\times 10^6)$	Competitor Trading
A (LP)	45.391	500,000
A (GP1)	45 062	500,000
A (GP2)	46 883	212,000
B (LP)	49.392	87,500
B (GP1)	45.297	500,000
C (LP)	56.703	500,000
C (GP1)	56 593	500,000
D (LP)	INFEASIBLE	
D (GP1)	37.000	500,000
E (LP)	58 269	500,000
E (GP1)	56 930	500,000

Table 2: Model Optimum Costs and Trading Levels

Model	Refinery Production			Depot Capacity					
	1	2	3	1	2	3	4	5	6
A (LP)	33.33%	50.74%	66.66%	·20%	20%	20%	20%	20%	55.81%
A (GP1)	33.33%	50.74%	66.66%	20%	19.76%	10.5%	10.08%	29.03%	64.81%
A (GP2)	33.33%	60.44%	66.66%	30.59%	19.76%	15.83%	10.08%	47.22%	70.49%
B (LP)	64.15%	66.66%	66.66%	35%	35%	35%	35%	35%	35%
B (GP1)	47.22%	57.07%	87.66%	21.55%	19.76%	27.67%	10.08%	29.03%	59.69%
C (LP)	50.49%	60%	60%	61.73%	20%	20%	20%	44.87%	80%
C (GP1)	48.1%	80%	63.22%	57.14%	10.29%	20%	7.77%	67.03%	87.7%
D (LP)	INFEASIBLE								
D (GP1)	31.12%	33.33%	33.33%	17.24%	15.81%	8 4%	8.06%	23.33%	40.93%
E (LP)	66.66%	66.66%	66.66%	80%	29.32%	28.06%	20%	70.84%	80%
E (GP1)	56.41%	71.09%	81 82%	80%	17.82%	28.5%	11.32%	80%	100%

Table 3: Refinery and Depot Operating Levels

7 Conclusions and Further Research

The results in model A show that depots are generally running at their lower limits which implies that an over-supply of depot capacity exists. The LP shows that most depots are operating at their lower bound capacities, due to the hard constraints. Both GP1 and GP2, however, do not reach these goals thus showing which depots are expensive to maintain at their lower level of production and providing more information as to their least-cost production level in the absence of under-production regulations. Relaxing the constraints on depot operating levels, in model A, achieved a cash saving of 329,000 units (0.87% of the total cost) in GP1. Restricting the competitor trading levels, however, in GP2 proved more costly eventhough less risk was taken, which would suggest that such trading produces considerable cash benefits. The refineries in model A were running within the given operating level bounds, suggesting that they are maintaining an appropriate level of production.

The refineries in the LP of model B are running at their upper bounds, and all of the depots are operating at their lower bounds of 35%. The GP, however, shows that strategically breaking these bounds can offer a substantial cash saving, 4,095,000 units (8.29% of the total cost). This relaxation also allows maximum competitor trading.

Model C offers the least cash benefit, comparing the LP against the GP, in all of the situations tested. This is due to the fact that the LP solution falls comfortably between the refinery and depot operating limits, and so the GP can only improve the cost slightly, but it still gives further information as to the scenario's desired working levels.

Combining the low customer demand with the high supply required at both refineries and depots in model D, the LP proved infeasible as neither the refineries or depots could operate within the hard constrained levels. The GP, however, could find a solution to this problem by relaxing the bound restrictions. The lower bounds which could not be achieved are immediately noticeable. The GP meets all customer demand. A solution therefore exists, but is not obtainable by the LP of model D, so LP and GP total costs cannot be compared.

Again, the GP of model E shows a more favourable cost than the LP - a total saving of about 2 million units (6.98% of the total cost). This is because in the GP, the more attractive refineries can be given a higher operating level, instead of forcing them all to work at their upper bound levels. The same is true of the depots, but there is one depot operating at 100% which in practice is probably not desirable. It is possible to use preference modelling techniques in the GP [6, 2] to overcome this difficulty.

Further investigation in this area could include: the effects of closing a depot, refinery or competitor trading link; the introduction of penalty functions for modelling the effects of different dissatisfaction levels for refinery production levels and depot operating levels; the introduction of binary (0-1) variables to LP and GP models preventing a customer zone receiving fuel from more than one depot.

References

1 Charnes A., and Collomb B. (1972); 'Optimal economic stabilization policy: linear goal-interval programming models', *Socio- economic Planning Sciences*, **6**, 431-435.

2 Jones D.F. and Tamiz M. (1995); 'Expanding the Flexibility of Goal Programming via Preference Modelling Techniques', *Omega, Int. J. Mgmt Sci.*, **23**, No. 1, 41-48.

3 Maximal Software (1991); 'MPL modelling system release 2.0'.

4 Mitra G., and Tamiz M. (1988, rev. 1990); 'FortLP manual', NAG Ltd.

5 Sear T.N. (1993); 'Logistics planning in the downstream oil industry', *Journal of the Operational Research Society*, **44**, No. 1, 9-17.

6 Tamiz M., and Jones D.F. (1993); 'Preference modelling in goal programming', School of Mathematical Studies, University of Portsmouth.

7 Tamiz M., Jones D.F., and El-Darzi E. (1993); 'A review of goal programming and its applications', presented to APMOD93, Budapest, Hungary.

Appendix

(A) Linear Programming Model

$$Min\ cost = \sum_i p_i \sum_j x_{ij} + \sum_i \sum_j c_{ij} x_{ij} + \sum_j \sum_k c'_{jk} y_{jk} + \sum_j \sum_l \sum_k c''_{jlk} z_{jlk} +$$

$$\sum_l \sum_k c'''_{lk} \sum_j z_{jlk} + \sum_k \sum_l \sum_m c''''_{klm} w_{klm} + 0.5 \left(\sum_j \sum_k T_{jk} t_{jk} + \sum_l \sum_k F_{lk} f_{lk} \right)$$

subject to,

$$\sum_j x_{ij} \leq MAXSITE_i \quad , i = 1 \ldots 3$$

$$\sum_j x_{ij} \geq MINSITE_i \quad , i = 1 \ldots 3$$

$$\sum_i x_{ij} - (0.002 y_{j1} + 0.0025 y_{j2} + 0.004 y_{j3}) = 0 \quad , j = 1 \ldots 3$$

$$y_{jk} \leq MAXPROD_{jk} \quad , j = 1 \ldots 3, k = 1 \ldots 3$$

$$\sum_k y_{jk} \leq 0.66 \sum_k MAXPROD_{jk} \quad , j = 1 \ldots 3$$

$$\sum_k y_{jk} \geq 0.33 \sum_k MAXPROD_{jk} \quad , j = 1 \ldots 3$$

$$\sum_j z_{jlk} + f_{lk} - \sum_m w_{klm} = 0 \quad , k = 1 \ldots 3, l = 1 \ldots 6$$

$$\sum_j t_{j1} + 0.8 \sum_j t_{j2} + 1.5 \sum_j t_{j3} - \sum_k f_{1k} + 0.8 \sum_k f_{2k} + 1.5 \sum_k f_{3k} = 0$$

$$\sum_l z_{jlk} + t_{jk} - y_{jk} = 0 \quad , j = 1 \ldots 3, k = 1 \ldots 3$$

$$\sum_l w_{klm} - DEMAND_{km} = 0 \quad , k = 1 \ldots 3, m = 1 \ldots 20$$

$$\sum_j z_{jlk} \leq DEPOTMAX_{lk} \quad , k = 1\ldots3, l = 1\ldots6$$

$$\sum_j \sum_l z_{jlk} \leq 0.8 \sum_k DEPOTMAX_{lk} \quad , l = 1\ldots6$$

$$\sum_j \sum_l z_{jlk} \geq 0.2 \sum_k DEPOTMAX_{lk} \quad , l = 1\ldots6$$

(B) Goal Programming Model

$$Min\ z = PCOST + \sum_j PUPREF_j + \sum_j NLOREF_j +$$

$$\sum_k PUPDEP_k + \sum_k NLODEP_k + PTRADE$$

subject to,

$$\sum_i p_i \sum_j x_{ij} + \sum_i \sum_j c_{ij} x_{ij} + \sum_j \sum_k c'_{jk} y_{jk} + \sum_j \sum_l \sum_k c''_{jlk} z_{jlk} +$$

$$\sum_l \sum_k c'''_{lk} \sum_j z_{jlk} + \sum_k \sum_l \sum_m c''''_{klm} w_{klm} +$$

$$0.5(\sum_j \sum_l T_{jk} t_{jk} + \sum_l \sum_k F_{lk} f_{lk}) + NCOST - PCOST = 0$$

$$\sum_j x_{ij} \leq MAXSITE_i \quad , i = 1\ldots3$$

$$\sum_j x_{ij} \geq MINSITE_i \quad , i = 1\ldots3$$

$$\sum_i x_{ij} - (0.002y_{j1} + 0.0025y_{j2} + 0.004y_{j3}) = 0 \quad , j = 1\ldots3$$

$$y_{jk} \leq MAXPROD_{jk} \quad , j = 1\ldots3, k = 1\ldots3$$

$$\sum_k y_{jk} + NUPREF_j - PUPREF_j = 0.66 \sum_k MAXPROD_{jk} \quad , j = 1\ldots3$$

$$\sum_k y_{jk} + NLOREF_j - PLOREF_j = 0.33 \sum_k MAXPROD_{jk} \quad , j = 1 \dots 3$$

$$\sum_j z_{jlk} + f_{lk} - \sum_m w_{klm} = 0 \quad , k = 1 \dots 3, l = 1 \dots 6$$

$$\sum_j t_{j1} + 0.8 \sum_j t_{j2} + 1.5 \sum_j t_{j3} - \sum_k f_{1k} + 0.8 \sum_k f_{2k} + 1.5 \sum_k f_{3k} = 0$$

$$\sum_l z_{jlk} + t_{jk} - y_{jk} = 0 \quad , j = 1 \dots 3, k = 1 \dots 3$$

$$\sum_l w_{klm} - DEMAND_{km} = 0 \quad , k = 1 \dots 3, m = 1 \dots 20$$

$$\sum_j z_{jlk} \leq DEPOTMAX_{lk} \quad , k = 1 \dots 3, l = 1 \dots 6$$

$$\sum_j \sum_l z_{jlk} + NUPDEP_j - PUPDEP_j = 0.8 \sum_k DEPOTMAX_{lk} \quad , l = 1 \dots 6$$

$$\sum_j \sum_l z_{jlk} + NLODEP_j - PLODEP_j = 0.2 \sum_k DEPOTMAX_{lk} \quad , l = 1 \dots 6$$

$$\sum_j t_{j1} + 0.8 \sum_j t_{j2} + 1.5 \sum_j t_{j3} + NTRADE - PTRADE = 0$$

(C) Variable Definitions

x_{ij}	=	amount (in tonnes) distributed from site i to refinery j
y_{jk}	=	amount of fuel k (in gallons) produced at refinery j
z_{jlk}	=	amount of fuel k transported from refinery j to depot l
w_{klm}	=	amount of fuel k received by zone m from depot l
t_{jk}	=	amount of fuel k sent to competitor from refinery j
f_{lk}	=	amount of fuel k received from competitor at depot l

$PCOST$ =	positive deviation from the cost objective
$NCOST$ =	negative deviation from the cost objective
$PUPREF_j$ =	positive deviation from a refinery's upper production level bound
$NUPREF_j$ =	negative deviation from a refinery's upper production level bound
$PLOREF_j$ =	positive deviation from a refinery's lower production level bound
$NLOREF_j$ =	negative deviation from a refinery's lower production level bound
$PUPDEP_j$ =	positive deviation from a depot's upper operating level bound
$NUPDEP_j$ =	negative deviation from a depot's upper operating level bound
$PLODEP_j$ =	positive deviation from a depot's lower operating level bound
$NLODEP_j$ =	negative deviation from a depot's lower operating level bound
$PTRADE$ =	positive deviation of the trade with competition
$NTRADE$ =	negative deviation of the trade with competition

(D)   Model Data

(D).1   Site to Refinery Costs and Site Limits and Prices

c_{ij}	Canvey	Rotterdam	Aberdeen
Essex	10	50	70
Birkenhead	50	80	35
Fawley	30	60	85
$MINSITE_i$	666	2000	666
$MAXSITE_i$	1000	3000	1000
Price p_i	50	49	51

(D).2   Refinery Production Costs

c'_{jk}	Essex	Birkenhead	Fawley
Petrol	20	19	17
Diesel	10	9	9
Fuel Oils	20	18	16

(D).3   Refinery Oil Production Limit

$MAXPROD_{jl}$	Essex	Birkenhead	Fawley
Petrol	600000	450000	300000
Diesel	300000	27000	24000
Fuel Oil	600000	27000	300000

(D).4 Refinery to Depot Costs

c''_{jlk} Petrol/Diesel/Fuel Oil	Essex	Birkenhead	Fawley
Grays	4/3/6	12/9/3	7/5/9
Coventry	8/5/10	8/4/9	9/5/11
Port Talbot	8/7/12	6/6/9	11/6/9
Peterlee	12/9/17	6/6/10	14/10/20
Preston	10/8/17	4/3/6	12/8/18
Southampton	7/5/10	11/7/16	4/2/5

(D).5 Depot Operating Costs per Fuel

c'''_{lk}	Grays	Coventry	Port Talbot	Peterlee	Preston	Southampton
Petrol	0.9	0.9	1.1	1.1	1.0	1.1
Diesel	0.7	0.7	0.9	0.9	0.8	0.9
Fuel Oils	1.5	1.5	1.9	1.9	1.7	1.8

(D).6 Depot Maximum Capacities

$DEPOTMAX_{lk}$	Grays	Coventry	Port Talbot	Peterlee	Preston	Southampton
Petrol	360000	360000	240000	270000	300000	270000
Diesel	300000	300000	225000	240000	270000	240000
Fuel Oils	180000	180000	135000	135000	150000	135000

(D).7 Depot to Customer Zone Costs and Demand

c''''_{klm} P/D/F	Grays	Coventry	Port Talbot	Peterlee	Preston	Southampton	$DEMAND_{km}$ (,000's)
1. Kent	8/7/12	-	-	-	-	10/9/16	35/28/18
2. London	8/7/12	-	-	-	-	9/8/15	50/31/19
3. Sussex	9/8/16	-	-	-	-	8/7/14	30/27/17
4. Wessex	-	-	11/10/19	-	-	7/5/9	30/27/17
5. Thames Valley	-	10/9/16	9/8/15	-	-	9/8/15	40/29/18
6. Home Counties	9/7/12	9/7/12	-	-	-	-	30/26/16
7. Norfolk	11/9/17	12/10/19	-	-	-	-	30/27/16
8. West Country	-	-	10/9/16	-	-	10/9/16	35/27/17
9. Devon/Cornwall	-	-	12/11/20	-	-	12/11/20	28/26/17
10. South Wales	-	-	7/5/9	-	-	-	32/28/18
11. North Wales	-	13/10/21	13/10/21	-	12/9/19	-	26/27/17
12. West Midlands	-	7/5/9	12/9/18	-	13/10/19	-	40/30/19
13. Herefordshire	-	9/7/12	9/7/12	-	-	-	25/27/17
14. East Midlands	14/11/23	8/6/11	-	-	12/9/18	-	37/28/18
15 Potteries	-	9/7/12	14/11/20	-	10/8/14	-	35/27/17
16. Lancashire	-	-	-	13/10/18	7/5/9	-	40/29/18
17. Yorkshire	-	-	-	11/9/15	12/10/17	-	38/29/18
18. Tyneside	-	-	-	7/5/9	13/11/19	-	40/30/18
19. Cumbria	-	-	-	13/11/20	12/10/18	-	30/27/17
20. Northumberland	-	-	-	12/10/18	15/11/24	-	32/27/17

(D).8 Fuel Costs in Trade with Competition

T_{jk}	Essex	Birkenhead	Fawley
Petrol	7	10	6
Diesel	9	5	6
Fuel Oils	4	6	11

F_{lk}	Grays	Coventry	Port Talbot	Peterlee	Preston	Southampton
Petrol	9	7	10	6	6	10
Diesel	8	4	8	5	5	7
Fuel Oils	11	9	15	8	10	16

(E) Diagramatic View of Distribution Area

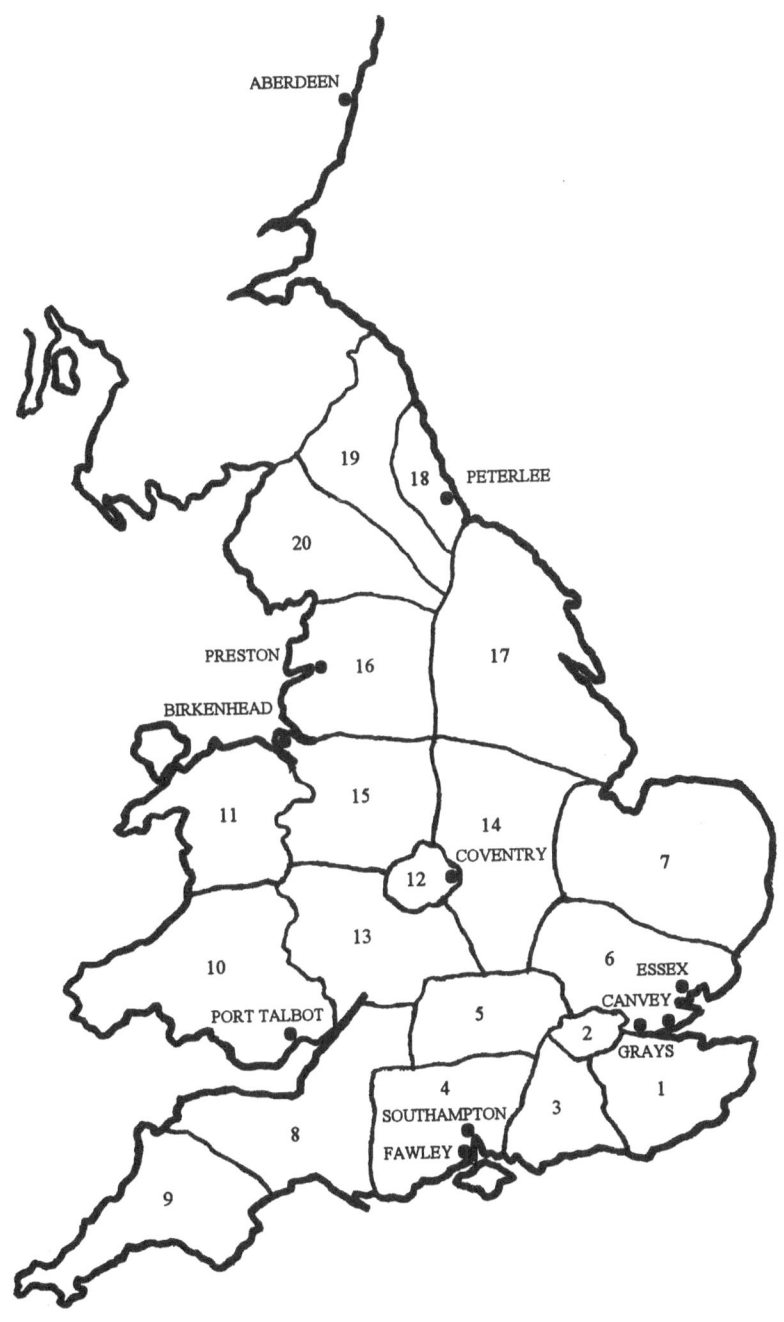